Sustainable Agriculture Reviews

Volume 42

Other Publications by Dr. Eric Lichtfouse

Books
Scientific Writing for Impact Factor Journals
https://www.novapublishers.com/catalog/product_info.php?products_id=42242

Environmental Chemistry
http://www.springer.com/978-3-540-22860-8

Sustainable Agriculture
Volume 1: http://www.springer.com/978-90-481-2665-1
Volume 2: http://www.springer.com/978-94-007-0393-3

Book series
Environmental Chemistry for a Sustainable World
http://www.springer.com/series/11480

Sustainable Agriculture Reviews
http://www.springer.com/series/8380

Journal
Environmental Chemistry Letters
http://www.springer.com/10311

Sustainable agriculture is a rapidly growing field aiming at producing food and energy in a sustainable way for humans and their children. Sustainable agriculture is a discipline that addresses current issues such as climate change, increasing food and fuel prices, poor-nation starvation, rich-nation obesity, water pollution, soil erosion, fertility loss, pest control, and biodiversity depletion.

Novel, environmentally-friendly solutions are proposed based on integrated knowledge from sciences as diverse as agronomy, soil science, molecular biology, chemistry, toxicology, ecology, economy, and social sciences. Indeed, sustainable agriculture decipher mechanisms of processes that occur from the molecular level to the farming system to the global level at time scales ranging from seconds to centuries. For that, scientists use the system approach that involves studying components and interactions of a whole system to address scientific, economic and social issues. In that respect, sustainable agriculture is not a classical, narrow science. Instead of solving problems using the classical painkiller approach that treats only negative impacts, sustainable agriculture treats problem sources.

Because most actual society issues are now intertwined, global, and fast-developing, sustainable agriculture will bring solutions to build a safer world. This book series gathers review articles that analyze current agricultural issues and knowledge, then propose alternative solutions. It will therefore help all scientists, decision-makers, professors, farmers and politicians who wish to build a safe agriculture, energy and food system for future generations.

More information about this series at http://www.springer.com/series/8380

Grégorio Crini • Eric Lichtfouse
Editors

Sustainable Agriculture Reviews 42

Hemp Production and Applications

Springer

Editors
Grégorio Crini
Chrono-environnement, UMR 6249
Université Bourgogne Franche-Comté
Besançon, France

Eric Lichtfouse
Aix-Marseille Univ, CNRS, IRD, INRAE, Coll France, CEREGE
Aix-en-Provence, France

ISSN 2210-4410 ISSN 2210-4429 (electronic)
Sustainable Agriculture Reviews
ISBN 978-3-030-41383-5 ISBN 978-3-030-41384-2 (eBook)
https://doi.org/10.1007/978-3-030-41384-2

This Springer imprint is published by the registered company Springer Nature Switzerland AG
The registered company address is: Gewerbestrasse 11, 6330 Cham, Switzerland

Preface

Why use up the forests which were centuries in the making and the mines which required ages to lay down, if we can get the equivalent of forest and mineral products in the annual growth of the hemp fields (Henry Ford, 1941)

Cannabis sativa – Male flower

Climate change will undoubtedly threaten the survival of life beings on Earth, including *Homo sapiens* who has so far not been so *sapiens* – wise in Latin – to cure this global issue. Our survival will depend on how and how fast we will be able to tune the climate globally, e.g., by geoengineering. A feasible solution is to replace fossil fuels by modern biomass to set up a carbon neutral economy that does not increase and preferably decrease atmospheric CO_2 levels. For instance, in the last two decades, the increasing interest in natural resources, energy conservation, and biomass conversion to chemicals, bioproducts, and biofuels has renewed the interest for hemp as a "new" low-cost, sustainable, ecological, biodegradable, recyclable, and multipurpose material. Hemp-based materials are suitable replacements for a variety of fossil-based products or applications.

The hemp plant *Cannabis sativa* Linn, referred to as industrial hemp, is a high-yielding annual industrial crop grown for fibers from hemp stalk and for seed oil. Hemp has many potential applications in textile, food, beverages, buildings, automotive, furniture, luxury market, cosmetics, and personal care. Actually, about 25,000 products are derived from industrial hemp.

This book, *Hemp Production and Applications*, is published in the series Sustainable Agriculture Reviews and written by 35 international contributors from 8 countries. The chapters review applications, developments, research trends, methods, and issues related to the applications of industrial hemp for fundamental research and technology.

The first chapter by Krystyna Żuk-Gołaszewska and Janusz Gołaszewski summarizes the current state of knowledge on hemp production with a particular focus on current trends, research challenges, and opportunities. The second chapter by Grégorio Crini et al. provides an overview of traditional and news uses of industrial hemp. In Chap. 3, John H. Fike et al. give a roundup of hemp activities in the United States Biljana Pejić et al. discuss in Chap. 4 the cultivation, processing, and applications of *Cannabis sativa* L. in Serbia. Applications of industrial hemp in construction sector are discussed in two chapters: the physicochemical characterization and development of hemp aggregates for highly insulating construction building materials by Yunhong Jiang et al. in Chap. 5 and the modeling of the hygrothermal behavior of hemp concrete by Yacine Aït Oumeziane et al. in Chap. 6. Three chapters focus on recent applications of hemp proteins. In Chap. 7, Anne Pihlanto et al. discuss the current knowledge of the hempseed proteins extractions and the functional properties of the enriched protein fractions. The processing, functional, and bioactive properties of hempseed proteins are detailed in Chap. 8 by Tamara Dapčević-Hadnađev et al. In Chap. 9, Saurel et al. describe the use of hemp proteins as new sources of plant protein for human consumption. The final chapter, Chap. 10, by Lavinia Tofan et al. summarizes recent applications of hemp fibers in wastewater treatment.

The editors extend their thanks to all the authors who contributed to this book for their efforts in producing timely and high-quality chapters. The creation of this book would not have been possible without the assistance of several colleagues and friends deserving acknowledgment. They have helped by choosing contributors and reviewing chapters and in many other ways. Finally, we would like to thank the staff of Springer Nature for their highly professional editing of the publication.

Besançon, France — Grégorio Crini
Aix-en-Provence, France — Eric Lichtfouse

Keywords

Adsorption
Agro-technical factors
Applications
Beverages
Bioactive peptides
Biocomposites
Biofuel
Biological activity
Building
Building materials
Cannabinoids
Carbonization
Carbonized hemp fibers
Chemically modified hemp fibers
Composites
Construction
Cosmetics
Cultivation
Digestibility
Energy production
Environmental impact
Environmental uses
Extraction
Fibers
Fibrous sorbents
Food
Functional properties
Gelling
Grain
HAM transfer
Heavy metal ions

Hemp aggregates
Hemp concrete
Hemp protein
Hemp seed protein
Hydration
Hygrothermal modeling
Hygrothermal properties
Interfacial properties
Isolation
Isolation technique
Life cycle assessment
Lime mortar
Loaded hemp fibers
Markets
Modification
Nanotechnology
Organic binder
Organic pollutants
Paper
Physiological parameters
Pollutants
Production
Protein
Removal
Sorbent
Sorption
Techno-functional properties
Textiles
Thermal properties
Wastewater
Water purification

Contents

Contributors

Yacine Aït Oumeziane FEMTO-ST Institute, University Bourgogne Franche-Comté, CNRS, Belfort, France

Martin Ansell BRE Centre for Innovative Construction Materials, Department of Architecture and Civil Engineering, University of Bath, Bath, UK

Gilles Chanet Eurochanvre, Arc-les-Gray, France

Florence Collet University of Rennes, LGCGM EA 3913, Rennes, France

Grégorio Crini Chrono-environnement, UMR 6249, Université Bourgogne Franche-Comté, Besançon, France

Tamara Dapčević-Hadnađev University of Novi Sad, Institute of Food Technology, Novi Sad, Serbia
University Business Academy in Novi Sad, Faculty of Pharmacy, Novi Sad, Serbia

Heather Darby Department of Plant and Soil Science, University of Vermont, Burlington, VT, USA

Manda Dizdar University Business Academy in Novi Sad, Faculty of Pharmacy, Novi Sad, Serbia

John H. Fike School of Plant and Environmental Sciences, Virginia Tech, Blacksburg, VA, USA

Janusz Gołaszewski Department of Plant Breeding and Seed Production, Faculty of Environmental Management and Agriculture, Center for Bioeconomy and Renewable Energies (Inter-faculty Unit), University of Warmia and Mazury in Olsztyn, Olsztyn, Poland

Miroslav Hadnađev University of Novi Sad, Institute of Food Technology, Novi Sad, Serbia
University Business Academy in Novi Sad, Faculty of Pharmacy, Novi Sad, Serbia

Davoud M. Heidari Interdisciplinary Research Laboratory on Sustainable Engineering and Eco-design (LIRIDE), Faculty of Engineering, Department of Civil and Building Engineering, Université de Sherbrooke, Sherbrooke, QC, Canada

Atif Hussain BRE Centre for Innovative Construction Materials, Department of Architecture and Civil Engineering, University of Bath, Bath, UK

Yunhong Jiang BRE Centre for Innovative Construction Materials, Department of Architecture and Civil Engineering, University of Bath, Bath, UK

Burton L. Johnson Department of Plant Sciences, North Dakota State University, Fargo, ND, USA

Mirjana Kostić Faculty of Technology and Metallurgy, University of Belgrade, Belgrade, Serbia

Christophe Lanos University of Rennes, LGCGM EA 3913, Rennes, France

Michael Lawrence BRE Centre for Innovative Construction Materials, Department of Architecture and Civil Engineering, University of Bath, Bath, UK

Eric Lichtfouse Aix-Marseille Univ, CNRS, IRD, INRAE, Coll France, CEREGE, Aix-en-Provence, France

Nataša Jovanović Lješković University Business Academy in Novi Sad, Faculty of Pharmacy, Novi Sad, Serbia

Sari Mäkinen Natural Resources Institute Finland, Jokioinen, Finland

Nadia Morin-Crini Laboratoire Chrono-environnement, UMR 6249, UFR Sciences et Techniques, Université Bourgogne Franche-Comté, Besançon, France

Bassam Moujalled Cerema, Equipe-projet BPE, L'Isle d'Abeau, France

Markus Nurmi Natural Resources Institute Finland, Jokioinen, Finland

Carmen Paduraru Department of Environmental Engineering and Management, "Cristofor Simionescu" Faculty of Chemical Engineering and Environmental Protection, "Gheorghe Asachi" Technical University of Iasi, Iasi, Romania

Biljana Pejić Faculty of Technology and Metallurgy, University of Belgrade, Belgrade, Serbia

Anne Pihlanto Natural Resources Institute Finland, Jokioinen, Finland

François Potin Procédés Alimentaires et Microbiologiques, UMR A 02.102, University of Bourgogne Franche-Comté, AgroSup Dijon, Dijon, France

Rémi Saurel Procédés Alimentaires et Microbiologiques, UMR A 02.102, University of Bourgogne Franche-Comté, AgroSup Dijon, Dijon, France

Larry Smart School of Integrative Plant Science, Horticulture Section, Cornell University, Geneva, NY, USA

Carmen Teodosiu Department of Environmental Engineering and Management, "Cristofor Simionescu" Faculty of Chemical Engineering and Environmental Protection, "Gheorghe Asachi" Technical University of Iasi, Iasi, Romania

Lavinia Tofan Department of Environmental Engineering and Management, "Cristofor Simionescu" Faculty of Chemical Engineering and Environmental Protection, "Gheorghe Asachi" Technical University of Iasi, Iasi, Romania

Marija Vukčević Faculty of Technology and Metallurgy, University of Belgrade, Belgrade, Serbia

Krystyna Żuk-Gołaszewska Department of Agrotechnology, Crop Management and Agribusiness, Faculty of Environmental Management and Agriculture, University of Warmia and Mazury in Olsztyn, Olsztyn, Poland

About the Editors

Grégorio Crini is environmental polymerist at the University of Bourgogne Franche-Comté, Besançon, France. His research activities focus on the design of new polymer networks and the environmental aspects of polysaccharide chemistry. He published more than 200 papers and edited 12 books. He is a highly cited researcher with a total number of citations exceeding 10,000 (h-index: 35).

Eric Lichtfouse is geochemist and professor of scientific writing at Aix-Marseille University, France, and visiting professor at Xi'an Jiaotong University, China. He has discovered temporal pools of molecular substances in soils, invented carbon-13 dating, and published the book *Scientific Writing for Impact factor Journals*. He is chief editor and founder of the journal *Environmental Chemistry Letters* and the book series Sustainable Agriculture Reviews and Environmental Chemistry for a Sustainable World. He has awards in analytical chemistry and scientific editing. He is World XTerra vice-champion.

Chapter 1
Hemp Production

Krystyna Żuk-Gołaszewska and Janusz Gołaszewski

Abstract This chapter reviews hemp production with emphasis on research challenges and opportunities. *Cannabis sativa* L. is an annual, cosmopolitan plant belonging to the genus *Cannabis*. Among a number of varieties, the economic significance have *Cannabis sativa* var. *sativa* and *Cannabis sativa* var. *indica*, commonly referred to as industrial cannabis/hemp and medical cannabis/medical marijuana, respectively. The in-between varieties are differentiated by around one hundred organic chemical compounds known as cannabinoids. The crucial distinction between industrial and medical cannabis is the content of the principal psychoactive cannabinoid – tetrahydrocannabinol (THC) in relation to non-psychoactive ingredient – cannabidiol (CBD).

Hemp is cultivated for biomass and fiber that constitute feedstock for industrial uses such as energy, construction and automotive markets and for hempseeds that are components of functional foods, animal feeds and medicinal products. The productivity of hemp is affected by both environmental and agronomic factors and their interaction. The hemp plants can act as retardants inhibiting weed growth, as natural insect repellents, and as a limiting factor of nematode growth in soil. The plants are capable of reducing greenhouse gas emissions by binding approximately 2.5 tons of CO_2 per ha and improving soil quality by phytoremediation of heavy metals. Under international and many national laws, industrial cultivation and provision is strictly regulated while medicinal plants are allowed only for medical and scientific purposes. For instance, according to Polish legislation on counteracting drug addiction all the hemp plantations can be established in delimited

K. Żuk-Gołaszewska (✉)
Department of Agrotechnology, Crop Management and Agribusiness, Faculty of Environmental Management and Agriculture, University of Warmia and Mazury in Olsztyn, Olsztyn, Poland
e-mail: kzg@uwm.edu.pl

J. Gołaszewski
Department of Plant Breeding and Seed Production, Faculty of Environmental Management and Agriculture, Center for Bioeconomy and Renewable Energies (Inter-faculty Unit), University of Warmia and Mazury in Olsztyn, Olsztyn, Poland
e-mail: janusz.golaszewski@uwm.edu.pl

G. Crini, E. Lichtfouse (eds.), *Sustainable Agriculture Reviews 42*, Sustainable Agriculture Reviews 42, https://doi.org/10.1007/978-3-030-41384-2_1

areas upon permission and when the content of psychoactive constituents does not exceed 0.2% on a dry matter basis while the other specific legislation is intended for particular nutritional uses and medical devices.

The future research on hemp production should be orientated towards breeding new varieties resistant to environmental stresses, optimization of agricultural treatments to improve effective uptake of nutrients, and modeling canopy structure in order to adapt the production to a changing environment.

Keywords Physiological parameters · Agrotechnical factors · Environmental impact

1.1 Introduction: Historical Background/Origin/Ancestry

Cannabis sativa L. (2n = 20) is a common plant that has been known since the dawn of humanity (Schultes 1970). This annual plant belongs to the botanical family *Cannabaceae,* and it is known under various names around the world: *konopie siewne* in Poland, *hemp* in English-speaking countries, *chanvre* in France, *konopi* in the Czech Republic, *konoplja* in Croatia, *hanf* in Germany, *kender* in Hungary, *cânepă* in Romania, *canapa* in Italy, *hennep* in the Netherlands, *maconha* in Portugal, *cannabis* in Spain, and *konopli* in Ukraine.

Paleobotanical and archaeological research indicates that cannabis has been known already in 6.000–8.000 BC, and its cultivation dates back to around 10.000–12. 000 BC. According to Li (1974), cannabis stems were used in the production of fibre and paper. Seeds with both psychoactive (*ma fen*) and non-psychoactive (*ma tze*) properties were grown in traditional Chinese medicine. Three geographic origins of cannabis have been postulated. The first independent trail begins in the Caucasus and the Altai Mountains in Central Asia (De Candolle 1884), from where cannabis spread east to China, south to India and west to Europe. The second trail leads from northern Afghanistan and the Hindu Kush mountains in Pakistan to India or China (Vavilov 1926). The third place of origin of *C. sativa* were the Himalayan foothills, the Hindu Kush mountains in South Asia (Sharma 1979) or India. These finding indicate that *C. sativa* originated in Asia, in particular in China and India where it was first cultivated by humans. Cannabis subsequently spread through the Middle East and Russia to reach Africa and Europe. The fact that the first paper mill in Europe was built by Muslims in 1150 provides evidence for the above. In the following 850 years, paper was manufactured mainly from hemp. In the sixteenth century, cannabis reached South America with Spanish explorers, and it was first introduced to Chile and Peru. The species then spread to North America, including Canada and the United States.

According to historical sources, hemp was classified by Linnaeus as *Cannabis sativa* L. in 1753. The second species, *Cannabis indica,* was identified by Lamarck in 1783, and the third species, *Cannabis ruderalis,* was classified by Janischevisky in 1924. Since then, the taxonomy of the species has been relatively well explored.

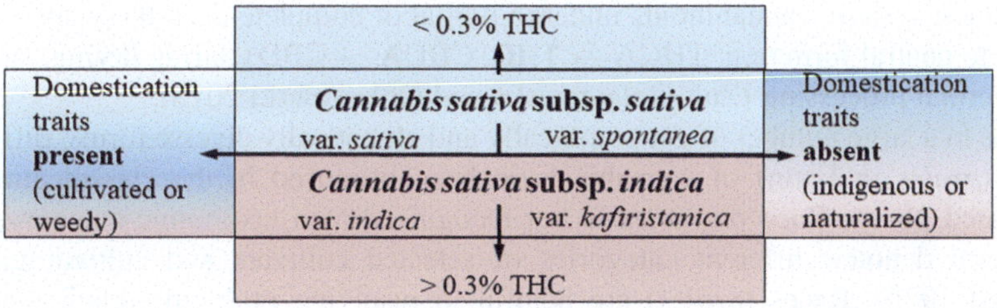

Fig. 1.1 Classification of *Cannabis sativa* according to Small and Cronquist (1976), authors' modification. THC – tetrahydrocannabinol. The two-step hierarchical approach: first step discriminates horizontally between subsp. *sativa* vs. subsp. *indica* on the basis of THC content under and over 0.3%, respectively*;* second step discriminates vertically between varieties with different THC content on the basis of the presence *vs.* the absence of domesticated traits, respectively

Small and Cronquist (1976) analyzed 2500 cannabis plants grown under standard conditions and found that the analyzed population belonged to the same species of *Cannabis sativa* L. regardless of the content of biologically active substances in non-intoxicant, semi-intoxicant and wild plants. The controversy surrounding the contemporary taxonomy and nomenclature of cannabis, including its scientific and vernacular names, has been discussed by McPartland and Guy (2017). This species has different subpopulations or botanical varieties, two of which have economic significance: *Cannabis sativa* subsp. (var.) *sativa* and *Cannabis sativa* subsp. (var.) *indica*. These varieties are commonly referred to as industrial hemp and medicinal hemp/medicinal marijuana, respectively (Small and Cronquist 1976; Hill 1983; Gigliano 2001). McPartland and Guy (2017) on the basis of Small's taxonomic concept (Small and Cronquist 1976) provide an indication on the cannabis classification system assuming two-step hierarchy approach (Fig. 1.1). The first step discriminates between two subspecies with 0.3% THC content in dried female flowering tops while the second step discriminates between varieties within subspecies *sativa* and *indica* on the basis of domestication phase. The authors pointed out that the content of THC and CBD can differentiate cannabis variety identities through three types of chemovars: Type I: THC > 0.3%, CBD < 0.5%; Type II: THC > 0.3%, CBD > 0.5%; or Type III: THC < 0.3%, CBD > 0.5%.

Genetic studies of *Cannabis sativa* germplasm based on Random Amplified Polymorphic DNA RAPD markers not only demonstrated the separation of Italian, Hungarian and Korean germplasm groups, but also revealed variability in a mixture of genotypes from different locations and within the analysed cultivars of Carmagnola and Fibranova (Faeti et al. 1996).

Botanical varieties of *Cannabis sativa* L. differ in chemical composition, plant habit, agronomic requirements and suitability for processing (Datwyler and Weiblen 2006). Industrial hemp is classified as an agricultural crop. Industrial cannabis is generally characterized by a low (below 1%) content of THCA and a CBD:THC ratio higher than 1. In turn, hemp used in the pharmaceutical industry is classified as a horticultural crop. Medicinal marijuana is characterized by a high content of THCA and a low content of cannabidiolic acid (CBDA). Regardless of the cultivation system, cannabinoids undergo partial or complete decarboxylation from acidic to neutral form (e.g. THCA → THC, CBDA → CBD) during drying, storage and thermal processing (Żuk-Gołaszewska and Gołaszewski 2018).

Due to a large number of geographically and climatically diverse forms, different races (*proles* in Latin) of cannabis have been proposed in the classification of cultivated plants. Race or *proles* is not recognized as a taxonomic category, and this term denotes different categories of selected cultivars and cultivar groups (Fryxell 1976). Races (*proles*) are helpful in resolving problems relating to the classification and terminology of cultivated plants. These taxa are given Latin or botanical names. Hemp races differ in the length of the growing season, plant height and yield. Northern races have a growing period of 50–60 days with cumulative temperatures of 800–900 °C and a considerably high seed yield. Plant height ranges from 30 cm to 50 cm. In turn, southern races have a growing period of 140–160 days with cumulative temperatures of 3500–4000 °C. These plants have a height of up to 4 m, and they are characterized by a relatively low seed yield (Hoffman 1961). Transitional races represent intermediate characteristics between northern and southern forms.

1.1.1 Botanical Description

The morphology of *Cannabis sativa* plants is determined by sex (male and female, dioecious and monoecious) and growing conditions. Female plants, known as głowacze in Polish, develop more leaves than male plants (płaskonie in Polish). Cannabis plants have straight non-branching stems that are covered with short, curved and oval hairs. Stems have 7 to 10 internodes between the root crown and inflorescences, and the longest internodes are found in the central part of the stem. Stem length ranges from 30 cm to 400 cm. Palmate leaves have petioles and 3–11 leaflets with an opposite arrangement in the lower and central parts of the stem and an alternate arrangement at the level of inflorescences. Leaflets are lanceolate, regularly serrate and covered with glandular hairs. Female flowers are arranged in panicles at the top of the stem. Flowers are green, without a perianth, with one pistil and two stigmata per flower. Male flowers are small and arranged in loose panicles at the top of the stem. Hemp plants grown in Marian Lewicki's farm in Linowiec are presented in Photo 1.1. The farm's scientific consultant, Krystyna Żuk-Gołaszewska, assists the farm in selecting the optimal methods of cannabis cultivation and disseminates knowledge on the practical applications of this plant species.

Cannabis plants have a taproot and numerous side roots with a length of up to 2 m. The main portion of the root system is located at a depth of 20–50 cm. In a study by Amaducci et al. (2008a), root biomass ranged from 2.41 to 3.21 t ha^{-1} and varied across the experimental years. The ratio of above-ground to below-ground biomass was determined at 5.46. The roots of hemp plants grown in a field in north-eastern Poland are presented in Photo 1.2.

Photo 1.1 Hemp plants var. Białobrzeskie grown in Marian Lewicki's farm in Linowiec, Poland

Root systems are characterized by considerable plasticity, but root growth is conditioned by numerous environmental factors, including metal and water levels in soil (Fitter 1991; Elisa et al. 2007). Exposure to copper led to a significant decrease in total root length, mean root diameter, root surface area and root volume. Less branched root systems absorb nutrients less efficiently (Fitter 1991), which is why copper treatments reduced root absorbing surface.

1.1.2 Phenological Growth Stages

Every living organism is characterized by specific stages of growth and development. The growth stages of plants are described based on their unique agrophenophases as well as with the use of a decimal coding system developed by Zadoks et al. (1974). The first plants to have been described by the proposed decimal code were cereals of the class Monocotyledoneae. Many scientists relied on the Zadoks scale to develop codes for plant species of the class Dicotyledoneae, including *Cannabis sativa* L. (Mediavilla et al. 1998; Mishchenko et al. 2017). Cereal growth stages are described on the BBCH scale which is derived from the

Photo 1.2 Roots of *Cannabis sativa* L. (Marian Lewicki)

names of the originally participating stakeholders: Biologische Bundesantalt, Bundessortenamt and CHemische Industrie. The abbreviation is said to unofficially represent the four companies that initially financed the scale's development: Bayer, BASAF, Ciba-Geigy and Hoehst. The BBCH scale is a system for the uniform coding of the phenological stages of cultivated plants. This standardized scale for describing the growth stages of agricultural plants is applied in different scientific disciplines, including agronomy, physiology, pathology, plant breeding and meteorology, and it is also used by farmers. The BBCH scale also eliminates linguistic barriers which still exist in science.

Cannabis sativa L. is usually a dioecious (unisexual) species, where individual plants develop only male or only female reproductive organs (flowers). Hermaphrodites are sporadically encountered in successive generations. *Cannabis sativa* is characterized by distinctive stages of growth and development, and four principal stages representing the life cycle of this plant have been described by Mediavilla et al. (1998). In secondary growth stages, the second digit of the code denotes the plant's sex, whereas the third and fourth digits indicate the developmental stage (Table 1.1).

Germination and Emergence (Principal Stage Code 1) Various processes take place inside seeds during germination and lead to the activation of the embryo. Seeds imbibe water, and the radicle emerges from the seed coat. *C. sativa* is characterized by epigeal germination, where the hypocotyl emerges and the cotyledons unfold

Table 1.1 Principal and secondary growth stages of hemp (Mediavilla et al. 1998)

Code	Description
0	**Germination and emergence**
0000	Dry seed
0001	Radicle apparent
0002	Emergence of hypocotyl
0003	Cotyledons unfolded
1	**Vegetative stage**
1002	1st leaf pair
1004	2nd leaf pair
1006	3rd leaf pair
1008	4th leaf pair
1010	5th leaf pair
10xx	Until last leaf pair
2	**Flowering and seed formation**
2000	Change of phyllotaxis on the main stem
2001	Flower primordia
Dioecious plant – male	
2100	Flower formation
2101	Beginning of flowering
2102	Full flowering
2103	End of flowering
Dioecious plant – female	
2200	Flower formation
2201	Beginning of flowering
2202	Full flowering
2203	Beginning of seed maturity
2204	Seed maturity
2205	End of seed maturity
Monoecious plant	
2300	Female flower formation
2301	Beginning of female flowering
2302	Full flowering
2303	Male flower formation
2304	Male flowering
2305	Beginning of seed maturity
2306	Seed maturity
2307	End of seed maturity
3	**Senescence**
3001	Leaf desiccation
3002	Stem desiccation
3003	Stem decomposition

above the soil to form the first assimilative leaves. The cotyledons are large, sessile and smooth with nonserrate edges. The optimal temperature for germination is 24 °C (Elisa et al. 2007). The minimal temperature for germination is 0 °C (van der Werf et al. 1995a). Germination usually takes 3–7 days (Clarke 1977).

Vegetative Stage (Principal Stage Code 1) The vegetative stage occurs between emergence and generative development, and it is characterized by steam and leaf growth. In hemp plants, the stem is hollow and unbranched, and it is grooved or furrowed to varying degrees (Heslop-Harrison and Heslop-Harrison 1958). All leaves have a petiole. True leaves have a single narrowly elliptic blade with serrate margins (Stearn 1970). During the vegetative stage, the plant forms up to five true leaf pairs and short internodes. The first leaf pair consists of a single leaflet. Internodes grow during dynamic stem elongation (Bócsa and Karus 1998). The number of leaflets increases from the second leaf, and every leaf pair has an odd number of leaflets, usually three to eleven (the first leaf pair has three leaflets, the third leaf pair has five leaflets, etc.) (Clarke 1977). Seven to twelve leaf pairs develop during the vegetative stage, and the relevant code is 10xx for the nth leaf pair (xx = 2n). Fibre hemp plants generally have unbranched stems.

Flowering and Seed Formation (Principal Stage Code 2) In male and female plants, the first flower primordia are indistinguishable and difficult to identify. Male primordia can be identified by their curved shape and round, pointed flower buds with five radial segments. In turn, female primordia are identified based on an enlarged symmetrical tubular calyx (Clarke 1977).

At present, the generative phase of male, female and monoecious plants can be easily identified. Flower formation is characterized by the appearance of the first closed male flowers (code 2100) and the opening of the first staminate flowers (code 2101). Female inflorescences are leafy and compact. Individual female flowers are small, green and inconspicuous, and they are hidden inside the perigonal bract (Pate 1994).

In dioecious plants, male and female plants generally occur in roughly similar proportions (Hoffman 1961). The inflorescences of male plants are strongly branched, with few or no leaves. Female inflorescences are leafy, stocky and unbranched. Female plants live 3–5 weeks longer than male plants, until the seeds are ripe. Monoecious plants produce male and female flowers. The flowering stage refers to female flowers. Male flowers usually appear when female flowers bloom on the tips of female branches. Flowering is uneven. The male flowering stage is reached when around 50% of staminate flowers in one plant are open and pollen is released, and flowering ends when 90% of all flowers are open (Table 1.1). In this phase, stem elongation slows down, and the number of leaflets per leaf decreases (Heslop-Harrison and Heslop-Harrison 1958).

Cannabis plants are anemophilous. Mature pollen grains are released by thecae and reach the stigmata where the male gametophyte develops. Pollen grains germinate on stigmata and produce pollen tubes. Generative cells divide to form two male gametes (generative and vegetative), and sperm cells reach pollen tubes. Pollination takes place when pollen grains are transferred from the male anther of a flower to the female stigma. The pollen tube penetrates the stigma and the pistil, and it enters the ovary and the ovule through the micropyle opening. The embryo sac is initially composed of a single cell containing eight haploid nuclei. Two polar nuclei are positioned in the centre of the embryo sac, and they fuse to form a diploid nucleus which is the secondary nucellus of the embryo sac. The secondary nucleus is diploid. The remaining six nuclei move to opposite poles of the embryo sac, and three nuclei are positioned at each pole. They are surrounded by cytoplasm to form three naked cells. One of the three cells on the side of the micropyle opening is an egg cell, and the remaining two cells are synergids. The three cells at the opposing pole are the antipodal cells. Pollination is followed by fertilization when the male gamete (sperm) fuses with the female gamete (egg cell). The pollen tube with two sperm cells penetrates the stigma and the pistil, and it enters the ovary and the ovule through the micropyle opening. Inside the ovule, sperm cells are released by the pollen tube and fertilization takes place. One of the sperm cells fuses with the egg to form a diploid zygote which further develops into an embryo. The second sperm cell fuses with the secondary diploid nucleus of the embryo sac to form a triploid cell (3n) that gives rise to triploid nutritive tissue – the endosperm. The ovule develops into a seed.

Senescence (Principal Stage Code 3) The seed maturation stage is defined for dioecious female plants (code 2306). The female flower has a small green organ, the bract, which completely encloses the ovary from which two stigmas protrude. This sheath is covered with slender hairs and stalked circular glands secreting resin containing cannabinoids (Stearn 1970). The seeds (achenes) turn hard, and seed shedding begins. Single seeds take 3–5 weeks to mature. Seed maturity is reached when 50% of the seeds are hard and have a characteristic shape and size. Cannabis seeds, referred to as achenes, are oval or ellipsoid in shape, somewhat flattened and grey-green in colour. The seed coat is shiny, and it takes on a matte appearance as the seed grows older. Cannabis seeds are 2–6 mm long, with a diameter of 2–4 mm and 1000 seed weight of 13–25 g. The germination capacity of cannabis seeds declines rapidly, which is why plants should be grown from the most recently harvested seeds (Heslop-Harrison and Heslop-Harrison 1958; Sacilik et al. 2003).

The phenological phases of *C. sativa* described by Mishchenko et al. (2017) are largely based on the decimal coding system developed by Zadoks et al. (1974). According to the authors, stage 4 does not occur in hemp plants (Table 1.2).

Table 1.2 Growth stages of *Cannabis sativa* L.

Code	Description
0002	Emergence of expected plants
2000	Flower induction of all plants
2102	Dioecious male flowering of all male plants
2202	Dioecious female flowering of all female plants
2302	Monoecious flowering of all monoecious plants
2204 or 2306	Seed maturity of female or monoecious plants

Mishchenko et al. (2017) relied on the research conducted by de Meijer et al. (1992) and de Meijer and Keizer (1994) to describe growth stages in 50% of the population in a given phenological phase. Code 2102 denotes dioecious male flowering in male plants, code 2202 represents dioecious female flowering in female plants, code 2302 indicates monoecious flowering in monoecious plants, and codes 2204 or 2306 denote seed maturity in female and monoecious plants. The proposed approach has useful implications for the cultivation of *C. sativa* plants.

The proposed code is a useful tool for identifying growth stages in hemp plants. Harvest date significantly influences the yield and quality of hemp plants used for various purposes. Plants grown for fibre should be harvested during male flowering (code 2012) or monoecious flowering (code 2302). In turn, plants cultivated for seeds should be harvested during seed maturity of female (code 2204) or monoecious plants (code 2306) (Bócsa and Karus 1998). In plants used in the production of essential oil, the recommended harvest date is 1–3 weeks before seed maturity (code 2203 and 2305) (Meier and Mediavilla 1998). However, regardless of the intended use (fibre, seeds), harvest should generally begin after seed formation (code 2204 or 2306) (Hennink 1997).

The decimal code supports the precise determination of phenological phases and facilitates cultivation. The described system promotes the application of fertilizers in the optimal growth stages (Mediavilla et al. 1998) and supports the identification of diseases that typically occur in a given stage (McPartland 1996). The decimal code is used in growth analyses, computer storage of data and statistical analyses. De Meijer et al. (1992) relied on the decimal code to determine phenological stages characterized by the highest cannabinoid content of inflorescences (code 2203 and 2305).

1.2 Seed Production

The seeds of *Cannabis sativa* have numerous applications, which creates new opportunities for cultivating this versatile crop. The area under cannabis continues to increase due to higher demand for cannabis seeds, preferably seeds from the last harvest.

1.2.1 Field Production Area and Certified Seeds

In 2017, the area under agricultural crops grown for seeds in Europe increased by 1.3% to 2,029,930 ha. France is Europe's largest producer of seeds (354,120 ha), including hemp seeds (www.escaa.org). In France, the area under hemp increased from 634.2 ha in 2010 to 1676.8 ha in 2017 (Table 1.3). The acreage of planted hemp was also high in Poland and Germany.

Table 1.3 Field area dedicated to the production of hemp seeds (ha) (www.escaa.org, accessed on 5 February 2019)

	Year								
Country	2010	2011	2012	2013	2014	2015	2016	2017	2018
France	634.2	384.4	398.9	534.7	791.64	1094.9	1450.1	1676.8	1307.9
Germany	39.3	68.8	66.3	71.6	122.2	154.4	255.8	300.9	324.4
Hungary	101.5	8.8	54.6	28.3	35.5	87.5	93.5	111.2	36.2
Italy	8.71	10.58	16	22.5	36.4	8.13	138.4	157.1	195.5
Lithuania	nd	nd	nd	nd	55.5	139.5	332.4	174.2	129.8
Netherlands	1.0	nd	1.0	6.0	1.0	12	17	14	7
Poland	227.48	56.8	188.6	25.9	19.8	104.1	116.3	113.1	664
Romania	8.9	1.6	4.0	12.0	19.0	124.6	138.3	559.7	nd

nd no data

Table 1.4 Production of certified hemp seeds (t) (www.escaa.org, accessed on 5 February 2019)

	Year								
Country	2010	2011	2012	2013	2014	2015	2016	2017	2018
France	668.8	479.5	500	481.3	821.5	1030.9	1131.8	1225.9	nd
Germany	33	70.4	72.2	50.5	109	250.8	193	295.5	nd
Hungary	8.1	9.4	8.1	15.2	12	11.2	16.0	18.0	nd
Italy	nd	1.7	1.3	5.4	3.6	10.5	34.0	0.0	nd
Poland	37	53	0.0	0.0	0.0	0.0	11.9	102.7	nd

nd no data

According to the Main Inspectorate of Plant Health and Seed Inspection (2019), Poland, the area under hemp crops cultivated for seeds was small and varied in 2010–2014. In the following 3 years (2015, 2016 and 2017), the area of seed plantations was similar, and it only marginally exceeded 100 ha. In 2018, field production increased sixfold to 664 ha. The above could be attributed to changes in legal regulations that facilitated the cultivation of hemp crops as well as increased supply of hemp seeds on the European market. In Poland, field area under hemp reached 6777.5 ha in 2017.

Certified hemp seeds are purchased mainly from France and Germany (Table 1.4), but in 2017, domestic production of certified seeds increased to 102.7 t. However, to the absence of local certified seeds, foreign varieties continue to be grown in Poland. An example can be HemPoland, a producer of hemp seed oil which relies on French varieties Futura and Felina. The company conducts research and shares its experience and knowledge with farmers and businesses. HemPoland also cooperates with international research institutions.

The European Industrial Hemp Association (EIHA) was founded in 2005, and it brings together 31 European countries that produce and process hemp seeds with various cannabinoid content. In 2011, the EIHA became a member of Technical

Committee 411 for Bio-Based Products of the European Committee for Standardization (CEN/TC 411). The EIHA is also a member of the ASTM D37 Committee on Cannabis, the Expert Group on Bio-Based Products of the European Commission, the Biomass Supply Thematic Working Group of the European Bioeconomy Panel, and the Sustainable Bioresources for a Growing Bioeconomy Strategic Working Group of the Standing Committee on Agricultural Research (SCAR) (www.eiha.org).

1.2.2 Varieties

The development of monoecious varieties of *C. sativa* has contributed to an increase in seed yield and improvements in biomass (fiber) quality. This is a considerable breeding achievement because the existing knowledge on inheritable agronomic traits in hemp is still limited. However, hemp varieties that adapt well to a broad range of agro-ecological conditions are still in short supply. At present, hemp breeding programs are initiated to pursue two major goals. The first is the industrial use of hemp, including hemp fiber which is qualitatively superior to woody core fiber for the production of paper pulp, and the production of oil from hemp seeds (Hennink 1994). Hemp biomass is also an excellent renewable source for energy generation. The second goal is the extraction of biologically active substances (cannabinoids) for medicinal use.

The majority of hemp varieties grown for industrial purposes are monoecious varieties, characterized by high yields, high processing suitability, early flowering and tolerance to a wide range of agro-economic conditions. The common catalogue of varieties of agricultural plant species has been published since 1975 in accordance with the provisions of Article 17 of Council Directive 2002/53/EC of 13 June 2002 on the common catalogue of varieties of agricultural plant species. The 37th complete edition of the catalogue, published in the Official Journal of the European Union 2019/C13/01, lists 68 varieties, including 8 Polish varieties. French, Dutch, Italian, Spanish, Hungarian varieties as well as conservation varieties from Latvia are also listed. The oldest Polish variety is Białobrzeskie, which was included in the catalogue in 1967, whereas varieties Glyana and Henola were registered in 2017 (Table 1.5).

1.3 Environmental Requirements

Cannabis sativa is very sensitive to environmental factors, including soil conditions, temperature and water availability. Malceva et al. (2011) and Stafecka et al. (2016) demonstrated that cannabis yields were more influenced by weather conditions during the growing season than by agronomic factors.

Table 1.5 Characteristics of selected varieties of *Cannabis sativa* L.

Białobrzeskie (Poland) – cultivated mainly for fiber. Plant height – 220 cm; straw yield – 14.9 t ha^{-1}; seed yield – 690 kg ha^{-1}; 1000-seed weight – 14.6 g; cellulose content – 48.2% NF DM; fat content of seeds – 34.8% DM; field certification – 116% (COBORU[a] 2017).
Glyana (Poland) – cultivated mainly for fiber. Straw yield – 12.6 t ha^{-1}; seed yield – 719 kg ha^{-1}; 1000-seed weight – 16 g; fat content of seeds – 33.8% DM; average fiber content of straw; relatively high content of long fiber; cellulose content of straw – 43% NF DM; content of Δ^9THC in the dry weight of panicles – 0.006%; satisfactory plant health (COBORU 2017).
Henola (Poland) – cultivated for seeds; seed yield – 1620 kg ha^{-1}; straw yield – 8.1 t ha^{-1}; 1000-seed weight – 15.9 g; fat content of seeds – 34.5% DM, trace amounts of Δ^9THC in the dry weight of panicles (0.01%); early to very early flowering and seed maturation; very short plants; resistant to lodging; satisfactory plant health (COBORU 2017).
Futura (France) – late maturity; inflorescence yield – 3.0 t ha^{-1}; stem yield at seed maturity – 8.34 t ha^{-1}; essential oil yield – 9 L ha^{-1} (Baldini et al. 2018).
Finola (Finland) – dioecious variety; short plants; resistant to lodging; branching absent or weak; photo-inhibition of inflorescence effectively prevents the production of mature seeds; fine fiber is more comparable to flax than hemp; low seed and biomass yield; growing season – 115–120 days (Callaway 2004a).

[a]COBORU = Research Center for Cultivar Testing

1.3.1 Soil

Cannabis thrives on fertile soils that are weeded and maintained in good condition. It can be farmed on both mineral and peat soils that are abundant in humic compounds and lime. Soils belonging to a very good and good wheat complex and a very good and good rye complex (according to the Polish soil quality classification system) are most suitable for cannabis cultivation. The highest yields are obtained on deep and porous soils with water table depth of more than 80 cm. The seed yield of cannabis cultivated on podzolic sandy loam ranged from 1050 to 2557 kg ha^{-1} seeds (Malceva et al. 2011). The most important factor is soil pH which should be close to neutral or slightly alkaline. In a study by Stafecka et al. (2016), hemp was grown on humic-podzolic gley soil with 6.5% organic matter content, pH 7.0, phosphorus (P_2O_5) content of 145 mg kg^{-1} soil, and potassium (K_2O) content of 118 mg kg^{-1} soil. García-Tejero et al. (2014) cultivated hemp on soil with the granulometric composition of clay loam containing 300 g kg^{-1} of sand, 310 g kg^{-1} of clay and 390 g kg^{-1} of silt. Effective depth was 2.5 m, organic matter content was below 15 g kg^{-1}, and water-holding capacity was 170 mm m^{-1}.

1.3.2 Temperature

Cannabis is a thermophilic crop. Total temperature during the growing season ranges from 800 to 3500 °C, depending on the region of cultivation (northern, southern and intermediate). In the temperate climate, total temperature ranges from 2000 to 3000 °C. The optimal temperature for photosynthesis is 25–30 °C, subject to genotype (Chandra et al. 2011). Hemp seeds can germinate already at 0 °C, and emerged seedlings are resistant to temperatures as low as −8 °C to −10 °C (Bócsa and Karus 1998). In a study by Sarsenbaev et al. (2013), spring temperatures ranged from 3.3 °C in March to 12 °C in April. The average temperature in June was 23 °C, and maximum temperature reached 43–44 °C. In non-irrigated/irrigated plants exposed to high temperatures, seed yield was determined at 93.9/330.4 kg ha^{-1} (var. Fedora) and 22.6/324.7 (var. Felina).

1.3.3 Water

Water is one of the key determinants of crop viability, including in cannabis (Cosentino et al. 2013). Hemp is generally regarded as a resource-efficient crop. However, Tang et al. (2018) demonstrated that long-term stress increased leaf senescence, decreased LAI, but retained total canopy N content. Cannabis is a water-intensive crop, and daily water requirements are estimated at 22.7 l per plant. Field-grown marijuana requires nearly 430 million L km^2 of water. During the growing season (May–September), total precipitation approximated 700 mm, and daily evapotranspiration was determined at 6–7 mm in May–July and 4 mm in September (García-Tejero et al. 2014). Irrigation compensating for 80–100% of evapotranspiration loss had no significant influence on yields which were more dependent on genotype. Lisson and Mendham (1998) found that irrigation in the amount of 30 mm and 60 mm of water had a significant effect on fiber hemp yield. In turn, in the study Grabowska and Koziara (2001) the hemp plants need a total rainfall of 200–300 mm. Additionally, the due to the well-developed root system, the plant is tolerant to short periods of drought.

1.4 Agricultural Aspects

Agronomic factors, including preceding crop, tillage, nutrient availability, seeding and plant protection, and the key limiting factors to fiber and biomass yield and seed oil composition.

1.4.1 Farming System

Cannabis plants can be grown indoors with artificial light, in greenhouses with natural light, artificial light or both, as well as outdoors under exposure to natural light (van Butsic and Brenner 2016). *Cannabis sativa* is grown mainly in the field around the world. The selection of the optimal growing site is a very important consideration which influences the growth, development, yield and health of plants.

Protein crops, root crops and cereals fertilized with manure are the most suitable preceding crops for cannabis.

Cannabis should not be cultivated in long-term monoculture. A 4–5 year rotation is recommended (Kilanowski 1974). Cannabis plants leave the soil in good condition and free of weeds such as couch grass. Cannabis plants have a well-developed taproot which decomposes after harvest and is an excellent source of humic compounds for the cultivation of cereals and root crops. Gorchs et al. (2017) demonstrated that cannabis was a highly suitable preceding crop for wheat that contributed to the sustainability of cereal-based cropping systems under humid rainfed Mediterranean conditions.

Cannabis is usually farmed in conventional tillage systems. After the preceding crop is harvested, soil should be skimmed to a depth of 8–12 cm and harrowed to eliminate weeds. Mineral soils are ploughed to a depth of 20–30 m, and peat soils – to a depth of 30–35 cm (Campiglia et al. 2017). In fields where cannabis is grown after perennial plants (grass, red clover), disking is the first treatment to remove the sod. In conventional tillage systems, soil is dragged and harrowed in spring before seeding (Kilanowski 1974). Strip tillage also contributes to rapid and uniform seedling emergence, and it enables cannabis plants to effectively compete with weeds.

1.4.2 Fertilization

Nutrients influence the growth and development of plants and determine crop yields and yield quality (van der Werf et al. 1995b; Papastylianou et al. 2018). Cannabis plants have a particularly high demand for nitrogen in early stages of development and growth. In general, hemp has high nitrogen requirements (Coffman and Gentner 1975; Amaducci et al. 2002). In hemp cultivated for fiber, nitrogen fertilization at a rate of 240 kg ha^{-1} increased biomass yield, stem dry weight and inflorescence weight by 37.3%, 48.2% and 16%, respectively. In addition, plant height and inflorescence length increased from 1.66 to 1.76 m and from 66.2 to 82.9 cm, respectively (Papastylianou et al. 2018). To minimize nitrogen loss, nitrogen fertilizers should be applied in split doses during seedling emergence and upon the formation of the third leaf pair (Mediavilla et al. 1998). According to Stafecka et al. (2016), the nitrogen rate that optimizes seed yield ranges from 60 to 90 kg ha^{-1}. The seed yield of hemp var. Purini with 41.2% oil content was influenced by the rate of N fertilization and ranged from 0.83 to 2.39 t ha^{-1}. Nitrogen fertilization generally influences plant habit, leaf surface area and the dry matter yield of stems (Struik et al. 2000). Nitrogen also determines the THC content of cannabis leaves and their position on the plant (Coffman and Gentner 1975). In a study by Hemphill et al. (1980), the THC content of leaves decreased from the top to the bottom of the plant. Bócsa et al. (1997) demonstrated that high nitrogen levels promoted a greater reduction in the THC content of older than younger cannabis leaves.

Potassium and phosphorus are also essential for the growth and development of plants. Potassium plays a major role in the regulation of water transport in plants. Finnan and Burke (2013) demonstrated that hemp has lower potassium requirements than other agricultural plants and that the optimal potassium fertilization rate for soils with moderate to high levels of potassium is >70 mg/L. In treatments supplied with 60, 90, 120, and 150 kg K ha^{-1}, significant relationships were not observed between hemp yields, the potassium rate or soil potassium levels relative to control (0 kg K ha^{-1}). Around 70–75% of the absorbed potassium was concentrated in the stem. Varied doses of K were applied in other studies. For example, Iványi and Izsáki (1996) found that 234 kg K ha^{-1} of potassium was incorporated into plants. Iványi et al. (1997) demonstrated that potassium fertilizer applied at 249 kg K ha^{-1} during the growing season increased stem yields by around 1 t ha^{-1}. Similarly to nitrogen, phosphorus plays an important role in plant function. Phosphorus is a structural component of nucleic acids and lipids, and it participates in protein synthesis. The demand for phosphorus and potassium is particularly high during flowering.

1.4.3 Seeding

Plant propagation material should be obtained from certified sources where seeds are bred specifically to produce only female flowers. Male plants should be identified in the flowering stage and eliminated (Żuk-Gołaszewska and Gołaszewski 2018). The length of the growing season of cannabis cultivars differs across climate zones and is closely correlated with flowering and harvesting dates. Early-sown cannabis is characterized by prolonged vegetative development and relatively longer stems (Sengloung et al. 2009). In the Polish climate, the optimal period for sowing cannabis is late April to early May (Burczyk et al. 2009). Similar sowing dates were recommended by Papastylianou et al. (2018). Hemp seeds should be sown at a depth of 2–3 cm. Medicinal cannabis should have a bushy growth habit with a large number of side shoots, numerous leaves and inflorescences. A low seeding rate leading to relatively low plant density per m^2 promotes a bushy habit. A higher seeding rate produces taller plants with fewer and more evenly distributed side shoots. The optimal density is 20,000 plants per hectare, but in some production systems, it can be increased to 30,000 or even 40,000 plants per hectare to maximize the quality of raw materials for pharmaceutical processing (García-Tejero et al. 2014; Hall et al. 2014). Stem biomass decreases with an increase in plant density. The self-thinning effect of cultivation in rows with a width of 12.5, 25 and 50 cm was relatively small. Generally, row width did not affect stem yield or quality. However, in early-grown plants, stem yields tended to decrease with an increase in row width (van der Werf et al. 1995b). Stafecka et al. (2016) analyzed four seeding rates (50, 70, 90 and 110 kg ha^{-1}) and found that 50 kg ha^{-1} was the optimal seeding rate. In a study by Burczyk et al. (2009), the optimal seeding rate was 30 kg ha^{-1}. The highest biomass yield was 13.3 t ha^{-1}. In turn, in hemp grown for seeds or panicles, the highest yields were obtained when hemp was sown at 10–20 kg ha^{-1} and harvested at full panicle maturity.

1.4.4 Weeds, Diseases and Pest Control

Successful cultivation of hemp is determined mainly by agronomic factors. Tillage is an important farming operation which reduces weed infestation and decreases the prevalence of diseases and pests. In the production of hemp, the optimal preceding crops are plants that effectively eliminate weeds. Hemp plantations are infested with weeds that are ubiquitous on agricultural land. The predominant weed species include *Chenopodium album* L., *Lamium amplexicaule* L., *Viola arvensis* Murr., *Phallopia convolvulus* L., *Stellaria media* Vill., *Thlaspi arvense* L., *Polygonum nodosum* Pers., *Galinsoga parviflora* Cav. and *Elymus repens* (L.) (Heller et al. 2007). Hemp plants that are adequately supplied with nutrients effectively compete with weeds. Hemp should be cultivated in wide rows to facilitate mechanical weeding before plants are sufficiently tall to suppress weeds (Grabowska 2005). Weed infestation is reduced in dense stands. According to Campiglia et al. (2017), weed infestation was affected by nitrogen rate and seeding rate. Weed biomass was determined at 73 g per m^2, and it was reduced by 31.5% in stands with a density of 120 hemp plants per m^2.

Plant diseases are caused mainly by fungi which thrive in environments characterized by high moisture levels, high temperature and lack of airflow. Crop yields are most severely compromised by the fungal species of *Botrytis cinerea* and *Sclerolinia sclerotiorum*, especially in wet years (van der Werf et al. 1995a). In hemp growing regions, diseases of hemp are rare. However, Agrios (1988) estimated that 11% of fiber crops are lost to diseases. The common diseases affecting cannabis plants are presented in Table 1.6.

Diseases can be controlled by keeping the field idle for 4–5 years, destroying post-harvest residues, implementing crop rotation schemes and introducing deep ploughing. Hemp plants are generally grown without herbicides, fungicides or insecticides (Struik et al. 2000; Jankauskiene and Gruzdeviene 2010). According to the guidelines of the Institute of Plant Protection (2019), the range of plant protection agents is limited, and preventive application of fungicides in early stages of development appears to be an effective method of disease control. The plants grown for medicinal purposes is protected with biological agents to guarantee the safety and high quality of raw material. Typical hemp pests include the hemp flea beetle (*Psylliodes attenuata*) and aphids (*Aphis fabae*). Bird predation can also compromise hemp seed yields (so called "seed disappearance" problem).

Hemp plants are environmentally-friendly crops that act as retardants suppressing weed growth. They also repel insects and pests, and inhibit the growth of nematodes in soil.

Table 1.6 Common diseases affecting cannabis plants

Gray mold – caused by *Botrytis cinerea*, thrives in temperate regions with high humidity and cool to moderate temperatures. Seedlings succumb to damping off. It arises as a gray-brown mat of mycelium which becomes covered by masses of conidia (fungal spores). Stems become chlorotic. Enzymes released by *B. cinerea* reduce stems to soft shredded cankers. Stems often snap at canker sites. Gray mold may encircle and girdle stems, wilting everything above the canker. Fiber varieties become more susceptible after canopy shadows (McPartland 1996).
Hemp canker – caused by *Sclerotinia sclerotiorum*. Symptoms begin as water soaked lesions on stems and branches of plants in the early maturity. The lesions collapse into cankers and become darkly discolored. Affected areas take on a shredded appearance, and the pith becomes filled with a white cottony mycelium. Plants remain in this condition or wilt and fall over. By September, large black sclerotia develop on the stem surface or within the pith of dead stalks. Crop losses can reach 40% (McPartland 1996).
Stem cankers – caused by *F. graminearum* and *F. avenaceum* in cooler climates, and by *F. sulphureum* and *F. sambucinum* in warmer climates (McPartland 1996).
Downy mildew – caused by two *Pseudoperonospora* species that affect fiber varieties
Root rot – caused by *Sclerotium rolfsii*, mostly in southern temperate zones and the tropics. The disease affects both fiber and drug varieties (Ferri 1961).
Premature wilt (charcoal rot) – caused by *Macrophomina phaseolina*. Infected plants prematurely wilt, turn yellow, brown, and die. The disease reduces yields (McPartland 1996).

1.5 Physiological Parameters

In crop production, the term "efficiency" is used synonymously with "performance", and it is defined in thermodynamic terms as the ratio of energy output (carbohydrates) to energy input (solar radiation) (Monteith 1977). The efficiency of plants is determined by agronomic (variety/cultivar, production system, plant protection agents, fertilization) and environmental factors, mainly temperature and water supply. In most agricultural crops, total dry matter yield is strongly correlated with the availability of solar energy. It was found that the photosynthetic efficiency of plants producing around 1.4 g of carbohydrates per 1 MJ of solar energy was 2.4% higher (Monteith 1977). The amount of light that is intercepted during the growing season and the efficiency with which captured light is utilized are important considerations. In general, plants intercept only around 40% of solar radiation, and they utilize 0.3% of that energy. The intercepted amount is determined by the growth stage and leaf area.

1.5.1 Photosynthesis

The photosynthetic process is a valuable physiological tool for evaluating plant responses to environmental stressors, and it facilitates the selection of plants for cultivation under varied environmental conditions (Żuk-Gołaszewska 2008; Chandra et al. 2015). Plant productivity is the rate at which dry matter (tissues, organs or the entire plant after dehydration) is produced and accumulated per unit of time (day, growing season, year) as a result of carbon assimilation (photosynthesis) and dissimilation, namely mitochondrial (dark) respiration or photorespiration. This dynamic concept is expressed by the amount of dry matter produced by plants per unit of area per unit of time (g cm^{-2} h^{-1}) (Pessarakli 2014). In the light of the above definition of plant productivity, Nalborczyk (1996) demonstrated that under optimal conditions, crop yield is determined by the rate of photosynthesis minus the biomass loss resulting from respiration and the quantity of assimilates transferred from plant organs which had developed before the emergence of plant organs that constitute the agricultural yield. In Cannabis plants of two chemotypes (drug and fiber) exposed to UVB radiation (daily dose of 6.7 or 13.4 kJ m^{-1}), the rate of photosynthesis was determined at 25.2 (drug) and 24.2 (fiber) μmol CO_2 m^{-2} s^{-1}, whereas transpiration reached 3.29 and 3.38 mmol H_2O m^{-2} s^{-1}, respectively (Lydon et al. 1987). In a study by Bazzaz et al. (1975), the maximum net photosynthetic rate ranged from 7.89 to 9.72 μmol CO_2 m^{-2} s^{-1}. This parameter was influenced by geographic location and temperature. In Cannabis plants grown in Nepal and Illinois, the rate of photosynthesis increased with a rise in air temperature from 20 °C to 25 °C. In turn, the photosynthetic rate of Cannabis plants grown in Jamaica and Panama increased with a rise in temperature from 20 °C to 30 °C. In a study by Tang et al. (2017), the net photosynthetic rate increased with a rise in leaf nitrogen content up to 31.2 μmol CO_2 m^{-2} s^{-1} at 25 °C. The analyzed parameter increased with a rise in leaf temperature to 25–35 °C, and it decreased when leaf temperature exceeded 35 °C. In high 9-THC yielding varieties of *Cannabis sativa* L., the highest rate of photosynthesis was determined at 19.82 to 25.54 μmol CO_2 m^{-2} s^{-1} (Chandra et al. 2015).

1.5.2 Chlorophyll Content

Nitrogen fertilizer rates should be optimized based on an accurate assessment of plants' nutritional status. Chlorophyll levels during growth and development can be used to determine the nitrogen content of plants, and to predict yields (Żuk-Gołaszewska 2008; Grove and Navarro 2013, Zuk-Golaszewska et al. 2015). In a study by Malceva et al. (2011), *Cannabis sativa* leaves with higher nitrogen content tended to be more abundant in chlorophyll ($R^2 = 0.774$). Exposure to heavy metals decreased the chlorophyll content of leaves by 30% relative to control plants and led to visible leaf chlorosis (Linger et al. 2005). Zielonka et al. (2017) demonstrated that chlorophyll concentration was determined by the growth stage, variety and fertilization treatment. In cannabis var. Białobrzeskie, chlorophyll content was highest at approximately 40–45 SPAD units in early stages of growth in treatments fertilized with sewage sludge and phosphogypsum at 100 kg ha^{-1}. Greater variations in the chlorophyll content of *Cannabis sativa* leaves (38–47 SPAD) were observed at full maturity. In the last stage of plant development, the chlorophyll content of senescing leaves decreased to 11–35 SPAD.

1.5.3 Chlorophyll Fluorescence

Photosynthetic efficiency can also be evaluated by measuring chlorophyll fluorescence emissions. Chlorophyll fluorescence is a robust indicator of photosynthetic capacity and the vitality of the photosynthetic apparatus in plants growing both in nature and under controlled conditions. Measurements of chlorophyll fluorescence support rapid and sensitive analyses of plant responses to disruptions in photosynthesis caused by multiple stressors as well as evaluations of the effectiveness of repair mechanisms and plants' ability to maintain homeostasis under adverse environmental conditions. Stress factors such as cold, drought, herbicide damage, atmospheric pollution and nutrient deficiency decrease the photosynthetic rate, compromise plants' photosynthetic efficiency and increase chlorophyll fluorescence (Kalaji et al. 2017). A chlorophyll fluorescence analysis carried out by Malceva et al. (2011) demonstrated that high rates of nitrogen fertilizer (100 kg N ha^{-1}) stimulated the photosynthetic performance of hemp plants. In a study by Linger et al. (2005), the photosynthetic rate of *Cannabis sativa* plants exposed to different concentrations of cadmium in soil (17 mg kg^{-1} and 72 mg kg^{-1} soil) decreased in treatments with higher Cd content. Similar results were reported by He et al. (2008) in *Oryza sativa* L.

1.5.4 Growth Parameters

The relative growth rate of plants, expressed as the ratio of leaf area to whole plant dry mass, can be interpreted as: (1) an indicator for comparing growth between different species (varieties) under identical environmental conditions or within species (varieties) under different environmental conditions; and (2) as an indicator of dry matter accumulation in the plant, characterizing the size and structure of the plant and its assimilating organs (Pietkiewicz 1985). Canopy structure parameters include: leaf area, leaf growth curve, pathogen infestation, leaf area per unit ground surface area, foliage surface area of a canopy, leaf angle and the degree of canopy shadows. Plant growth analysis can be performed to evaluate the growth rate of plants as well as biomass production and distribution, which are determined by growing conditions and agronomic factors such as mineral fertilization, seeding rate, plant protection, light conditions, water stress, salinity stress and plant response to UVB radiation, expressed with multiple indicators (Żuk-Gołaszewska et al. 2003; Hyer and Gotem 2004). Well-known and reliable growth and structural parameters of plants and canopies are presented in Tables 1.7a and 1.7b.

The above indicators are used for plant growth modeling, including the distribution and translocation of dry matter in plants during ontogenesis. They may also serve as criteria for evaluating the morphological plasticity of biomass-related parameters, stress resistance and adaptability to adverse environmental conditions.

Table 1.7a Plant growth indicators

LAR (Leaf Area Ratio) – the total area of assimilating organs (usually leaves) of a plant *A* divided by the dry mass of the entire plant. It is equal to the product of: • SLA (Specific Leaf Area) – the ratio of leaf area *A* to leaf dry weight W_L • LWR (Leaf Weight Ratio) – the ratio of leaf dry weight W_L to whole plant dry weight *W*. In a study by Żuk-Gołaszewska et al. (2003), which investigated the effect of UVB radiation on agricultural crops, LAR values were determined by plant species rather than UVB radiation. An analysis of common weeds revealed that LAR ranged from 440 to 700 in *Avena fatua*, and from 460 to 580 in *Setaria viridis*, which points to higher or lower environmental adaptability.	$LAR = \frac{A}{W}$
ULR (Unit Leaf Rate) – the rate of photosynthesis per unit area of leaf, the rate of increase in net dry weight of the whole plant (difference between total dry weight production and loss) per unit leaf area over minimum of 24 h. Depending on the rate of photosynthesis, the same leaf area or weight can produce a different amount of biomass, which is related to daytime/nighttime. ULR is positively correlated with the rate of photosynthesis, and negatively correlated with respiration. During plant development, ULR decreases due to leaf senescence. ULR increases with an increase in N content, and it is correlated with the protein content of plants, including the enzyme Rubisco (ribulose-1,5-bisphosphate carboxylase) involved in the dark cycle reactions of photosynthesis. A decrease in soil N content is followed by a decrease in ULR within a few days. Average ULR value in C3 plants is 2–10 g DM per m^2 per day, and it is fivefold higher in C4 plants (Kopcewicz and Lewak 2005).	$ULR = \frac{1}{A} \cdot \frac{dW}{dt}$
RGR (Relative Growth Rate) – used to quantify an increase in whole plant mass over a given period of time. RGR is the only indicator that can also be applied to non-photosynthetic biomass, i.e. to determine the rates of breakdown and utilization of storage compounds during seed germination (Kopcewicz and Lewak 2005).	$RGR = \frac{1}{A} \cdot \frac{dW}{dt}$

1.6 Harvest, Drying and Storage

Hemp harvesting dates are determined mainly by the production profile and differ in plants that are grown for medicinal purposes (inflorescences), fiber (stems) and oil (seeds). Inflorescences are harvested from August to September (Grabowska 2005). Plants grown for medicinal purposes should be harvested manually, stems should be separated from inflorescences which should be dried at a temperature of 40 °C for 15 h to obtain high-quality material. Dried material for pharmaceutical processing should have a moisture content of approximately 8% and should be stored at a temperature of 18–20 °C. Inflorescences should be free of leaves (which are usually removed manually), they should be of high quality and should be certified as normalized medical cannabis. The plants for medicinal use are evaluated for moisture content, health, the presence of heavy metals, pesticide residues and cannabinoid levels (McPartland et al. 2000; Upton et al. 2013).

In conventional production systems, stems and seeds are harvested simultaneously with a combine harvested. After harvest, the remaining straw is baled. Harvest poses the greatest problem in seed production. When seeds mature, hemp stems are already lignified, and they become entwined around machine parts during

Table 1.7b Canopy indicators

LAI (Leaf Area Index) – leaf area of the entire canopy A_c per unit ground surface area *P*. LAI indicates how many times the leaf area contains the surface area occupied by canopy, and it is used to calculate seeding rate. Too high plant density (number of plants per unit area) decreases leaf area. In a study by Daughtry and Walthall (1998), the LAI of cannabis plants was 8. According to Tang (2018), the LAI of hemp plants is determined by N fertilization, and it reached 6.4 at N fertilizer rate of 120 kg ha^{-1}, and 2.3 in the unfertilized treatment. Additionally, long-term water stress enhanced leaf senescence and reduced LAI. Zielonka et al. (2017) reported that the maximal efficiency of the photosynthetic apparatus in *C. sativa*, expressed as LAI, was 3.4 m^{-2}. This physiological parameter was determined by growth stage, agronomic factors (fertilization) and breeding progress (varieties Białobrzeskie, Tygra, Beniko). A close correlation was also found between chlorophyll concentration and LAI (Haboudane et al. 2002). Differences in LAI result from leaf shape and angle, the rate of photosynthesis and the light requirements of plants. In most monocotyledonous plants (cereals, grasses), leaf inclination angle is acute, which facilitates light penetration through the canopy. In shadow-loving plants, LAI may range from 5–8 to even 10.	$LAI = \frac{A_c}{P}$
CGR (Crop Growth Rate) – is a measure of the increase in size, mass or number of crops over a period of time. It is the product of ULR and LAI. An increase in LAI can compensate for a decrease in ULR, thus maintaining CGR at a relatively stable level (Kopcewicz and Lewak 2005).	$CGR = ULR \cdot LAI$

harvest. As a result, harvesting equipment has to be frequently cleaned, and it is often damaged. For this reason, hemp plants grown for fiber should be harvested before seed maturation (Čeh 2018). Harvested hemp seeds should be dried at low temperature to a moisture content of around 9–10%. Fast drying at high temperatures can crack, burn seed and damage seeds, thus compromising their nutritional value and germination rate (Callaway 2004b).

1.7 Raw Material and Uses

Today, monoecious cannabis plants are more widely cultivated due to their higher processing suitability, i.e. higher fiber and seed yields, and medical use (Small and Marcus 2002; Amaducci et al. 2008b). Hemp fiber, located in stem tissues, is characterized by high strength and resistance to putrefactive changes. The length of hem fibers is determined by the technical length of stems (referred to as hemp straw), which is measured from the root crown to the mid-inflorescence. Heuser (1927), who investigated *C. sativa* grown in the temperate zone, found that the cross-sectional shape of fibers was grape-like and deeply wrinkled at the top, hexagonal in the middle and circular at the stem base. The stem is covered with epidermis. Collenchyma, support tissue, is located next to the epidermis (Herse 1979). The next layer is made up of primary and secondary fiber bundles. Highly valuable primary fibers are located in the middle and at the top of hemp stems. Useless secondary fibers, found at the bottom of the stem, are left in the field after harvest. During plant growth and development, secondary fibers form mechanical tissue, which keeps the plant upright and contribute to its stiffness. Individual fibers are 10–25 mm long, forked, with bluntly pointed tips. In a study by Thomsen et al. (2005), hemp fibers contained 73–77% w/w cellulose, 7–9% w/w hemicellulose and 4–6% w/w lignin. Hemp shives contained 48% w/w cellulose, 21–25% w/w hemicellulose and 17–19% w/w lignin. The content of the above components varied across varieties – Felina contained least lignin whereas Futura and Fasamo contained least cellulose. The color of fibers was determined by pretreatment, and it was brighter after wet oxidation and hydrothermal treatment compared with steam explosion.

Hemp fiber constitutes valuable raw material for textile fibers, it has excellent moisture absorption and release properties, air permeability, warmth retention, cold and warm sense, high temperature resistance, insulation, anti-ultraviolet, anti-radiation qualities, anti-mildew and antibacterial properties, and sound-deadening properties. Hemp fiber has been widely used in many fields and products, such as clothing, sails, rope, paper and medical supplies (van der Werf and van Geel 1994; Zhang et al. 2016). Whole plant biomass (20–30 t ha^{-1}) is used in the power industry and sustainable housing. Hemp-lime composite known as hempcrete is newly-discovered construction material (Gołębiewski 2017) obtained by mixing hemp shiv, water, lime–based binder and/or sand in appropriate proportions. Hempcrete is characterized by high strength, therefore it can be used in sustainable architecture for building insulation, including filling walls with wooden or steel construction frames. Hemp composite material is non-flammable and has desirable acoustic properties (Bevan and Woolley 2008). Hempcrete production does not generate any waste, and the material is 100% recyclable. Other advantages of hempcrete include naturalness, environmental potential and a healthy indoor environment in buildings where hemp shiv was used as construction material. Hemp can also be a suitable crop for the bio-economy as it requires low inputs for producing high biomass yields.

Hemp seeds have a high content of fat (25–35%), protein (20–25%), phytic acid, choline, trigonelline, lecithin, chlorophyll, vitamins K and E, iron, calcium, zinc, phosphorus and magnesium (Deferne and Pate 1996; Główczewska-Siedlecka et al. 2016). They have medicinal properties and are used in dietary supplements and herbal preparations. Additionally, Ferenczy (1956) demonstrated the antibiotic properties of cannabis seeds, a factor that may aid its winter survival. Adherent resin on the seed surface, as well as a surrounding mulch of spent cannabis leaves, may serve in this regard.

Dried flower buds, leaf and flower extracts are used in the treatment of many diseases, in particular chronic pain. Medicinal cannabis is generally characterized by a high content of tetrahydrocannabinolic acid (THCA) and a low content of cannabidiolic acid (CBDA). The psychotropic threshold for cannabis plants is generally estimated at 1% THCA. Varieties of *C. sativa* L. have different predominant traits that make them suitable for pharmaceutical applications. They differ in chemical composition which is determined by the proportions of around

100 identified organic chemical compounds – cannabinoids, natural metabolites produced by cannabis plants.

Medical cannabis has analgesic properties, and it alleviates the suffering of patients who experience chronic pain (Murnion 2015). Aizpurua-Olaizola et al. (2016) identified at least 113 different cannabinoids isolated from cannabis plants. Psychoactive Δ9-trans-tetrahydrocannabinol (THC or Δ9-THC) and non-psychoactive cannabidiol (CBD) are the most known and most therapeutically valuable cannabinoids. Cannabinoid extracts with high levels of THC and low levels of CBD are most widely used for therapeutic purposes. Cannabis plants exert psychotropic effects when their THC content approximates 1% (Small and Marcus 2002). Small and Beckstead (1973) identified 3 chemical phenotypes (chemotypes) of cannabis: chemotype I – THC > 0.3% and CBD < 0.5%, chemotype II – predominance of CBD and various concentrations of THC, and chemotype III – marginal levels of THC.

1.8 Markets and Marketing

Due to their multiple uses, *C. sativa* plants play an important role in the economy. An increase in hemp production has been noted recently, mainly in Canada, the USA, China, Australia, and many other countries (www.faostat.fao.org). The number of cannabis farms increased by 58% and the total area under cannabis cultivation increased by 91% in 2012–2016. Hemp production increased by 40% on steep slopes, in the vicinity of public lands and in areas of high environmental sensitivity, including an 80–116% increase in cultivation sites near high-quality habitats for threatened and endangered salmonid species (Van Butsic et al. 2018). The food processing sector is making efforts to minimize the environmental impact of production, reduce the levels of toxic residues, use by-products more efficiently and manufacture high-quality products with desirable nutritional and organoleptic properties.

The market value of hemp-based products is difficult to estimate accurately because it is a nascent industry in terms of regulations, and cannabis plants have a variety of applications. Canada, the largest country in the world to legalize cannabis, has dramatically reshaped the cannabis investment landscape (www.bdsanalytics.com). The market of hemp-based products has been growing rapidly. According to forecasts, legal cannabis revenue is projected to reach $ 23.4 billion in the USA, $ 5.5 billion in Canada and $ 3.1 billion in the rest of the world by 2022 (Fig. 1.2) (Arcview Market Research and BDS Analytics 2018).

By 2022, the European market of CBD extracted from industrial hemp and medicinal cannabis will exceed $ 4.2 billion, whereas the segment of industrial hemp products will reach the value of $ 1.9 billion. *Cannabis sativa* is a niche crop, cultivated on more than 33,000 ha in the European Union (2016). France is the Europe's biggest producer of industrial hemp, and the world leader in hemp-seed production accounting for 59% of the global total (www.escaa.org). In Poland, according to the Statistical Yearbook of Agriculture (2017), crop production in the

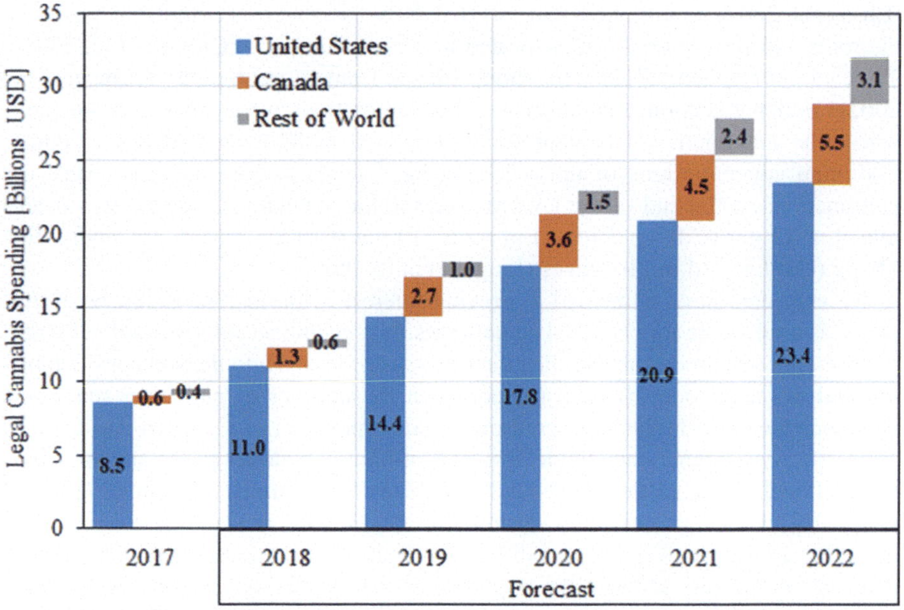

Fig. 1.2 Cannabis spending (Arcview Market Research and BDS Analytics 2018)

area of industrial plants including hemp is financed mostly under the EU Rural Development Program 2014–2020 where subsidies for cultivating fiber hemp reached 368,55 (PLN thous.) in 2016.

1.9 Environmental Performance and Impact

Hemp (*Cannabis sativa* L.) is a cosmopolitan species that is widely distributed around the world. Cannabis plants are eco-friendly crops that do not require extensive pesticide treatments, especially for weed suppression, and they can be cultivated in organic farming systems. Cannabis plants are tall and have a bushy growth habit, therefore they can protect other crops from wind in open spaces. *Cannabis sativa* is a good plant for soil phytoremediation (Linger et al. 2002). The species is capable of purifying soil and air, and it is resistant to environmental pollution. Therefore, cannabis farms can be established on fallow land and in areas degraded by industrial operations. Cannabis plants are characterized by high biomass yield and an extensive root system with numerous side roots. Although this trait is determined genetically, cannabis plants are highly adaptable and capable of absorbing and accumulating heavy metals in shoots and roots (Fitter 1991; Chiatante et al. 2005). Cannabis roots usually occupy the upper soil layer to a depth of 20 cm, but they can penetrate soil to a depth of up to 2 m (Zatta et al. 2012). These properties enhance the accumulation of Cu in cannabis roots and shoots (Angelova et al. 2004; Bona et al. 2007). A study of root morphology and protein expression revealed that *C. sativa* is tolerant to copper

(Elisa et al. 2007). In turn, heavy metals (Cd, Ni and Cr) were accumulated mainly in the roots and were partially translocated to aerial plant parts (Citterio et al. 2003). The cadmium content of cannabis shoots ranged from 14 to 66 μg g^{-1} (Linger et al. 2002). Under greenhouse conditions, Cd concentrations were highest in the roots where the maximum Cd content was determined at 830 mg (Cd) kg^{-1} (d.m.). Cadmium concentrations of up to 72 mg kg^{-1} (soil) had no negative effect on germination. At the end of the growing season, the average Cd content was determined at 42 mg (Cd) kg^{-1} (d.m.) in roots, 20 mg (Cd) kg^{-1} (d.m.) in stems, and 15 mg (Cd) kg^{-1} (d.m.) in leaves (Linger et al. 2005).

Recent years have witnessed a growing interest in the use of *Cannabis sativa* plants as a surrogate for chemical nematicides. In a cucumber farm severely affected by root-knot nematodes, the application of *C. sativa* leaves into the soil considerably reduced the nematode invasion (reduction in the number of galls, egg mass and nematode fecundity) relative to control. The incorporation of *C. sativa* into the soil as an effective organic nematicide against root-knot nematodes poses an alternative to traditional chemical treatments and minimizes environmental pollution (Kayani et al. 2012). Cannabis sativa has several advantages over conventional nematicides, including low cost, wide availability as well as environmental safety. Cannabis extracts significantly reduced the populations of phytopathogenic nematodes, bacteria, fungi, protozoans, insects, mites and weeds. Pure cannabinoids extracted from cannabis demonstrate similar properties (McPartland 1997).

In turn, hempseed oil showed absorbance in the UVC (100–290 nm) and UVB (290–320 nm) ranges and exhibited unique transmittance properties in the UVB and UVA (320–400 nm) spectrum. These findings indicate that hempseed oil has potential for use as a protectant against both UVA (source of oxidative stress in skin) and UVB radiation (Oomah et al. 2002). Hempseed oil containing CBD also exerts an indirect positive impact on the environment because cannabinol extracts have been shown to be helpful in smoking cessation. Hempseed oil contributes to an improvement in air quality by reducing the amount of tobacco smoke containing toxic tar compounds and heavy metals which are detrimental not only to human health, but also to the ecosystem.

Cannabis cultivation is eco-friendly and does not require plant protection agents. During growth and development, cannabis plants are capable of absorbing atmospheric carbon dioxide which is responsible for the greenhouse effect. Crops that absorb carbon dioxide, including cannabis, contribute to a reduction in greenhouse gas levels. A cannabis plantation with an area of 1 hectare is capable of absorbing 2.5 tons of atmospheric CO_2. Cannabis plants are also an abundant source of biomass for the generation of renewable energy, and the energy efficiency of cannabis can exceed that of wood and coal. This alternative source of energy minimizes tree felling and, consequently, prevents ecosystem degradation.

The area under cannabis has increased dynamically due to the numerous potential uses of the p, in particular in medicine. In the United States, the number of cannabis farms increased by 58%, the production of cannabis plants increased by 183%, and the area under cannabis increased by 91% between 2012 and 2016. An increase in the number of sites where cannabis is grown (80%) as well as their size (56% per site) has also contributed to the observed expansion. Cannabis production soared in areas of high environmental sensitivity, and a 80%–116% increase in cannabis cultivation was reported in the proximity of high-quality habitats for threatened and endangered salmonid fish species (van Butsic et al. 2018). However, intensive production and transportation of cannabis is also energy intensive, and it has contributed to air pollution through higher emissions of carbon dioxide, nitrous oxides and airborne fungal spores (Mills 2012). Intensive cannabis cultivation, in particular in small farms, also exerts a negative impact on the local landscape.

The environmental impact of production is an important consideration. The life cycle assessment is a new approach to analyzing the environmental impact of a process or a product. In food production, this technique is used to analyze the authenticity of food products, their place of provenance, safety and nutritional value as well as environmental impacts of production and distribution systems; LCA contributes to improving environmental and economic efficiency in the food supply chain, taking into the account the effects exerted by raw material, production, distribution, packaging, sale and consumption. The method can also be applied to hemp plants which produce significant quantities of useful straw co-products that could act as temporary CO_2 storage (O'Mahony 2011). The LCA carried out by Casas et al. (2005) included the absorption of CO_2 during the growth of hemp plants, which conventionally is excluded from LCAs of biodiesels under the assumption that CO_2 absorption will be balanced out by the corresponding CO_2 emission during fuel combustion.

1.10 Legal Regulations

The production and processing of cannabis is governed by strict laws in many countries. In Poland, a state license is required to establish a cannabis plantation in a delimited area. Only certified or elite seeds can be sown, and growers can supply cannabis only to legitimate contractors who are registered by the voivodeship marshal. Cannabis production is regulated by the provisions of the Act of 5 July 2005 on counteracting drug addiction. In the above act, hemp is defined as fibrous-class *Cannabis sativa* L. plants where the total content of Δ-9-THC, Δ-9-tetrahydrocannabinolic acid and Δ-9-THC-2-carboxylic acid in flower buds and fruit buds containing resin does not exceed 0.20% on a dry matter basis.

Cannabis cultivation is also regulated by the Act of 7 July 2017 amending the provisions of the Act on counteracting drug addiction and the Act on the reimbursement of medicines, foodstuffs intended for particular nutritional uses and medical devices. Pursuant to Article 1 of the Act of 7 July 2017, the provisions of Article 33 the Act of 29 July 2005 on counteracting drug addiction (Journal of Laws, 2017, item 783) have been amended by inserting Articles 33a–33d: "Article 33a: 1. Cannabis plants other than fiber hemp, pharmaceutical extracts and tinctures, extracts from cannabis plants other than fiber hemp as well as cannabis resin other than fiber hemp resin, as stipulated in Annex 1 to the Act, may constitute pharmaceutical raw materials, as stipulated in Article 2, point 40 of the Pharmaceutical Law of 6 September 2001, for the production of prescription drugs, as stipulated in Article

3, section 4, point 1 of the Pharmaceutical Law of 6 September 2001, that may be placed on the market under a license granted by the President of the Office for the Registration of Medicinal Products, Medical Devices and Biocidal Products, hereinafter referred to as the President of the Office. 2. The license stipulated in section 1 hereinabove can be issued or denied, the data and documents which constitute a basis for granting the license can be modified, the term of the license can be prolonged and shortened, and the license can be revoked by a decision of the President of the Office. 3. The license stipulated in section 1 hereinabove shall be issued for a period of 5 years. Article 33b. 1. The application for: 1) granting the license stipulated in Article 33a, section 1 hereinabove, 2) modifying the data that constitute the basis for granting the license, 3) modifying the validity of the license shall be submitted by the responsible entity, as defined by Article 2, point 24 of the Pharmaceutical Law of 6 September 2001, to the President of the Office. 2. The provisions of Article 10, Article 18, Article 23, section 1, Article 29, section 1, Article 30, section 1, Article 31, section 1 and Article 33, section 1 of the Pharmaceutical Law of 2001, and the provisions of Article 36 of the Pharmaceutical Law of 6 September 2001 applicable to the licensing requirements stipulated in Article 20 of the said law shall apply to the application and the license stipulated in section 1 hereinabove. 3. The provisions of chapters 3 and 5 of the Pharmaceutical Law of 6 September 2001 shall apply to the pharmaceutical raw materials stipulated in section 1 hereinabove. 4. Prescription drugs stipulated in Article 33a, section 1 hereinabove shall be classified into schedules, as indicated by Article 23a, section 1, point 4 of the Pharmaceutical Law of 6 September 2001. Prescription drugs indicated in Article 33a, section 1 hereinabove shall not be prescribed by veterinary physicians. Article 33c. 1. The application for the license stipulated in Article 33a, section 1 shall indicate: 1) the name of the pharmaceutical raw material and the active ingredient; 2) packaging size (Journal of Laws, item 1458); 3) the name and permanent address of the responsible entity, as defined by the Pharmaceutical Law of 6 September 2001, submitting the application as well as the data pertaining to the manufacturer or manufacturers if the pharmaceutical raw material is not manufactured by the responsible entity; 4) list of documents attached to the application. 2. A copy of the license stipulated in Article 38, section 1 of the Pharmaceutical Law of 6 September 2001 shall be attached to the application indicated in section 1 hereinabove. 3. The specimen application and the detailed scope of the data and documents that constitute the basis for licensing pharmaceutical raw products for the production of prescription drugs containing cannabis plants other than fiber hemp, pharmaceutical extracts and tinctures, extracts from cannabis plants other than fiber hemp as well as cannabis resin other than fiber hemp resin, as stipulated in Annex 1 to the Act, shall be defined by a regulation of the minister competent for public health, taking into consideration special requirements regarding the proper administration of the drug, the patient's safety and the protection of public health. Article 33d. 1. The production of an active ingredient for the manufacture of pharmaceutical raw materials consisting of cannabis plants other than fiber hemp, pharmaceutical extracts and tinctures, extracts from cannabis plants other than fiber hemp as well as cannabis resin other than fiber hemp resin, as stipulated in Annex 1 to the Act, shall involve the comminution of dried plant parts, physicochemical processes that lead to the production of the above substances, including extraction, and packing into bulk packaging, and it shall conform to the Good Manufacturing Practice for active pharmaceutical ingredients. 2. The provisions of chapter 3a of the Pharmaceutical Law of 6 September 2001 shall apply to the operations stipulated in section 1 hereinabove. 3. The production of pharmaceutical raw materials stipulated in section 1 hereinafter shall involve repackaging of the active ingredient from bulk packaging into primary and secondary packaging that will be distributed to pharmacies, and it shall conform to the Good Manufacturing Practice for active pharmaceutical ingredients. 4. The provisions of chapter 3 of the Pharmaceutical Law of 6 September 2001 shall apply to the operations stipulated in section 3 hereinabove." Article 2. The provisions of Article 6 of the Act of 12 May 2011 on the reimbursement of medicines, foodstuffs intended for particular nutritional uses and medical devices (Journal of Laws, 2016, items 1536 and 1579; Journal of Laws, 2017, item 1200) are hereby amended by inserting section 5a after section 5: "5a. The provisions of section 5 shall not apply to prescription drugs manufactured from pharmaceutical raw materials that have been licensed pursuant to the provisions of Article 33a, section 1 of the Act of 29 July 2005 on counteracting drug addiction (Journal of Laws, 2017, items 783 and 1458)". Article 3. This Act shall come into force 3 months after notification.

In the Netherlands, medicinal cannabis is available on prescription in pharmacies. Research and production of medicinal cannabis are monopolized by Bedrocan International under government supervision (Law on opium 2003). In turn, Canada is a pioneer in medicinal cannabis legislation. The Medical Cannabis Access Program came into force in 1999. Personal cultivation and commercial production of cannabis for medicinal purposes were legalized in 2016. In Canada, licensed producers who hold a valid permit may import cannabis and its derivatives for commercial use.

The next example describes the production and use of *Cannabis sativa* in Colombia. In Colombia, medicinal cannabis had been legally available on prescription already since 1986, but Law 1787/2016, Decree 613/2017 and Resolutions 577, 578 and 579/2017 created a legal framework for the production and sale of cannabis for medicinal purposes. Four types of licenses for the legal production of marijuana can be obtained from the government: 1. License for the manufacture of cannabis derivatives, 2. License for the cultivation of non-psychoactive cannabis, 3. License for the cultivation of psychoactive cannabis, and 4. License for the use of cannabis seeds. These licenses have different modalities, and they can be obtained for research, national use or export. Colombian law sets an example for the legalization of cannabis in Latin American countries such as Ecuador, Peru and Mexico. The newly adopted legal framework will place Colombia at the forefront of cannabis production and research, and it will boost revenues from the export of cannabis derivatives (cannabis flowers may not be exported) (information obtained privately).

1.11 Challenges

Hemp is the crop of the future because it has numerous industrial applications – it can be used as fiber, building material, composite material in the automotive industry, a source of oil and essential oils, animal feed and litter, briquettes and fuel. The medicinal and therapeutic use of cannabis poses a challenge. The yield and pharmaceutical quality of cannabis plants are determined mainly by plant variety and agricultural treatments, in particular sowing date, fertilization, seeding rate and watering. Cannabinoids, including tetrahydrocannabinol (THC), identified in cannabis exert analgestic effects, stimulate appetite and protect neurons. Cannabidiol (CBD) alleviates some symptoms of depression and mood-related disorders, and may help combat alcohol addiction.

Cannabis sativa is an eco-friendly plant species that requires few crop protection chemicals, suppresses weeds and inhibits the growth of selected pests such as the large white and the Colorado potato beetle because limonene and pinene – compounds responsible for its aroma – are potent insect repellents.

There is a need for standards and regulations on the cultivation and medicinal use of cannabis across the EU member states and in the world in order to breed varieties with desirable qualitative parameters and offer alternative treatment options.

Besides, industrial hemp is a promising energy crop that can be converted to bioethanol and biogas, and used as biomass for combustion. In open spaces, hemp plants perform esthetic functions and add diversity to monotonous landscapes.

It appears that future research into hemp cultivation should focus on:

- breeding new varieties that would convert solar energy into biomass more effectively, use water more efficiently, and be more resistant to multiple environmental stressors,
- optimization of agricultural treatments, including their effects on canopy structure, in view of more effective use of substrates (light, solar energy, carbon dioxide and water) to increase photosynthetic productivity and yields,
- modeling canopy structure in different varieties in view of crop yield forecasting and management under changing environmental conditions due to global warming, soil contamination, etc.

References

Agrios G (ed) (1988) Plant pathology, 3rd edn. Academic, New York, 803 p

Aizpurua-Olaizola O, Soydaner U, Öztürke SD, Simsir Y, Navarro P, Etxebarria N, Usobiaga A (2016) Evolution of cannabinoid and terpene content during the growth of *Cannabis sativa* plants from different chemotypes. J Nat Prod 79:324–331. https://doi.org/10.1021/acs.jnatprod.5b00949

Amaducci S, Zatta A, Raffanini M, Venturi G (2008a) Characterisation of hemp (*Cannabis sativa* L.) roots under different growing conditions. Plant Soil 313:227–235. https://doi.org/10.1007/s11104-008-9695-0

Amaducci S, Zatta A, Pelatti F, Venturi G (2008b) Influence of agronomic factors on yield and quality of hemp (*Cannabis sativa* L.) fibre and implication for an innovative production system. Field Crop Res 107:161–169. https://doi.org/10.1016/j.fcr.2008.02.002

Amaducci S, Errani M, Venturi G (2002) Response of hemp to plant population and nitrogen fertilization. Ital J Agron 6:103–111

Angelova VR, Ivanova R, Delibaltova V, Ivanov K (2004) Bio-accumulation and distribution of heavy metals in fibre crops (flax, cotton and hemp). Ind Crop Prod 19:197–205. https://doi.org/10.1016/j.indcrop.2003.10.001

Arcview Market Research and BDS Analytics (2018)

Baldini M, Ferfuia C, Piani B, Sepulcri A, Dorigo G, Zuliani F, Danuso F, Cattivello C (2018) The performance and potentiality of monoecious hemp (*Cannabis sativa* L.) cultivars as a multipurpose crop. Agronomy 8:162–178. https://doi.org/10.3390/agronomy8090162

Bazzaz FA, Dusek D, Seigler DS, Haney AW (1975) Photosynthesis and cannabinoid content of temperate and tropical populations of *Cannabis sativa*. Biochem Syst Ecol 3(1):15–18. https://doi.org/10.1016/0305-1978(75)90036-8

Bevan R, Woolley T (2008) Hemp lime construction: a guide to building with hemp lime. HIS BRE Press, Watford

Bócsa I, Karus M (1998) The cultivation of hemp: botany, varieties, cultivation and harvesting. Hemptech, Sebastopol, p 184

Bócsa I, Máthé P, Hangyel L (1997) Effect of nitrogen on tetrahydrocannabinol (THC) content in hemp (*Cannabis sativa* L.) leaves at different positions. J Int Hemp Assoc 4:78–79

Bona E, Marsano F, Cavaletto M, Berta G (2007) Proteomic characterization of copper stress response in *Cannabis sativa* roots. Proteomics 7:1121–1130. https://doi.org/10.1002/pmic.200600712

Burczyk H, Grabowska L, Strybe M, Konczewicz W (2009) Effect of sowing density and date of harvest on yields of industrial hemp. J Nat Fibers 6:204–218. https://doi.org/10.1080/15440470902972588

Callaway JC (2004a) Hemp seed production in Finland. J Ind Hemp 9:97–103. https://doi.org/10.1300/J237v09n0111

Callaway JC (2004b) Hemp seed as a nutritional resource: an overview. Euphytica 140:65–72

Campiglia E, Radicetti E, Mancinelli R (2017) Plant density and nitrogen fertilization affect agronomic performance of industrial hemp (*Cannabis sativa* L.) in Mediterranean environment. Ind Crop Prod 100:246–254. https://doi.org/10.1016/j.indcrop.2017.02.022

Casas XA, Rieradevall J, Pons I (2005) Environmental analysis of the energy use of hemp analysis of the comparative life cycle: diesel oil vs. hemp diesel. Int J Agric Res Gover Ecol 4:133–139

Čeh B (2018) Hemp in Slovenia – the area, varieties and its response to some agrotechnical arrangements. Agric Food 6:219–226

Chandra LS, Lata H, Khan IA, ElSohly MA (2011) Temperature response of photosynthesis in different drug and fiber varieties of *Cannabis sativa* L. Physiol Mol Biol Plants 17:297–303. https://doi.org/10.1007/s12298-011-0068-4

Chandra LS, Lata H, Mehmedic Z, Khan IA, ElSohly MA (2015) Light dependence of photosynthesis and water vapor exchange characteristics in different high 9-THC yielding varieties of *Cannabis sativa* L. J Appl Res Med Aromat Plants 2:39–47. https://doi.org/10.1016/j.jarmap.2015.03.002

Chiatante D, Di Iorio A, Scippa GS (2005) Root responses of Quercus ilex L. seedlings to drought and fire. Plant Biosyst 138:124–139. https://doi.org/10.1080/11263500500160591

Citterio S, Santagostino A, Fumagalli P, Prato N, Ranalli P, Sgorbati S (2003) Heavy metal tolerance and accumulation of Cd, Cr and Ni by *Cannabis sativa* L. Plant Soil 256:243–252

Clarke RC (1977) The botany and ecology of Cannabis. Pods Press, Ben Lomond

Coffman CB, Gentner WA (1975) Cannabinoid profile and elemental uptake of *Cannabis sativa* L. as influenced by soil characteristics. Agron J 67:491–497

Cosentino SL, Riggi E, Testa G, Scordia D, Copani V (2013) Evaluation of European developed fibre hemp genotypes (*Cannabis sativa* L.) in semi-arid Mediterranean environment. Ind Crop Prod 50:312–324. https://doi.org/10.1016/j.indcrop.2013.07.059

Datwyler SL, Weiblen GD (2006) Genetic variation in hemp and marijuana (*Cannabis sativa* L.) according to amplified fragment length polymorphisms. J Forensic Sci 51:371–375. https://doi.org/10.1111/j.1556-4029.2006.00061.x

Daughtry CST, Walthall CL (1998) Spectral discrimination of *Cannabis sativa* L. leaves and canopies. Remote Sens Environ 64:192–201. https://doi.org/10.1016/S0034-4257(98)00002-9

De Candolle A (1884) The origin of cultivated Plants. K. Paul, London

De Meijer EPM, van der Kamp HJ, van Eeuwijk FA (1992) Characterisation of *Cannabis* accessions with regard to cannabinoid content in relation to other plant characters. Euphytica 62:187–200

De Meijer EPM, Keizer LCP (1994) Variation of *Cannabis* for phenological development and stem elongation in relation to stem production. Field Crop Res 38:37–46

Deferne JL, Pate DW (1996) Hemp seed oil: a source of valuable essential fatty acids. J Int Hemp Assoc 3:4–7. https://doi.org/10.1007/s11104-008-9695-0

Elisa B, Marsano F, Cavaletto M, Berta G (2007) Copper stress in *Cannabis sativa* roots: morphological and proteomic analysis. Caryologia 60:96–101

Faeti V, Mandolino G, Ranalli P (1996) Genetic diversity of *Cannabis sativa* germplasm based on RAPD markers. Plant Breed 115(367):370. https://doi.org/10.1111/j.1439-0523.1996.tb00935.x

Ferenczy L (1956) Antibacterial substances in seeds of *Cannabis*. Nature 178:639–640

Ferri F (1961) Sensibilitá di Sclerotium rolfsii a vari funghicidi. Phytopathol Mediterr 3:139–140

Finnan J, Burke B (2013) Potassium fertilization of hemp (*Cannabis sativa*). Ind Crop Prod 41:419–422. https://doi.org/10.1016/j.indcrop.2012.04.055

Fitter AH (1991) Characteristics and functions of root systems. In: Waisel Y, Eshel A, Kafkafi U (eds) Plant roots. The hidden half. Marcel Dekker, New York, pp 3–24

Fryxell AP (1976) A nomenclator of Gossypium. The botanical name of cotton. Technical Bulletin 1491, U.S. Department of Agriculture, Washington, DC, March 1976, 114 p

García-Tejero F, Durán-Zuazo VH, Pérez-Álvarez R, Hernández A, Casano SM, Morón M, Muriel-Fernández JL (2014) Impact of plant density and irrigation on yield of hemp (*Cannabis sativa* L.) in a Mediterranean semi-arid environment. J Agric Sci Technol 16:887–895

Gigliano GS (2001) *Cannabis sativa* L. – botanical problems and molecular approaches in forensic investigation. Forensic Sci Rev 13:1–17

Główczewska-Siedlecka E, Madra-Gackowska K, Nowacka K, Kędziora-Kornatowska K (2016) Konopie siewne i ich zastosowanie ze współczesnej geriatrii. (*Cannabis sativa* and their use of modern Geriatrics). J Edu Health Sport 6:34–39

Gołębiewski M (2017) Architektura i renesans konopi. (Architecture and renaissance of hemp). ZNUV 53:154–165

Gorchs G, Lloveras J, Serrano L, Cela S (2017) Hemp yields and its rotation effects on wheat under rainfed Mediterranean conditions. Agron J 109:1551–1560. https://doi.org/10.2134/agronj2016.11.0676

Grabowska L (2005) Perspektywy uprawy konopi przemysłowych w Polsce. Rola infrastruktury i techniki w zrównoważonym rozwoju rolnictwa (Prospects for industrial hemp cultivation in Poland. Role of infrastructure and technology in sustainable agriculture). IBMiER, Warszawa

Grabowska L, Koziara W (2001) Wpływ temperatury i opadów na plonowanie konopi włóknistych odm. Białobrzeskie. (Effect on temperature and precipitation on yields of the *Cannabis sativa* variety Białobrzeskie). Natural Fibres 45:32–46

Grove JH, Navarro MM (2013) The problem is not N deficiency: active canopy sensors and chlorophyll meters detect P stress in corn and soybean. Precision agriculture '13, 9th European conference on Precision Agriculture, Lleida, Catalonia, Spain 7–11 July, pp 137–144. ISBN:978-90-8686-224-5

Haboudane D, Miller JR, Tremblay N, Zarco-Tejada PJ, Dextraze L (2002) Integrated narrow-band vegetation indices for prediction of crop chlorophyll content for application to precision agriculture. Remote Sens Environ 81:416–426

Hall J, Bhattarai SP, Midmore DJ (2014) Effect of industrial hemp (*Cannabis sativa* L.) planting density on weed suppression, crop growth, physiological responses, and fibre yield in the subtropics. Renewable Biores 2:1–7

He JY, Ren YF, Zhu C, Yan YP, Jiang DA (2008) Effect of Cd on growth, photosynthetic gas exchange, and chlorophyll fluorescence of wild and Cd-sensitive mutant rice. Photosynthetica 46:466–470

Heller K, Adamczewski K, Baraniecki P (2007) Zbiorowiska chwastów segetalnych w uprawach konopi (Segetal weed communities in hemp stands). Pam Puł 145:121–128

Hemphill JK, Turner JC, Mahlberg PG (1980) Cannabinoid content of individual plant organs from different geographical strains of *Cannabis sativa* L. (*Cannabinaceae*). J Nat Prod 43:112–122

Hennink S (1994) Optimization of breeding for agronomic traits in fiber hemp (*Cannabis sativa* L.) by study of parent off spring relationships. Euphytica 78:69–76

Hennink S (1997) EU regulations on hemp cultivation. J Int Hemp Assoc 4:38–39

Herse J (ed) (1979) Szczegółowa uprawa roślin (Crop production). PWN, Warszawa, 609 p

Heslop-Harrison J, Heslop-Harrison Y (1958) Studies on flowering-plant growth and organogenesis III. Leaf shape changes associated with flowering and sex differentiation in Cannabis sativa. Proc Roy Irish Acad Sect B 59:257–283

Heuser O (1927) Hanfund hartfaser. Julius Springer Verlag, Berlin, p 156

Hill RJ (1983) Marijuana, *Cannabis sativa* L. regulatory horticulture. Weed Circular 5(9):57–66

Hoffman W (1961) *Cannabis sativa* L. Hdbch d Pflanzenz V:204–264

Hyer EJ, Gotem SJ (2004) Comparison and sensitivity analysis of instruments and radiometric methods for LAI estimation: assessments from a Boral forest site. Agric Forest Meteo 122:15–174

Institute of Plant Protection (2019) National Research Institute, Poznań, Poland

Iványi I, Izsáki Z (1996) Effect of nutrient supply on nutrient uptake of fibre hemp during the growing season. Növénytermelés 45:181–193

Iványi I, Izsáki Z, der Werf HMG V (1997) Influence of nitrogen supply and P and K levels of the soil on dry matter and nutrient accumulation of fibre hemp (*Cannabis sativa* L.). J Int Hemp Assoc 4:84–89

Jankauskiene Z, Gruzdeviene E (2010) Evaluation of *Cannabis sativa* cultivars in Lithuania. Zemdirbyste-Agriculture 97:87–93

Kalaji MH, Goltsev VN, Żuk-Gołaszewska K, Zivcak M, Brestic M (2017) Chlorophyll fluorescence: understanding crop performance – basics and applications. CRC Press/Taylor & Francis Group, LLC. 222. ISBN:9781498764490

Kayani MZ, Mukhtar T, Hussain MA (2012) Evaluation of nematicidal effects of *Cannabis sativa* L. and *Zanthoxylum alatum* Roxb. against root-knot nematodes, Meloidogyne incognita. Crop Prot 39:52–56. https://doi.org/10.1016/j.cropro.2012.04.005

Kilanowski W (1974) Rośliny włókniste. (Fibre crops). Państwowe Wydawnictwo Rolnicze i Leśne, Warszawa, pp 147–184

Kopcewicz J, Lewak S (eds) (2005) Fizjologia roślin (Plant Physiology). Wyd. Naukowe PWN 806. ISBN:83-01-14549-8

Li HL (1974) An archaeological and historical account of *Cannabis* in China. Econ Bot 28:437–448

Linger P, Müssig J, Fischer H, Kobert J (2002) Industrial hemp (*Cannabis sativa* L.) growing on heavy metal contaminated soil: fibre quality and phytoremediation potential. Ind Crop Prod 16:33–42. https://doi.org/10.1016/S0926-6690(02)00005-5

Linger PA, Ostwald A, Haensler J (2005) *Cannabis sativa* L. growing on heavy metal contaminated soil: growth, cadmium uptake and photosynthesis. Biol Plant 49:567–576

Lisson S, Mendham N (1998) Response of fiber hemp (*Cannabis sativa* L.) to varying irrigation regimes. J Int Hemp Assoc 5:9–15

Lydon J, Teramura AH, Ffman CB (1987) UV-B radiation effects on photosynthesis, growth and cannabinoid production of two *Cannabis sativa* chemotypes. Photochem Photobiol 46:201–206

Main Inspectorate of Plant Health and Seed Inspection (2019) National Plant Protection Organization, Warszawa, Poland (https://piorin.gov.pl/en/about-inspection/)

Malceva M, Vikmane M, Stramkale V (2011) Changes of photosynthesis-related parameters and productivity of *Cannabis sativa* under different nitrogen supply. Environ Exp Bot 9:61–69

McPartland JM (1996) A review of *Cannabis* diseases. J Int Hemp Assoc 3:19–23

McPartland JM (1997) Cannabis as repellent and pesticide. J Int Hemp Assoc 4:87–92

McPartland JM, Clarke RC, Watson DP (2000) Hemp diseases and pests: management and biological control. CABI, Wallingford, p 251

McPartland JM, Guy GW (2017) Models of cannabis taxonomy, cultural bias, and conflicts between scientific and vernacular names. Bot Rev 83:327–381. https://doi.org/10.1007/s12229-017-9187-0

Mediavilla V, Jonquera M, Schmid-Slembrouck I, Soldati A (1998) Decimal code for growth stages of hemp (*Cannabis sativa* L.). J Int Hemp Assoc 5:68–74

Meier C, Mediavilla V (1998) Factors influencing the yield and the quality of hemp (*Cannabis sativa* L.) essential oil. J Int Hemp Assoc 5:6–20

Mills E (2012) The carbon footprint of indoor cannabis production. Energy Policy 46:48–67. https://doi.org/10.1016/j.enpol.2012.03.023

Mishchenko S, Mokher J, Laiko I, Burbulis N, Kyrychenko H, Dudukova S (2017) Phenological growth stages of hemp (*Cannabis sativa* L.): codification and description according to the BBCH scale. Žemės Ūkio Mokslai 24:31–36

Monteith JL (1977) Climate and efficiency of crop production in Britain. Philos Trans R Soc Lond B 281:277–294

Murnion B (2015) Medicinal cannabis. Aust Prescr 38:212–215

Nalborczyk E (1996) Dobór i wykorzystanie nowoczesnej aparatury kontrolno-pomiarowej w doświadczalnictwie polowym. Cz. I. Oznaczanie wymiany gazowej, struktury przestrzennej i bilansu energii fotosyntetycznie czynnej radiacji roślin w łanie (Selection and use of modern control and measurement equipment in field experiments. Part I. Determination of gas exchange parameters, spatial structure and energy balance of photosynthetically active radiation in crop stands). Zesz Probl Postep Nauk Rol 447:81–90

O'Mahony SJ (2011) Life cycle assessment of biodiesel from *Cannabis sativa* L. for transport fuel in the UK. A thesis submitted to the school of environmental sciences at the University of East Anglia for the degree of Doctor of Philosophy, p 269

Oomah BD, Busson M, Godfrey DV, Drover JC (2002) Characteristics of hemp (*Cannabis sativa* L.) seed oil. Food Chem 76:33–43. https://doi.org/10.1016/S0308.8146(01)00245

Papastylianou P, Kakabouki I, Travlos I (2018) Effect of nitrogen fertilization on growth and yield of industrial hemp (*Cannabis sativa* L.). Not Bot Horti Agrobo 46:197–201. https://doi.org/10.15835/nbha46110862

Pate DW (1994) Chemical ecology of *Cannabis*. J Int Hemp Assoc 2:32–37

Pessarakli M (ed) (2014) Handbook plant and crop physiology. CRC Press/Taylor and Francis Group, Boca Raton/London/New York, p 993

Pietkiewicz S (1985) Metodyka prac doświadczalnych i technika obliczeń we wskaźnikowej analizie wzrostu roślin (Methodology of experimental research and calculation technique in the index analysis of plant growth). Wiad Bot 2:111–126

Research Center for Cultivar Testing (COBORU) (2017) Słupia Wielka, Poland

Sacilik K, Ozturk R, Keskin R (2003) Some physical properties of hemp seed. Biosyst Eng 86:191–198. https://doi.org/10.1016/S1537-5110(03)00130-2

Sarsenbaev KN, Kozhamzharova LS, Baytelieva AM (2013) Influence high temperature, drought and long vegetation period on phenology and seed productivity European hemp cultivars in Moinkum Desert. World Appl Sci J 23:638–643. https://doi.org/10.5829/idosi.wasj.2013.23.05.13095

Schultes RE (1970) Random thoughts and queries on the botany of Cannabis. In: Joyce CRB, Curry SH (eds) The botany and chemistry of Cannabis. J.A.Churchill, London, pp 11–38
Sengloung T, Kaveeta L, Nanakorn W (2009) Effect of sowing date on growth and development of Thai hemp (*Cannabis sativa* L.). Kasetsart J (Nat Sci) 43:423–431
Sharma GK (1979) A botanical survey of cannabis in the Himalayas. J Bombay Nat Hist Soc 76:17–20
Small E, Beckstead HD (1973) Common cannabinoid phenotypes in 350 stocks of Cannabis. Lloydia 36:144–165
Small E, Cronquist A (1976) A practical and natural taxonomy for *Cannabis*. Int Assoc Plant Taxon (IAPT) 25:405–435. https://doi.org/10.2307/1220524
Small E, Marcus D (2002) Hemp – a new crop with new uses for North America. In: Janicj J, Whipkey A (eds) Trends in new crops and new uses. ASHS Press, Alexandria, pp 284–326
Stafecka I, Stramkale V, Stramkalis A, Kroica I, Ivanovs S (2016) Impact of the agro- environmental factors on the seed yield and yields component productivity of Latvian original hemp. J Res Appl Agric Eng 61:164–167
Statistical Yearbook of Agriculture (2017) Poland, Warsaw
Stearn WT (1970) The Cannabis plant: botanical characteristics. In: CRB J, Curry SH (eds) The botany and chemistry of Cannabis. Churchill J, Churchill A, London, pp 1–10
Struik PC, Amaducci S, Bullard MJ, Stutterheim NC, Venturi G, Cromack HTH (2000) Agronomy of fibre hemp (*Cannabis sativa* L.) in Europe. Ind Crop Prod 11:107–118. https://doi.org/10.1016/S0926-6690(99)00048-5
Tang K (2018) Agronomy and photosynthesis physiology of hemp (*Cannabis sativa* L.). PhD thesis
Tang K, Struik PC, Amaducci S, Stomph TJ, Yin X (2017) Hemp (*Cannabis sativa* L.) leaf photosynthesis in relation to nitrogen content and temperature: implications for hemp as a bio-economically sustainable crop. GCB Bioenergy 9:1573–1587. https://doi.org/10.1111/gcbb.12451
Tang K, Fracasso A, Struik PC, Yin X, Amaducci S (2018) Water- and nitrogen-use efficiencies of hemp (*Cannabis sativa* L.) based on whole-canopy measurements and modeling. Front Plant Sci 9:951. https://doi.org/10.3389/fpls.2018.00951
Thomsen AB, Rasmussen S, Bohn V, Nielsen KV, Thygesen A (2005) Hemp raw materials: the effect of cultivar, growth conditions and pretreatment on the chemical composition of the fibres. Riso-R-1507(EN) 30. ISBN:87-550-3419-5
Upton R, Craker L, ElSohly M, Romm A, Russo E, Sexton M (2013) American herbal pharmacopeia: Cannabis inflorescence: Cannabis spp.: standards of identity, analysis, and quality control. American Herbal Pharmacopoeia, Scott's Valley, p 63
Butsic V, Brenner JC (2016) Cannabis (*Cannabis sativa* or *C. indica*) agriculture and the environment: a systematic, spatially-explicit survey and potential impacts. Environ Res Lett 11:1–11. https://doi.org/10.1088/1748-9326/11/4/044023
Butsic V, Carah JK, Baumann M, Stephens C, Brenner JC (2018) The emergence of cannabis agriculture frontiers as environmental threats. Environ Res Lett 13:124017. https://doi.org/10.1088/1748-9326/aaeade
Van der Werf HMG, Van Geel WCA (1994) Fiber hemp as a raw material for paper. Akkerbouwn en de Groenteteelt in de Vollegrond 177:1–68
Van der Werf HMG, van Geel WCA, Wijlhuizen M (1995a) Agronomic research on hemp (*Cannabis sativa* L.) in the Netherlands, 1987–1993. J Int Hemp Assoc 2:14–17
Van der Werf HMG, van Geel WCA, van Gils LJC, Haverkort AJ (1995b) Nitrogen fertilization and row width affect self-thinning and productivity of fibre hemp (*Cannabis sativa* L.). Field Crop Res 42:27–37. https://doi.org/10.1016/0378-4290(95)00017-K
Vavilov NI (1926) Studies on the origin of cultivated plants. Biull Appl Bot 16:139–248
Zadoks JC, Chang TT, Konzak CF (1974) A decimal code for the growth stages cereals. Weed Res 14:415–421
Zatta A, Monti A, Venturi G (2012) Eighty years of studies on industrial hemp in the Po Valley (1930–2010). J Nat Fibers 9:180–196. https://doi.org/10.1080/15440478.2012.706439

Zhang H, Zhong Z, Feng L (2016) Advances in the performance and application of hemp fiber. Int J Simul Syst Sci Technol 17. https://doi.org/10.5013/IJSSST.a.17.09.18

Zielonka D, Nierebiński M, Kalaji HM, Augustynowicz J, Predecka A, Russel S (2017) Efficiency of the photosynthetic apparatus in *Cannabis sativa* L. fertilized with sludge from a wastewater treatment plant and with phosphogypsum. Ecol Quest 28:55–61. https://doi.org/10.12775/EQ.2017.039

Zuk-Golaszewska K, Kurowski T, Załuski D, Sadowska M, Golaszewski J (2015) Physio-agronomic performance of spring cultivars *T. aestivum* and *T. spelta* grown in organic farming system. Int J Plant Prod 9:211–236

Żuk-Gołaszewska K (2008) Produkcyjność i produktywność jęczmienia jarego (*Hordeum vulgare* L.) uprawianego w różnych warunkach agrotechniki. (Productivity and productiveness of spring barley (*Hordeum vulgare* L.) cultivated in different agrotechnical conditions). Dissertations and Monographs, Wyd. UWM Olsztyn 136:1–110

Żuk-Gołaszewska K, Gołaszewski J (2018) *Cannabis sativa* L. – cultivation and quality of raw material. J Elem 23:971–984. https://doi.org/10.5601/jelem.2017.22

Żuk-Gołaszewska K, Upadhyaya MK, Gołaszewski J (2003) The effect of UV-B radiation on plant growth and development. Plant Soil Environ 49:135–140

Chapter 2
Traditional and New Applications of Hemp

Grégorio Crini, Eric Lichtfouse, Gilles Chanet, and Nadia Morin-Crini

Abstract Recent trends in natural resources, energy conservation, biomass conversion to chemicals, bioproducts and biofuels have renewed the interest on hemp as a new low-cost, sustainable, ecological, biodegradable, recyclable, and multi-purpose material. Hemp-based materials are indeed suitable substitutes for many fossil-based materials and applications.

The hemp plant *Cannabis sativa* Linn, referring to industrial hemp, is a high-yielding annual industrial crop grown, for the fibers from hemp stalk and for the oil from hemp seeds. Although hemp is a niche crop, hemp production is currently undergoing a renaissance. More than 30 countries grow hemp, with China being the largest hemp producing and exporting country. Europe and Canada are also important actors in the global hemp market.

Hemp has many better properties than other plant species, such as superior fiber length, strength and absorbency, interesting thermal and acoustic properties, excellent oil quality for both industrial and feed uses, environmental and economic benefits, and a myriad of applications. For millenia, hemp has been a source of fibers, proteins, and oilseed used worldwide to produce a variety of consumer products for the four fundamental human needs, i.e. food, shelter, clothes and energy.

G. Crini (✉)
Chrono-environnement, UMR 6249, Université Bourgogne Franche-Comté, Besançon, France
e-mail: gregorio.crini@univ-fcomte.fr

E. Lichtfouse
Aix-Marseille Univ, CNRS, IRD, INRAE, Coll France, CEREGE, Aix-en-Provence, France

G. Chanet
Eurochanvre, Arc-les-Gray, France
e-mail: gilles.chanet@interval.coop

N. Morin-Crini
Laboratoire Chrono-environnement, UMR 6249, UFR Sciences et Techniques, Université Bourgogne Franche-Comté, Besançon, France
e-mail: nadia.crini@univ-fcomte.fr

G. Crini, E. Lichtfouse (eds.), *Sustainable Agriculture Reviews 42*,
Sustainable Agriculture Reviews 42, https://doi.org/10.1007/978-3-030-41384-2_2

Traditionally, hemp as fiber plant has been used in the production of apparels, fabrics, papers, cordages, and building materials. For instance, hemp was found in cloth unearthed from the tomb of Pharaoh Alchanaten dating back to 1200 B.C. and the Guttenberg Bible was printed on hemp paper in 1456. The hurds, as waste byproduct of fiber production, were used for bedding of animals, the seeds to human nutrition, e.g. as flour, and the oil for a wide range of purposes, from cooking to cosmetics. Hemp has also been an important crop throughout human history for medicine, e.g. to treat various disorders for thousands of years in traditional Chinese medicine.

Other more recent applications include materials for insulation and furniture, automotive composites for interior applications and motor vehicle parts, bioplastics, jewelry and fashion sector, animal feed, animal bedding, e.g. mulch for horses and kitty litter, and energy and fuel production. Foods containing hemp seed and oil are currently marketed in worldwide for both animal and human nutrition. They also find applications in beverages, e.g. in beer and wine industries, and in neutraceutical products. Hemp oil is also used for cosmetics and personal care items, paints, printing inks, detergents, and solvents. It is estimated that the global market for hemp consists of more than 25,000 products.

Currently, the construction and insulation sector, paper and textile industries, and food and nutrition domains are the main markets while the cosmetics and automotive sector are growing markets. Innovative applications, e.g. in the medical and therapeutic domains, cosmeceuticals, phytoremediation, acoustic domain, wastewater treatment, biofuels, bio-pesticides and biotechnology, opens new challenges. Hemp is also the object of numerous fundamental studies.

This chapter provides an overview of the traditional and new uses of industrial hemp.

Keywords Hemp applications · Textiles · Paper · Building materials · Food · Beverages · Cosmetics · Biocomposites · Energy production · Biofuel · Environmental uses

2.1 Introduction

Hemp is a dicotyledonous plant from the order of Rosales, from the family of Cannabaceae, genus *Cannabis*, like marijuana (Bouloc 2013). Hemp originated from Asia and is considered one of the oldest domesticated crops known to man (De Candolle 1883; Vavilov 1992; Lu and Clarke 1995; Amaducci and Gusovious 2010; Allegret 2013; Amaducci et al. 2015).

Hemp-based products such as bits of hemp fabric have been discovered in tombs dating back to 8000 B.C. (Lu and Clarke 1995; Bouloc 2013). Hemp was also cultivated 4500 years ago in China for fiber, seeds and oil production for multiple applications, e.g. cords imprinted on ancient pottery shards containing hemp residues and piece of paper made from hemp have been discovered (Vavilov 1992;

Roulac 1997; Ranalli 1999; Bouloc 2013). According to ancient Chinese writings, extracts of hemp were also used to treat a wide range of diseases. In Ancient Egypt, the pharaohs also used hemp, e.g. hemp was found in cloth unearthed from the tomb of Pharaoh Alchanaten dating back to 1200 B.C.

In the 600 s, Germans and Vikings made paper, sails and rope from hemp. The Guttenberg Bible was printed on hemp paper in 1456 (Lu and Clarke 1995). In sixteenth to eighteenth century, hemp was a major fiber crop in Europe, Russia and North America (Vavilov 1992; Lu and Clarke 1995; Gibson 2006; Bouloc 2013).

President George Washington grew hemp (Fike 2016) and Americans were legally bound to grow hemp during the Colonial Era and Early Republic (source: Hemp Industries Association). In Colonial days, farmers indeed grew hemp and processed it to make paper, braid ropes, sew clothing, and manufacture canvas sacks. The first American flag was made from it and the Declaration of Independence drafted on hemp paper in 1776. Pionners used hemp covered wagons (Fike 2016).

For thousands of years, hemp fibers were used for clothing and shoes, cordages, carpets and tarps, maritime ropes, sails and nets, and paper production, the hurds for bedding of animals and as straw for agricultural purposes and other uses, e.g. as waste to be burned and more recently as fuel to power farm equipment, the seeds as nutrition, and the oils for a wide range of purposes, from cooking to personal care. The leaves were also used for mulch, composting and animal bedding (De Candolle 1883; Vavilov 1992; Roulac 1997; Ranalli 1999; Ranalli and Venturi 2004; Amaducci 2005; Gibson 2006; Milanovic et al. 2012; Allegret 2013; Bouloc 2013).

In the middle of nineteenth century, hemp cultivation decreased with the disappearance of the sailing navy and competition from other natural fibers such as cotton and jute for textile applications, and later due to intensive development of synthetic fibers (Small and Marcus 2002; Ranalli and Venturi 2004; Milanovic et al. 2012). In the 1930s, cultivation was forbidden in most Western countries and in North America due to the fact that both hemp and marijuana come from the same genus and species plant, and this introduced a lot of confusion and social, political and moral controversies (Amaducci 2005; Holler et al. 2008; Bouloc 2013; Sawler et al. 2015; Cherney and Small 2016; Chandra et al. 2017; Das et al. 2017; Johnson 2018).

Indeed, hemp, i.e. plant species *Cannabis sativa* L., became a controversial crop due to its genetic closeness to THC-producing plants. THC, delta-9-tetrahydrocannabinol, is the chemical most responsible for the phsychoactive properties in marijuana. THC is a psychoactive ingredient present in all hemp varieties to some extent. Marijuana, so-called medical cannabis, contains about 10–30% of THC while hemp refers to the non-psychoactive varieties of *Cannabis sativa* L. (Bouloc 2013; Sawler et al. 2015; Pacaphol and Aht-Ong 2017; Żuk-Gołaszewska and Gołaszewski 2018). Hemp, referred to industrial hemp, has less than 0.2–0.3% THC but contains high levels of cannabidiol (CBD). However, industrial hemp is often incorrectly associated with hemp for narcotics.

Schultes wrote: "*Hemp is a green, very abundant and ubiquitous plant, economically valuable, a versatile and multi-purpose product, possibly dangerous and certainly in many ways mysterious*" (Schultes 1970). This definition is pertinent in describing the valuable but contradictory nature of this industrial crop.

The 1990s marked the renewal of hemp cultivation from an agricultural, industrial and scientific point of view throughout the world (Bócsa and Karus 1998; Small and Marcus 2002; Ranalli and Venturi 2004; Sponner et al. 2005; Kozlowski et al. 2005; Figueiredo et al. 2010; Thomas et al. 2011; Fike 2016). Indeed, growing interests in the commercial cultivation of hemp and other "forgotten fibers" in Europe and North America have emerged since the 1990s, mainly due to the increasing consideration of natural resources, energy conservation, and biomass conversion to bioproducts and biofuels (Roulac 1997; Gibson 2006; Kostić et al. 2003; Ranalli and Venturi 2004; Sponner et al. 2005; Kozlowski et al. 2005; Figueiredo et al. 2010; Thomas et al. 2011; Faruk et al. 2012).

Since 1992, France, The Netherlands, England, Spain and Germany have passed legislation allowing for the commercial cultivation of low-THC hemp. Two years later, Canada also proposed regulations for hemp cultivation, about 60 years after its prohibition.

Hemp is then considered as a valuable crop with particular agronomic characteristics that provided raw materials, i.e. fibers and oil, suitable for multiple potential applications. It quickly found concrete industrial applications, e.g. hempcrete was becoming a viable alternative to traditional building materials, particularly for insulation, panels and roofs (Small and Marcus 2002; Amaducci 2005). Using environmental-friendly and sustainable hemp-based materials for building insulation and construction was a pertinent approach to address reduction in carbon dioxide emission.

To build cars, processing processes allowed hemp fibers to be incorporated into biocomposites that are lighter than steel and much stronger. Another valuable field was the manufacture of biodegradable and non-toxic plastics based on hemp, referred as bioplastics. Indeed, industrial hemp has become an attractive biomass for bioplastic production although this material is typically resistant to enzymatic hydrolysis and direct fermentation by microorganism, due to the complex interactions of cellulose, hemicellulose lignin, and pectin.

The hemp oil was also used for food uses and for applications in cosmetics and personal care. All these areas continue to be exploited today and there are also other very active and innovative industrial markets for hemp-based materials and products (Fig. 2.1).

Currently, more than 30 nations grow industrial hemp as an agricultural commodity with high potential (Food and Agriculture Organization of the United Nations, www.faostat.fao.org). This plant is being mainly cultivated in China, the largest supplier in the world, in North Korea, Chile, Russia, Canada and Europe (Bouloc 2013; Amaducci et al. 2015; Salentijn et al. 2015; Żuk-Gołaszewska and Gołaszewski 2018). France is the top European producer of hemp with an area of cultivation around 12,000 ha (Morin-Crini et al. 2018, 2019). Hemp cultivation and hemp industry are also important in Lithuania, Germany, Italy, Poland, The Netherlands, Romania, Estonia, Hungary, Spain, and England (Żuk-Gołaszewska and Gołaszewski 2018).

Its production is however strictly controlled under existing National and European laws. In most European countries, the current upper legal limit for cultivation of industrial hemp for fiber and seeds production is 0.2% THC on dry basis (Russo and Reggiani 2013; Salentijn et al. 2015; Frassinetti et al. 2018). This

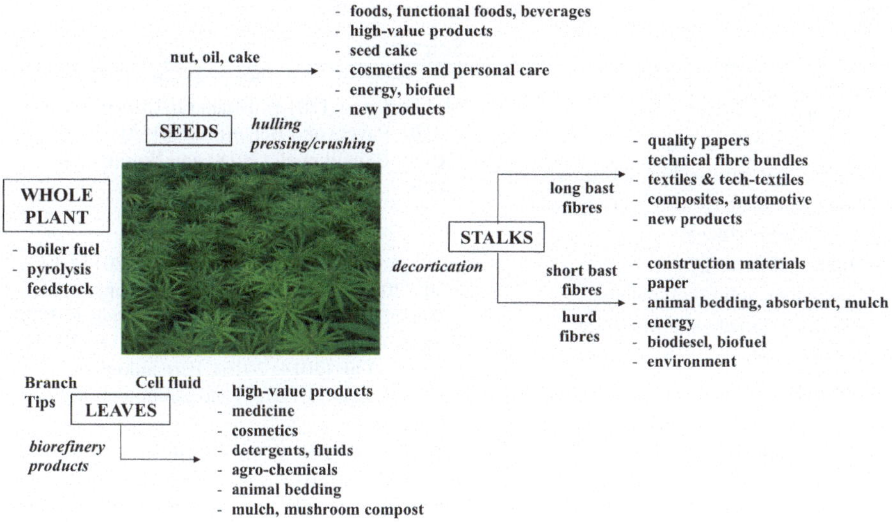

Fig. 2.1 Modern applications of hemp

restriction has reduced the number of varieties available for its cultivation. Hemp is currently subsidized by the European Union for non-food agriculture and research purposes, and a considerable initiative for its further development in Europe is underway.

Hemp is considered as a low-cost, ecological, sustainable and multi-use plant (Faruk et al. 2012; Shahzad 2012; Bouloc 2013; Bono et al. 2015; Johnson 2018). An important advantage is the fact that the entire plant, i.e. seeds and plant stem, is recoverable (Roulac 1997; Johnson 1999; Ranalli and Venturi 2004; Sadrmanesh and Chen 2019). Another advantage is hemp versatility, being useable in various forms, e.g. fibers, felts, powders, shives, and seed products including seeds themselves, oil and oil cake. All these products can be used for a myriad of applications as reported in Fig. 2.1.

The global market for industrial hemp has more than 25,000 products in nine main submarkets: textiles, agriculture, automotive, food and beverages, paper, furniture, construction, recycling, and personal care. Johnson (2018) recently reported that the sales of hemp based product in the U.S. alone is about $600 million dollars in 2017.

From a scientific point of view, hemp is also the object of numerous fundamental studies and a great deal of research is being conducted around the world. An indication of the widespread exploitation and constantly growing importance of this plant is the total of over 3750 scientific articles published between in the last 5 years (ISI Web of Science database). It is interesting to note that a majority of these investigations have been done in China, Korea, North America, Russia and Europe.

This chapter provides an overview of the traditional and new uses of industrial hemp, based on a large number of relevant references published over the last decade (Table 2.1).

Table 2.1 Applications of industrial hemp: selected publications during the last decade

Topics	References
Agriculture – Agrochemistry Mulch and animal bedding Animal feed, birdseeds Organic agriculture, bio-fertilizers Insect pest management, eco-friendly insecticide/herbicide	Kolodziejczyk et al. (2012), Bouloc (2013), Mukhtar et al. (2013), Isman (2015), Bedini et al. (2016), Pavela and Benelli (2016), Abé et al. (2018), Benelli et al. (2018) and Fiorini et al. (2019)
Biofuels – Bioenergy Boiler fuel, solid fuel, pellets Biodiesel, bioethanol, methanol Biogas: methane, biohydrogen Electricity	Burczyk et al. (2008), Ahmad et al. (2011), Li et al. (2010), Sipos et al. (2010), Kreuger et al. (2011), Prade et al. (2011, 2012), Gomes (2012), Finnan and Styles (2013), Rehman et al. (2013), Kuglarz et al. (2014, 2016), Fernando et al. (2015), Das et al. (2017) and Schluttenhofer and Yuan (2017)
Building materials Insulation, hemp wool, paneling, fiberboard Concrete, cement blocks, mortar	Elfordy et al. (2008), Arnaud and Gourlay (2012), Ip and Miller (2012), Amziane and Arnaud (2013), Bouloc (2013), Collet et al. (2013), Pretot et al. (2014), Walker and Pavía (2014), Walker et al. (2014), Abd Rashid and Yusoff (2015), Cigasova et al. (2015), Latif et al. (2015), Ingrao et al. (2015), Aït Oumeziane et al. (2016), Arizzi et al. (2016), Fangueiro and Rana (2016), George et al. (2016), Jonaitiene and Stuoge (2016), Kinnane et al. (2016), Niyigena et al. (2016), Fernea et al. (2017), Gourlay et al. (2017), Kiruthika (2017), Liuzzi et al. (2017), Mazhoud et al. (2017), Mirski et al. (2017), Chernova et al. (2018), Jiang et al. (2018), Kallakas et al. (2018), Moujalled et al. (2018), Nováková (2018), Pichardo et al. (2018), Pittau et al. (2018), Usmani and Anas (2018), Iucolano et al. (2019) and Jami et al. (2019)
Composites – Furniture Green composites, biocomposites Plastic composites Furniture industry Nanomaterials	Faruk et al. (2012), Shahzad (2012), Bhavani (2015), Bono et al. (2015), Cigasova et al. (2015), Fernando et al. (2015), Pil et al. (2016), Ummartyotin and Pechyen (2016) and Gallos et al. (2017), Lamberti and Sarkar (2017), Liu et al. (2017), Mirski et al. (2017), Nunes (2017), Nurazzi et al. (2017), Chernova et al. (2018), Karaduman et al. (2018), Musio et al. (2018), Sarasini and Fiore (2018), Sepe et al. (2018), Spierling et al. (2018), Usmani and Anas (2018) and Väisänen et al. (2018)
Cosmetology – Hygiene Oils, lotions, moisturizer, body care products Shampoos, bath gels, soaps, anti-microbe hand soap Essential oils, beauty products	Bertoli et al. (2010), Kolodziejczyk et al. (2012), Ionescu et al. (2015), Hartsel et al. (2016), Ligeza et al. (2016), Bonini et al. (2018)

(continued)

Table 2.1 (continued)

Topics	References
Environmental purposes	Kostić et al. (2008, 2010, 2014, 2018), Pejić et al. (2008, 2009, 2011), Rosas et al. (2009), Tofan et al. (2009, 2010a, b, c, 2013, 2015, 2016a, b), Yang et al. (2011, 2012), Vukčević et al. (2012, 2014a, b, 2015), Zou et al. (2012), Cassano et al. (2013), Rezić (2013), Sun et al. (2013), Balintova et al. (2014), Feng and Zhang (2015), Lupul et al. (2015a, b), Kyzas et al. (2015), Wang et al. (2015), Ahmad et al. (2016), Bugnet et al. (2017a, b), Loiacono et al. (2017a, b, c, 2018a, b) and Saxena et al. (2020)
Compost, growth medium	
Elimination of contaminants: metals	
Phytoremediation, phytoextraction	
Activated carbon production	
Water and wastewater treatment	
Air/oil filtration	
Food industry – Beverages	Yin et al. (2008, 2009), Kolodziejczyk et al. (2012), Bouloc (2013), Russo and Reggiani (2013), Girgih et al. (2014c), Malomo et al. (2014), The and Birch (2014), Dunford (2015), Malomo and Aluko (2015a, b), Andre et al. (2016), Fike (2016), Hartsel et al. (2016), The et al. (2016), Chandra et al. (2017), Korus et al. (2017), Mikulcová et al. (2017), Pihlanto et al. (2017), Frassinetti et al. (2018), Hadnađev et al. (2018), Johnson (2018), Devi and Khanam (2019a, b), Fathordoobady et al. (2019), Fiorini et al. (2019), King (2019), Mamone et al. (2019), Mikulec et al. (2019), Wang and Xiong (2019) and Zajac et al. (2019)
Nutrition and beverages: salad oil, margarine, granola, protein flour, beers, wines	
Functional food, nutritional supplement	
Nutraceutical products	
High-value products: fatty acids, nutrients, lecithin	
Animal nutrition, feed additives	
Paper	Harris et al. (2008), Barberà et al. (2011), Bouloc (2013), Miao et al. (2014), Feng and Zhang (2015), Danielewicz and Surma-Slusarska (2017), Przybysz Buzala et al. (2017), Yu et al. (2017) and Bajpai (2018)
Paper pulp, cardboard, packaging	
Cigarette paper	
Printing papers, newsprint	
Fine and specialty papers	
Technical filter paper, filters	
Medicine	Rodriguez-Leyva and Pierce (2010), Cassano et al. (2013), Richard and Dejean (2013), Russo and Reggiani (2013), Girgih et al. (2014a, b), Cherney and Small (2016), Hartsel et al. (2016), Parian and Limketkai (2016), Pathak et al. (2016), Bonini et al. (2018), Zhou et al. (2018), Devi and Khanam (2019a), Fathordoobady et al. (2019), Fiorini et al. (2019), Mamone et al. (2019) and VanDolah et al. (2019)
Active/psychoactive substances: cannabinoids, terpenes	
Drugs: glaucoma, vomiting, spasms	
Antibacterial products	
Textiles & Tech-Textiles – Ropes – Furniture	Amaducci and Gusovious (2010), Müssig (2010), Bouloc (2013), Kostić et al. (2014), Pil et al. (2016), Zhang et al. (2016), Bran et al. (2017), Lamberti and Sarkar (2017), Mirski et al. (2017), Rijavec et al. (2017), Gedik and Avinc (2018) and Miller (2018)
Fabrics, bacteria-fighting fabrics	
Clothes, sport clothing, military clothes, socks, knitted	
Underwear, T-shirt	
Bags, canvas bags, shoes	
Rops, twines, nets, carpets, nonwoven, geotextiles	

2.2 Hemp Products

Hemp production can provide social, economical, environmental, agronomic and industrial benefits.

From a social and economic point of view, hemp cultivation is a vector of development for local agricultural resources in emerging countries and an industrial output for crops in developed countries. Both on the production and demande sides, there are in each country an active movement towards the development of a hemp industry due to a multitude of potential benefits often cited (Bócsa and Karus 1998; Johnson 1999; Fike 2016; Żuk-Gołaszewska and Gołaszewski 2018), as an increase efficiency compared to others inputs for some industrial uses such as paper production, health benefits of both hemp seed and hemp oil consumption, and a competitive use as an input in textile manufacturing or furniture.

From an environmental point of view, hemp is one of the most efficient plants known for its ability to utilize sunlight and capture large quantities of CO_2 to photosynthesize, with an annual growth up to 5–6 m in height. Hemp is more environmentally friendly than traditional crops (Fike 2016; Żuk-Gołaszewska and Gołaszewski 2018).

From an agricultural point of view, hemp is considered as an excellent crop for sustainable agriculture (Żuk-Gołaszewska and Gołaszewski 2018). Indeed, the hemp cultivation has always been considered a good rotational crop because of its beneficial effect on following crops. Its cultivation helped to replenish the soil and did not require much weed control (Bócsa and Karus 1998). Gorchs et al. (2000) previously reported that an increased yield is obtained when wheat grown after hemp.

On the production side, hemp is also appreciated by organic growers for its adaptability to a wide range of agronomic conditions, ease of production, and rapid growth, e.g. after sowing, cultivation is not difficult. Hemp requires low fertilization and irrigation is not necessary. The crop requires only rarely the use of plant protection products. Herbicides are not considered necessary because hemp's fast growth, which results in weed suppression (Fike 2016; Sandler and Gibson 2019).

O'Brien and Arathi (2019) recently pointed out that, being wind pollinated, dioecious and staminate hemp plants produce large amounts of pollen that are attractive to bees. While hemp does not produce any nectar, the pollen rich nature of the flowers can make hemp an ecologically valuable crop.

Industrial hemp is cultivated to yield four products: the fiber, the seed and its extracted oil, and the bioactive substances from the seeds, seed cakes and leaves, the main product being the bast fiber (Sadrmanesh and Chen 2019). It is important to note that, for millennia, hemp was cultivated for (bast) fiber, mainly for textile and clothing applications while hurds were considered as byproducts of the fiber production. Indeed, hemp was traditionally known as a fiber plant and most historical cultivation of the plant was with fiber use in mind. Although consuming, the use of seeds and oil was not widespread.

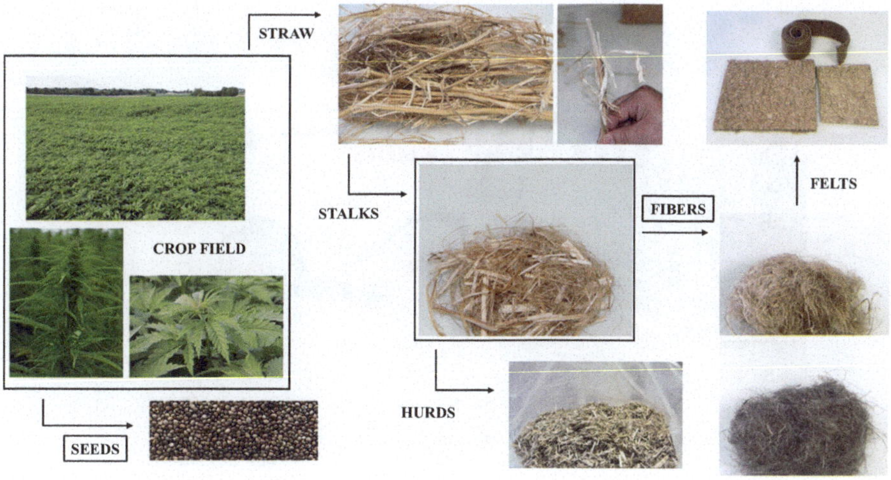

Fig. 2.2 Industrial hemp products; *Cannabis sativa* L. is the source of two types of natural fibers: long bast fibers in fibrous form and woody short core fibers, called hurds or shives, in granular form; the hemp stem photography shows bark, secondary bast fibers and woody core. (Source: G. Crini, Chrono-environnement, Besançon, France)

Nowadays, hemp can be grown for both fiber and seed harvesting (Fig. 2.2). Fiber is still the main product derived from hemp straw while the principal product made from hemp seeds today is undoubtedly the seed oil. Different hemp cultivars produce variable quantities and qualities of fiber, i.e. long bast fibers or short core fibers, as well as seed size and oil composition (Bócsa and Karus 1998; Karus and Vogt 2004). There are mainly two groups of Cannabis varieties being cultivated today (Żuk-Gołaszewska and Gołaszewski 2018): varieties primarily cultivated for their stalks, i.e. fibers used for construction material, clothing or animal related purposes, and varieties grown for seeds from which oil is extracted, e.g. for breeding and food.

Hemp can also be cultivated as dual-purpose crop that implies that both fibers and seeds can be processed. This practice has an impact on quality and quantity of fiber, e.g. a dedicated fiber crop yields the highest quality bast fiber for textiles and composites (Bouloc 2013). The yield and quality of hemp biomass are largely determined by the genetic background of the hemp cultivar. They are also strongly affected by environmental factors such as temperature and photoperiod (Salentijn et al. 2015, 2019).

Harvesting hemp fiber is however a bit tricky. Machinery has to be adapted to deal with the plant because the fibers can wrap among a harvester's moving parts, leading to possible mechanical failures. In addition, harvester knifes and blades must be kept sharp and in good condition in order to cut through the hemp stalk. Figure 2.3 shows a modern French forage chopper for cutting and windowing hemp. The preparation of the fiber is also a delicate activity that requires special know-how and machinery (Bócsa and Karus 1998; Karus and Vogt 2004; Münder et al. 2004; Fike 2016).

Fig. 2.3 A modern French forage chopper for cutting and windowing hemp. (Source: G. Chanet, Eurochanvre, Arc-les-Gray, France)

The fiber is divided into three fractions during mechanical processing: the short and long bast fiber bundles and the hurds/shives (Fig. 2.2). Hemp stem consists of approximately 20–40% of bast fibers and 60–80% of hurds, depending on variety or usage type, and planting density (Stevulova et al. 2014; Sadrmanesh and Chen 2019). The outside of the stem is covered with bark. Inside the stems are bast fibers and the woody core. The separation of the bast fiber is carried out through defiberization or decortication using specialized machineries, i.e. breaking the woody core of the stems into short pieces and separation of bast fiber from the hurds. The hemp fibers are situated in the bast of the hemp plant (Small and Marcus 2002; Hautala et al. 2004; Bouloc 2013).

The byproduct of the hemp stem obtained after the industrial defiberization process is the hurd, a short fibre which typically contains 20–30% lignin. In contrast to the high quality of bast fibers, the hurds or shives are the least valuable part of the plant, chemically close to wood. The industrial process also provides dust (10–15%) that can be find used as fuel and soil improver in powder form, or as granulated animal bedding.

There are two types of bast fibers: primary fibers, make up 70–90% of the bast, which extend nearly the entire length of the stalk and are coarser, and secondary fibers, 10–30% of the bast fibers, which are shorter, finer fibers that are found next to the wood core. The long, strong bast fibers are similar in length to soft wood fibers and are very low in lignin content while the short core fibers, more lignified, are more similar to hard wood fibers. Primary bast fibres are the most valuable part of the stalk. Low stand density of hemp crop and/or use of different genotypics favors development of secondary bast fibres.

Technical fibers are characterized with heterogeneous chemical composition and fairly complicated structure. The main constituents are cellulose, hemicelluloses and lignin, while minor components such as pectins and waxes are regarded as surface impurities (Kozlowski et al. 2005; Sponner et al. 2005; Ansell and Mwaikambo 2009; Thomas et al. 2011; Placet et al. 2014, 2017; Sadrmanesh and Chen 2019).

The fiber length and cellulose and lignin content are key quality parameters. As member of the bast fiber family, hemp contains over 75% of cellulose and less than 10 to 12% of lignin, compared with 60% and 30% respectively for wood. The bast fibers contain higher amounts of cellulose than the hurds. Given their differing physical properties, the two types of fibers have different ideal end uses (Bismarck et al. 2005; Thomas et al. 2011).

Most hemp oils available on the market are produced by cold pressing because of the simplicity of this method and its low serial production cost. However, only 60–80% of the oil can be recovered. There are other technologies such as soxhlet extraction using solvents, e.g. ethanol, n-hexane and petroleum ether, ultrasound-assisted extraction, techniques using supercritical carbon dioxide or microwave-assisted extraction (Da Porto et al. 2012; Chang et al. 2017; Subratti et al. 2019). In general, each extraction technique does not affect the fatty acids composition of the oils.

Da Porto et al. (2012), comparing the soxhlet extraction using n-hexane with supercritical carbon dioxide extraction, showed that a higher oil yield was achieved by the former method. Supercritical carbon dioxide extraction of oil depends on temperature, pressure and the size of hemp seed particles. Compared to cold pressing, the main disadvantages of this technique are high costs (expensive equipment) and high time costs. Solvent extraction is efficient but it requires longer extraction time compared to cold pressing and mostly it involves the use of organic solvents.

Subratti et al. (2019) recently proposed liquefied dimethyl ether as a green solvent for the extraction of hemp seed oil from hulled hemp seeds. The authors showed that the resulting yields were higher using liquefied dimethyl ether in comparison to conventional organic solvents. The oil was acquired in high purity. The process developed is simple to set up, affordable to use in any laboratory, does not require the use of a rotary evaporator, economical, and allows an easy recovery of the dimethyl ether without traces of solvent remaining in the crude. Indeed, dimethyl ether evaporates quickly at room temperature without leaving any residual solvent. The absence of applied heat is interesting because it preserves the integrity of the oil.

2.3 Hemp Fibers in Industrial Applications

From an industrial point of view, industrial hemp is a valuable raw material due to its natural origin, abundancy, renewable character, and mostly considered as a multiple-use plant with great potential (Fike 2016). Hemp fibers are suitable replacements for a variety of fossil-based products or applications. They are also valuable for various other reasons: their low cost, low carbon footprint, interesting life cycle analysis, low greenhouse gas emissions, etc.

All the hemp-based products mentioned above find a wide range of applications ranging from textiles and papers, to insulation and building materials, foods and beverages, cosmetics, livestock feed and bedding, and moulded plastics.

Traditionally, bast fibers were used to produce fabrics and textiles, bags and hard-wearing fabrics, twines, ropes, and paper. These fibers are now used in the manufacture of fine and specialty papers, insulation products, fiber-reinforced composites and bioplactics.

An important example is hempcrete/hemp-lime, perhaps the most researched bio-based building material. This material is a mixture of hurds and lime products used commonly in the construction sector, e.g. as a lightweight insulating material, hurd board, additive to bricks or loam construction, or as pour-in insulation.

Hemp is also used in plastics and composites for use as a fiberglass alternative by the automotive and aviation sectors. Hemp fibers, as eco-friendly material, are good candidates as a (partial) substitute for synthetic fibers such as glass, carbon or metallic fibers. Fine and specialty paper prepared from hemp can be used for making cigarettes. Hurds find applications as horse bedding or for bedding of small animals. Traditionally, the woody core was burnt for heating or to fuel the steam engines.

The oil extracted from the hemp seeds and their derivatives are most commonly used in various food-processing applications, nutritional supplements or as animal feed. The oil has also been considered for cosmetic applications, personal care products and skin care owing to its high polyunsaturated fatty acid content. Cannabinoid-based derivatives and other biological active substances are used by the pharmaceutical industry.

An interesting aspect of the cultivation of hemp is that hemp can be grown on polluted soil and can contribute to phytoremediation. In the last decade, hemp has been proposed as complexing material for pollutant removal. Hemp is also considered a suitable biomass crop for energy production or biodiesel.

All these applications are detailed in the following sections.

2.3.1 Textiles, Fabrics and Furniture

Hemp fiber has served humanity for thousands of years to create textiles, fabrics, ropes, yarns, rugs, and canvas. The fiber can be spun, then woven or knitted into many fabrics suitable for durable and comfortable clothing. Hemp fiber was indeed a common material for clothing because of its durability and versatility until the cotton industry grew stronger all over the world gained (Small and Marcus 2002; Liberalato 2003; Wang et al. 2003; Kozlowski et al. 2005; Sponner et al. 2005; Thomas et al. 2011; Bouloc 2013).

Hemp fiber is very strong compared with other natural fibers such as cotton, flax and nettle. So, historically, the fiber has also long been widely used in ropes, rigging, net, and sail production. Other advantages are its flexibility, strength, and resistance to water damage. Thus, in past centuries, hemp was extremely important to the Navy, the shipping trade, and also fishing. Christophe Colomb sailed to America on ships rigged with hemp (Bouloc 2013). The oldest known woven fabric was made from hemp, as were Levi Strauss' original denim jeans, and the first American flag.

Nowadays, hemp can be used to make a variety of similar but more durable fabrics than cotton (Amaducci and Gusovious 2010; Müssig 2010; Bouloc 2013). Numerous consumer brands such as Adidas, Patagonia, and commercial channels have added hemp products to their offer that help to popularized hemp as a clothing fibre or accessory.

Currently, clothing is a high profile market for hemp. Hemp fibers find numerous applications in textiles and the possibilities for hemp fabrics are immense: apparel, jeans, sport clothing, bags, hats, cushion covers, blankets, shoes, socks, accessories, ropes, yarns, rugs, furniture and home furnishings, and also hemp jewelry, e.g. bracelets, necklaces, anklets, rings, watches, and other adornments (Amaducci and Gusovious 2010; Müssig 2010; Bouloc 2013; Kostić et al. 2014; Bhavani 2015; Pil et al. 2016; Zhang et al. 2016; Bran et al. 2017; Lamberti and Sarkar 2017; Mirski et al. 2017; Rijavec et al. 2017; Gedik and Avinc 2018).

Textiles are easy to produce, durable, breathable, versatile, biodegradable, having strong thermal qualities. Fabrics have the best capacity ratio compared to other fibers, meaning they keep the wearer cool in the summer and warm in the winter. Preference for hemp textiles in summer is often associated with their excellent hygienic properties. However, natural fibers without modification do not provide good UV protection. The fabrics are also anti-microbial and hypoallergenic, as well as resistant to mold. Fibers are also more resistant to weather and ultraviolet rays than cotton and silk. Hemp clothing is stronger and more durable that cotton clothing and does not deform as easily. Apparel made from hemp merge easily with dyes and do not discolor easily. No difference between hemp and cotton fabrics in terms of colorfastness to crocking, oily stain release, flammability, tearing strength, breaking strength and elongation, as recently pointed out by Lamberti and Sarkar (2017). Fibers can also be mixed with other materials to create clothing hybrids, e.g. fibers can be blended with cotton or linen for specific textures and performance.

For all these reasons, hemp is also an eco-friendly fabric for upholstery and furniture: home furnishing textiles, seating, tables, fashion accessories, mirrors, wall decorations, decorative objects, etc. Finished goods such as clothing, shoes and hats are made from 100% hemp or combined with other natural or synthetic fibers. Figure 2.4 shows a hemp reusable bag made from 100% fibers. Hemp is also a viable and excellent fiber for making rugs, pure hemp carpets, and similar textiles. However, two main problems often cited are their primarily higher cost and the need for manufacturing machinery to be adapted.

Cordage is an age old use for hemp fiber. While its use in the marine world has largely being replaced by cheaper, long-lasting and lighter synthetics, hemp rope still has its uses. It is well-known that, due to its coarser texture, hemp rope can bind against itself for better knot stability, and this is useful in some situations. Hemp yarn is used for crafted jewelry because it is smooth, consistent, strong and comfortable in contact with the skin. Hemp twine is also interesting in crafting, gardening and landscaping.

Fig. 2.4 A hemp green bag. (Source: G. Chanet, Eurochanvre, Arc-les-Gray, France)

Hemp oil is also an interesting product for furniture, leather or wooden floor, e.g. Wise Owl Hemp. Pure oil, completely free of solvents or other chemical additives and preservatives, can be used to seal raw wood and furniture pieces. The advantages cited are: 100% natural, safe quality food, biodegradable, easy to use, both interior and exterior use, conservation of the initial color of wood, water and alcohol resistance, ideal for food surface, etc. Hemp oil for furniture is derived from the same pressed hemp seeds used for food grade edible oils, the only difference being the level of amino acids, which affect the drying time. The odor of the oil is much like crushed walnuts. Hemp oil may be used on finished wood to clean and protect or on unfinished wood to bring out the richness of the wood grain. It makes an excellent sealant over stained furniture.

However, although the hemp markets for clothing, furniture, luxury and fashion have seen a strong increase in popularity, the markets for specialty fibers tend to be cyclical. In addition, due to the rise in the industrial use of cotton followed by synthetic fibers in Europe, users had to turn to importing hemp, particularly from Asia. Nevertheless, several European companies and cooperatives, particularly in France, are carrying out several research and development programs for the reintroduction of European hemp into the textile industry because the possibilities for hemp fabrics are immense.

2.3.2 *Paper*

Historically, hemp has been used to make paper for thousands of years. In ancient China, hemp fiber was primarily produced for use in paper scrolls (Robinson 1996). As already mentioned, the first copies of the Bible were made of hemp paper and the American constitution was written on hemp paper (Ranalli and Venturi 2004). Hemp paper was more resistant to decomposition, stronger, especially when wet, less prone to yellowing than tree based paper (Harris et al. 2008; Ranalli and Venturi 2004; Bouloc 2013).

After the "rediscovery" of hemp in Europe in the 1990s, hemp fibers were mainly used for the manufacture of special pulp and paper. Hemp as non-wood fiber is indeed an ideal yield raw material for making specialty paper due to its high-quality physical properties of its pulp, and its tensile strength (Ranalli and Venturi 2004; Harris et al. 2008; Barberà et al. 2011; Bouloc 2013; Miao et al. 2014; Feng and Zhang 2015; Danielewicz and Surma-Slusarska 2017; Przybysz Buzala et al. 2017; Yu et al. 2017; Bajpai 2018).

Compared to conventional wood paper, hemp paper has superior qualities like higher strength, length and fineness. Hemp's long bast fibers makes a fine quality paper that is naturally acid free. Hemp paper uses less chemicals in production than tree based paper. It does not become yellow and brittle or disintegrate over time like conventional paper. It is a faster and more efficient way of growing fiber than the use of trees. The paper is also of a high-quality and durable for long term storage for forms, currency paper and cigarettes production. However, it is more expensive, i.e. the cost of hemp pulp is three to six times higher compared to conventional wood-based pulp production. This can be explained by the fact that the use of hemp in papermaking is mainly concentrated on bast fibers, the woody core of the plant being often considered as waste. The bast fibers are only a small part of the plant stem (hemp stalk averages around 20–30% bast fibers) and separation tends to lead to high production costs. Another difficulty which causes an increase of the hemp pulp price is that the hemp is harvested once and needs to be stored all year long.

The applications are limited to a very few applications such as technical filters, bank notes, bible paper, dielectric and medical paper, and cigarette paper, due to the high price of hemp pulp (Yu et al. 2017). Specialty papers also include teabag paper, coffee filter, speciality non wovens, greaseproof papers, carbon tissues and condensing tissues. Currently, the only well established market for hemp pulp is the cigarette paper market (European Industrial Hemp Association). Noteworthyy, in the production of cigarette paper, all the fiber in the stem can be used.

Nowadays, industrial hemp is not competitive on the paper market for the production of conventional papers, because the market is cyclical. For instance, kenaf has many economic advantages over hemp as non-wood fiber for the paper industry. Perhaps, the growing market for recycled pulp and paper may increase the demand for agricultural fibers to strengthen recycled papers. Similar to textiles, hemp fibers can be used as blends with other pulp fibers such as wheat straw or flax or even recycled wood, in order to increase paper performance, strength and recyclability. In addition, core and whole stalk are proposed to make lower end paper products, depending on available pulping technology that is tooled to process hemp efficiently. These markets can grow (Danielewicz and Surma-Slusarska 2017; Przybysz Buzala et al. 2017; Yu et al. 2017; Bajpai 2018).

Another promising market is filtration. Through comparison of oil filtration properties and air filtration properties with commonly used automobile engine oil/air cotton paper filtration materials, Feng and Zhang (2015) reported that hemp paper had the smaller thickness, weight, mean pore diameter, porosity and oil/air penetration, while the better oil/air filtration efficiency and higher pressure drop. Due to the higher filtration efficiency of hemp papers and green, biodegradable, sustainable resources of hemp plants in the case of environmental requirements, a pilot trial hemp paper automobile engine oil filter was successfully manufactured. The results indicated that hemp papers had better oil/air filtration properties than cotton paper in practical application.

2.3.3 Insulation and Building Materials

Compared to the traditional synthetic fiber materials, natural fibers such as hemp fibers represent a sustainable solution to be used for different applications in building construction, mainly because of their hygrothermal properties. Hemp-based materials can be manufactured into a variety of commercial products of various densities that resemble concrete, wood, and even plastic. In addition to the environmental value of using plant matter, these materials benefit from the mechanical strength of the hemp fibers.

Bio-based materials containing hemp also offer many other advantages over more established mineral and oil-based alternatives. These building materials are durable, light weight, affordable to produce, water-proof, fire-proof, self-insulating, resistant to mold, moisture-proof, highly breathable, and resistant to pests, and they have good heat resistance in winter time and cool in summer. The materials are also ideal for resisting damage caused by earthquakes, floods or other natural disasters (Shahzad 2012).

It is also claimed that hemp building materials trap carbon dioxide, making their use attractive from an environmental perspective (Zabalza Bribián et al. 2011; Ip and Miller 2012; Amziane and Arnaud 2013; Collet et al. 2013; Dubois et al. 2014; Latif et al. 2015; Jami et al. 2016, 2019; Jiang et al. 2018; Miller 2018; Moujalled et al. 2018; Pittau et al. 2018). However, the main disadvantage is the lack of guarantees in terms of durability (Walker et al. 2014; Arizzi et al. 2016).

As industrial products available on the market, hemp-based materials are subject to a number of National and European technical regulations and have been approved officially for use in the construction industry. Hemp fibers are mainly used for insulation, e.g. insulating material and insulation wool, and/or construction. Insulation is the second important application for hemp fibers today (European Industrial Hemp Association). These fibers can also be used for acoustic and soundproofing purposes. This is an area of research that is currently in full expansion.

With the exception of hemp seed, which has no significant use as a building material, hemp straw can provide two products/co-products for the construction industry: (1) the bast fibers and (2) the woody core of the stalk used to produce hurds. Due to their different properties, these two products are used to produce different construction and insulation materials (Figs. 2.5 and 2.6).

The materials can be made from the compressed inner short hemp fiber. The outer hemp fibers can also be used like straw in bale construction paired with mud for an old-style cob building. Hemp boon, obtained after crushing hemp woody core of the stem, is useful for thermal insulation composite production due to their small-pored structure.

High-quality building blocks are produced by mixing a binder, e.g. lime, clay or cement, with non-fibrous hurds particles (Elfordy et al. 2008; Pretot et al. 2014; Walker and Pavía 2014; Walker et al. 2014; Kinnane et al. 2016; Gourlay et al. 2017; Jami et al. 2019). These materials are bio-composites referred to hemp concrete, hempcrete or hemp-lime (Fig. 2.6). Hemp shives in combination with lime is currently another increasing market for construction, e.g. stucco/plaster production, caulking.

Fig. 2.5 Hemp panel wallboards made from 100% fibers. (Source: G. Chanet, Eurochanvre, Arc-les-Gray, France)

Fig. 2.6 Hemp-based products for insulation or building. (Source: G. Chanet, Eurochanvre, Arc-les-Gray, France)

Hemp provides all sorts of building materials: blocks and bricks, slabs and paneling, wallboard (Fig. 2.5), fiberboard, roofing tiles, and insulation products (Fig. 2.6). Hurds/shives are used to produce light concretes and mortars for different end uses, e.g. wall construction, insulation, underfloor, etc.

Due to their low density, elasticity, permeability, and insulating properties, hurds are also used for production of particle boards (Elfordy et al. 2008; Bouloc 2013; Walker et al. 2014; Cigasova et al. 2015; Ingrao et al. 2015; Fangueiro and Rana 2016; George et al. 2016; Jonaitiene and Stuoge 2016; Kinnane et al. 2016; Niyigena et al. 2016; Fernea et al. 2017; Gourlay et al. 2017; Kiruthika 2017; Liuzzi et al. 2017; Mirski et al. 2017; Chernova et al. 2018; Kallakas et al. 2018; Nováková 2018; Pichardo et al. 2018; Usmani and Anas 2018).

These environmentally friendly building materials are suitable for the construction of houses (e.g. Chanvribat®, Béton Chanvre Tradical® France). Hemp-based materials can indeed replace wood and other materials used to build homes and other structures including foundations, underfloor, walls, shingles, paneling, pipes, and paint.

Houses can be made nearly 100% out of hemp materials: walls can be hemp wallboard (Fig. 2.5) or hemp cement for walls (Fig. 2.6); insulation can be made of hemp, e.g. hemp cement to insulate under the floor boards; hemp bricks and hemp plaster; pipes can be made out of hemp plastic; paint made with hemp oil; hemp carpet for interiors, and even a hemp roofing material.

Hemp concrete was developed in France in the 1990s as an alternative to wattle and daub for the restauration of historic building (Bouloc 2013), in particular with the works of Charles Rasetti on the restauration of a mediaval timber framed building, "*Maison de la Turque*" at Nogent-sur-Seine in 1987. At that time, other projects on the renovation of historical buildings based on hemp-lime use, e.g. in Versailles, were also developed (Canosmose®, IsoChanvre®, ChanvriBloc®). However, Charles Rasetti was the first to patent hempcrete in 1986, Canobiote® and Canamose® (International Hemp Association). Canamose® was a light-weight concrete made with hemp hurds and natural lime, which might be used for all types of non load-bearing masonry, and was perfectly suited to walls sectioned with wooden supports. Canobiote® was hemp hurds coated with mineral salts, which was an effective process for the insulation of wood-framed, closed lofts and floors intended for regular use.

Currently, hemp concrete/hempcrete is commonly used in Europe, e.g. in France (Fig. 2.7), England, Germany, Ireland, Belgium, Luxembourg and Switzerland, as a multifunctional ecological material for building applications. Hemp-lime construction had also been adopted in North America (Quebec), South Africa, Israel, Australia and New Zealand.

The advantages often cited are: excellent thermal insulation properties, low thermal conductivity, high hygrothermal efficiency, breathability, low environmental impact, e.g. in terms of emissions of green-house gases, interesting acoustical properties, and particular hygrothermal behavior enabling a natural moisture regulation (Arnaud and Gourlay 2012; Mazhoud et al. 2017; Aït Oumeziane et al. 2016).

Hemp-lime is also a cheap and low density material with associated low thermal conductivity depending on the density and moisture level, and it presents an interesting balance between low mass and heat storage capacity compared with classical insulation materials. It is considered as a sustainable, carbon negative and low embodied energy construction material (Zabalza Bribián et al. 2011; Latif et al. 2015; Bouloc 2013; Kinnane et al. 2016).

2.3.4 Composites and Plastic Alternatives

Hemp fibers are used not only for insulation but also for the production of composites in the automotive, furniture and fashion industries where synthetic fibers are replaced by hemp fiber. Indeed, bast fibers are desirable for composite/biocomposite production due to their length and strength while hurds, the byproducts of extracting the bast fibers from the stalk, are interesting for particle board and biodegradable plastic production.

Fig. 2.7 A half-timbered building in Normandie, France, dating from the early nineteenth century. The walls and slabs were renovated using differents solutions based on hemp concrete. (Source: with the kind courtesy of Florence Collet (Rennes, France) and Yacine Aït Oumeziane (Belfort, France))

Active research continues in these areas (Faruk et al. 2012; Shahzad 2012; Bono et al. 2015; Cigasova et al. 2015; Fernando et al. 2015; Pil et al. 2016; Ummartyotin and Pechyen 2016; Gallos et al. 2017; Lamberti and Sarkar 2017; Liu et al. 2017; Mirski et al. 2017; Nunes 2017; Nurazzi et al. 2017; Chernova et al. 2018; Karaduman et al. 2018; Musio et al. 2018; Sarasini and Fiore 2018; Sepe et al. 2018; Spierling et al. 2018; Usmani and Anas 2018; Väisänen et al. 2018).

In 1941, Henry Ford used hemp-based plastics to build car doors and fenders. He produced a prototype, the Hemp Body Car, that showed the great potential of hemp used in combination with plastic technology; indeed 70% of straw mixed with a plastic composite. Ford demonstrated the strength of the hemp composite by hitting the car with a club and leaving no trace in the bodywork, a scene immortalized by a famous photograph (The Collections of the Henry Ford).

Ford predicted that his prototype "would be lighter, sager and less expensive" (New York Times, February 2, 1941). However, at that time, the technology was never put into mass production (Small and Marcus 2002). The car also was to run on hemp-ethanol. Ford was a visionary.

Currently, industrial hemp presents great opportunity for supplying a sustainable and carbon positive source of plasticizing material and several hemp-based plastics or bioplastics are on the market, e.g. in the automotive sector. Hemp can be processed into different forms used in automotive interior and exterior applications, e.g. carmakers are currently using hemp composite inserts, trunks, head-liners, spare wheel covers, parcel trays etc. An important market is the press moulding market for interior applications: door panels and car boot trims, rear shelf and roof liner panels, dashboards, pillar trims, seat shells, under bodies and other applications.

The biocomposites and bioplastics are price competitive in high-quality interior concepts although hemp fibers are also in competition with flax, jute and kenaf fibers. They are also considered less expensive than fiberglass counterparts and show the favourable mechanical properties of rigidity and strength in combination with low density.

Another interesting advantage can be cited: The material does not splinter and leaves no edges which is an important characteristic especially in the case of automobile accidents. It is also claimed that hemp fiber reinforced plastics show energy and greenhouse gas savings in comparison with their fossil-based counterparts. After use, their incineration with energy recovery shows higher energy and emissions of greenhouse gases savings compared to an analysis from cradle-to factory gate.

The bioplastics containing hemp are, for example, used to replace synthetic polymers and copolymers such as polyacrylonitrile-butadiene-styrene in dashboards. These products are hemp fibers reinforced polypropylene compound and are designed for automotive structural parts by injection process. They are mainly used in the German automotive industry (e.g. BMW, Mercedes, Volkswagen Golf), followed by the French (e.g. Peugeot 308, Megane, DS7 Crossback) and Italian (e.g. Alfa Roméo Giulia) industries. An example is Refine® Hemp-PP for the inserts and top roll of the door panels in the Peugeot 308. Another example is NAFILean™ used as biocomposite in the dashboard of the Alfa Roméo Giulia.

Hemp fiber reinforced plastics are also used in the production of furniture or other consumer products such as briefcases, glasses and fashion accessories. Biocomposites are proposed for high-performance products like yachts, composite sink basin, sunglasses and ski goggles (HempEyeWear®), or trays for grinding discs. Shives are proposed for insulation of airline seats. A hemp-plastic resin, HempStone™, is proposed for use in musical instruments, loudspeakers, and furniture, e.g. tables and chairs.

Khattab and Dahman (2019) recently reported the production and recovery of poly (3-hydrobutyrate) from agro-industrial residues of hemp hurd biomass, to prepare ecological bioplastics with low environmental impact. Their technology focused on converting wastes into useful biomaterials such as poly(3-hydrobutyrate), contributing to reduce the environmental footprint, through the concept of circular economy. Poly(3-hydrobutyrate) is a biodegrable and biocompatible material which stands out as replacement for fossil-derived plastics. However, much work is still needed to reduce the cost of this environmentally friendly biodegradable plastic. This polymer is also proposed for absorbable sutures, scaffolds, heart valves, and cardiovascular tissue supports. This research can lead to new markets.

2.3.5 Uses of Hemp for Domestic Animals

As already mentioned, fiber processing yields a significant quantity of hurds, corresponding to some 60% of the weight of the plant. Hurds are often considered as a byproduct with no worth (Small and Marcus 2002; Gibson 2006; Fike 2016). Indeed, they have a high bulk to price ratio, even when compressed, and cannot be economically transported very far. Today, special uses such as animal bedding and horticulture have become widespread and popular.

Due to its hydrophilic properties and its ability to absorb water, a valuable market for hurds is indeed in high quality bedding for horses, poultry and pet litter (Small and Marcus 2002). The hurds are more absorbent than straw, about four fold, and need to be changed less often. They have better odor suppressing abilities and are less allergenic than alternatives such as wood shavings, straw and hay. In addition, hemp, once on the ground, provides a stable surface that does not tend to move or slip. It is comfortable for horses, for it made up of small particles. However, the bedding is likely to ingested if the horses are hungry.

Cat litter made from granules of hemp powder absorbs urine and faeces and is particularly good at capturing odors. Cleaning of the litter tray is easy as the granules compact down after becoming soiled. They can then be used as a fertilizer of composted down. Hemp straw is used in cattle sheds as an alternative to barley or wheat straw. Small pieces of hurds are suitable as bedding for small mammals, e.g. hamsters, guinea pigs, chinchillas, mice, rats, and pigeons, as well as rabbits and snakes (Small and Marcus 2002; Bouloc 2013; Fike 2016; Johnson 2018).

2.3.6 Horticulture and Market Gardening

Hemp hurds are also used for horticultural mulch. Hemp mulch, like traditional mulch, is primarily used as a surface application for gardens that include vegetable, flower and even container plants such as shrubs. Hemp mulch is easy to handle and is not prone to caking on the surface. It does an excellent job of maintaining soil moisture, minimizing erosion, and suppressing weed growth and seed germination. Indeed, it eliminates the need for manual or chemical weeding control and act as a screen to keep seeds from germinating. Its water absorption capacity cuts down on the need for watering, as the shives keep the ground damper for longer. Hemp mulch is superior in insulating the soil from the hot sun and cold winter, especially frost protection. Product is also pH neutral, compatible with the pH of the soils, fully biodegradable and add humus to soils, i.e. as it decomposes, it helps to enrich the soil (Bouloc 2013; Cherney and Small 2016; Chandra et al. 2017).

Several manufacturers also offer hemp-based fleeces in the form of felt or mulch cloth, supplied mostly in rolls, e.g. Geochanvre® products, France. These materials are a biodegradable layer of fibers used to create a suitable planting environment that warms the soil, retains moisture, and deters weeds. It is also an ecological protection and an alternative to the use of herbicides. Figure 2.8 shows hemp-based products in felt form used in horticulture. They are considered as agricultural textiles or geotextiles (Small and Marcus 2002). For potted plants, they are also used for plant propagation and as a substrate (Kolodziejczyk et al. 2012; Jonaitiene and Stuoge 2016).

Fig. 2.8 Hemp-based products in felt form used in horticulture. (Source: G. Chanet, Eurochanvre, Arc-les-Gray, France)

Used as mulch mats for green areas, e.g. along highways or on roadsides against weed invasion, these biodegradable felts constitute an effective and environment-friendly alternative to plastic mulching. The plants are inserted through the mat, which promotes rapid growth. Once established, the constituent parts of the felt degrade, leaving natural organic matter for continued protection. Hemp fleeces can also be used in earthworks and water engineering, for example, to protect from erosion or to divide soil layers. These products show high saving potentials in terms of energy and emissions of greenhouse gases compared to polypropylene fleece (Kolodziejczyk et al. 2012; Fike 2016; Jonaitiene and Stuoge 2016; Johnson 2018)

2.4 Hemp Products in Animal Nutrition and Food Industry

Hemp seed is a valuable food for both human and animals. Since the 1990s, in several developed countries such as Canada, U.S, France, U.K. and Germany, the seed is gradually making a comeback as ingredients in food products, beverages or as nutritional supplements (Callaway 2004; Kolodziejczyk et al. 2012; Leson 2013; Russo and Reggiani 2013; Girgih et al. 2014c; The and Birch 2014; Bono et al. 2015; Dunford 2015; Andre et al. 2016; Fike 2016; Hartsel et al. 2016; The et al. 2016; Mikulcová et al. 2017; Pihlanto et al. 2017; Frassinetti et al. 2018; Jonhson 2018; Devi and Khanam 2019a, b; Fiorini et al. 2019; Mamone et al. 2019; Wang and Xiong 2019). Currently, there are a worldwide interest in health-promoting functional foods and dietary supplements. France is the world leader in hemp seed production (European Seed Certification Agencies Association, www.escaa.org).

Fig. 2.9 Hemp-based foods and beverages. (Source: G. Crini, Chrono-environnement, Besançon, France)

The renewal of hemp-based foods was due to the fact, in the 1990s, greater attention was paid to the nutritional composition of hemp seeds, particularly of their fatty acid spectrum, which have been found to have a unique and probably beneficial balance of so-called omega-3 and omega-6 fatty acids for health.

Other reasons can also be mentioned such as the growing interest in the valorization of agro-food byproducts, the search of new sources of proteins, the production of bioproducts, e.g. bioactive peptides, natural antioxidants and new natural beverages (Fig. 2.9), and different concerns such as food allergies, animal welfare, and the negative impact on the environment associated with animal-derived proteins (Pihlanto et al. 2017).

For instance, bioactivities claimed in numerous publications are antioxidative, antihypertensive, antimicrobial, anticholesterolemic, and also antitumoral activities, which have attracted growing attention not only from scientists but also from the food industry and consumers. However, more studies need to be done on these health benefits (Girgih et al. 2014a, b; Pihlanto et al. 2017).

Most hemp seeds are used as whole seeds, followed by hemp seeds for oil and dehulled seeds. The whole seed, the cheapest and less processed product, is mainly used for animal feed. In contrast, the dehulled hemp seed, first produced in quantity in Europe, is mainly used for human food (Small and Marcus 2002). The most expensive product, hemp oil, is almost entirely used for human food and cosmetics.

2.4.1 Hemp as a Source of Feed Additives

Currently, hemp is a feedstuff for animal diets (Bouloc 2013; Leson 2013; Fike 2016; Johnson 2018). Four main essentially different types of materials derived from the plant may be used: hemp seed, hemp seed cake/meal, hemp seed oil and whole plant. Further products such as hemp flour and protein isolate from seeds can also be used.

Seed and seed cake are particularly interesting as feed materials for all animal species, i.e. birds, ruminants, pigs, horses, poultry, pigeons, fishes (Bouloc 2013). Birds and fish feed are the main market for hemp seeds in animal feed. Seeds play an important role in the animal nutrition, providing mineral nutrients, vitamins, dietary fibers, as well as biologically-active compounds. Hemp seeds are also used by anglers as bait. The whole plant, including stalk and leaves, is also used for ruminants and horses. Cannabinoid-based formulations containing protein powders such as EliXinol™ are available as dogs treats.

Given the high value of the oils, future use of products in animal feeds may nevertheless be limited to the byproducts produced after the oils have been extruded (Leson 2013; Fike 2016). Indeed, the production of plant seed oils generates tons of processing wastes called seed cakes. These byproducts are then further processed into animal feed due to their high protein and energy contents.

In the last decade, the potential of hemp as an animal feed is dwarfed by its value to the foods, supplements, nutraceuticals, and cosmetics industries that create products and biologically-active compounds with expected health and diseases-prevention benefits for human use, nutrition or consumption. These markets are more lucrative (Johnson 2018).

2.4.2 Hemp Seed and Essential Oils as a Source of Human Food Additives

All hemp food products originate from hemp seeds, i.e. seeds themselves and its products such as meal, flour and protein powder, oil and bioactives substances (Callaway 2004). All these products are achieving a growing popularity in human nutrition as an important food resource (Andre et al. 2016; Frassinetti et al. 2018), the principal product made from hemp seeds today being undoubtedly the seed oil (Johnson 2018).

Demands for natural foods, functional foods, plant-derived proteins, gluten-free products, and bioproducts in worldwide markets have been a major driver of hemp production (Fike 2016). Growth in this market is not surprising given that hemp seeds have excellent fatty acid profiles and protein qualities, and have been used for centuries to treat various disorders (Leson 2013).

In the last years, an increasing use of hemp seeds was observed, due to their nutritional and beneficial properties among people interested in improving and maintaining their health status by changing dietary habits (Fike 2016; Pihlanto et al. 2017; Frassinetti et al. 2018; King 2019). Indeed, they play an important role in the human diet and are an excellent source of nutrients, containing all essential amino acids and fatty acids in sufficient amount and ratio to satisfy the dietary human demand.

Hemp seeds contain more than 30% of oil, i.e. 320–380 g/kg oil, 80% of which is polyunsaturated fatty acids (Callaway 2004; Rodriguez-Leyva and Pierce 2010; Russo and Reggiani 2013; The and Birch 2014; The et al. 2016; Pihlanto et al. 2017).

The oil as functional food is an exceptionally rich source of the two essential fatty acids, linoleic acid or omega-6 and alpha-linolenic acid or omega-3. The omega-6 to omega-3 ratio in hemp seed oil is between 2:1 and 3:1, which is considered to be optimal for human health. A 2.5 value is found in Mediterranean and Japanese diets where the incidence of heart diseases has been historically low (Russo and Reggiani 2013).

The metabolites of the two essential fatty acids, i.e. gamma-linolenic acid and stearidonic acid, are also present in oil. Hemp currently is the only known natural source of gamma-linolenic acid, a widely consumed supplement with numerous health benefits., e.g. to treat eczema and mastalgia.

A nut also contains 25–35% of lipids, 20–30% of carbohydrates, 10–15% of insoluble dietary fibers, and 20–25% of proteins with considerable amounts of vitamins, e.g. vitamin E (90 mg/100 g) and a rich array of minerals such as phosphorus, potassium, magnesium, sulfur, calcium, iron, and zinc (Callaway 2004; Rodriguez-Leyva and Pierce 2010). The seed, and thus oil, does not contain THC. Hemp seed are rich in proteins and contains absolutely no cholesterol. The two main proteins are globulin/edestin (60–80% of the total protein content) and albumin and both of the proteins are easily digested in the human gastrointestinal tract (Malomo and Aluko 2015a; Pihlanto et al. 2017).

Hemp seed protein is considered as a useful food ingredient and a suitable alternative source of functional proteins to traditional ingredients. However, there is scanty information on the structural and functional properties of the seed globulin and albumin fractions. Hemp seeds also contain nutritionally significant amounts of all essential amino acids, especially high levels of the amino acid arginine.

For all these reasons, hempseed oil is marketed as a nutritional additive and a health-promoting product. Oil from hemp seeds is far more valuable in terms of concentrated nutrients and proteins than soybean the nearest vegan alternative.

To retain its valuable constituents, hemp seed oil must be un-refined and cold pressed from non-heat treated seed. Before treatment, hemp seed oils are off-yellow to dark green color. Purified or refined, oil is clear and colorless and has a pleasant, nutty taste/flavor. It is a versatile product, used in liquid or capsule form (oil gel caps). It is best consumed raw, in salad dressings or as a garnish or mayonnaise, and it can be poured over pasta to give extra flavor. Indeed, due to its precious fragile essential fatty acids, it is better not to cook it, e.g. this can create toxic trans-fatty acids. Unsaturated oils also oxidize very easily, which is why such oil quickly becomes rancid on exposure to the air. Hemp seed oil is fairly unstable and becomes rancid rather quickly unless preserved.

Other products being produced today include hemp sauce, butter, hemp meal and gluten-free flour, protein powders (EliXinol™), pasta and spaghetti, sorghum and hemp cakes (Fiorentini®), snack foods, energy bars, hulled hemp seed, muesli, toasted hemp seeds, burger mix, crackers (FoodsAlive©), pancakes, porridge, fruit crumble, chocolate (Fig. 2.8), sweets (Plus™), sour hemp gum, frozen dessert and ice cream (SanMarco®), hemp cheese, etc. Hemp seeds are also incorporated into pizza or used as salt substitute.

Consumers are increasingly looking for products which are not only diverse in terms of taste, but also demonstrate improved nutritional properties, selecting foods with specific health-promoting properties (Fathordoobady et al. 2019; Mikulec et al. 2019). Over the recent years, it has become very popular to enrich food products, including bread, with functional additives.

Mikulec et al. (2019) recently reported hemp flour as a valuable component for enriching physicochemical and antioxidant properties of wheat bread. The aim of their study was to use hemp flour for the production of bread and to determine their impact on selected chemical, texture, organoleptic characteristics, the color of the crumb, changes in the crumb texture, polyphenol profile, the total polyphenol and furan derivatives content. There is a lack of data regarding the influence of hemp flour on the antioxidant potential of bread nor the formation of furan derivates.

The results showed that bread with hemp flour was characterized by significantly higher protein content, in comparison to wheat bread. The share of 30 and 50% of hemp flour contributed to the reduction of organoleptic assessment of the bread. The hemp flour content significantly inhibited the changes in the hardness of bread crumb by reducing bread stalling index from 1.12 (wheat bread) to 0.05 (50% of the additive). The share of hemp flour influenced the color of the crumb by increasing its browning index from 29.69 (standard bread) to 46.26 (50% of the additive). The share of hemp flour influenced the polyphenols content and the formation of furan derivatives, e.g. furfuryl alcohol, furfuryl aldehyde and hydroxymethylfurfural, was dependent on the participation of hemp flour. For industrial production, the share of hemp flour should not exceed 30% (Mikulec et al. 2019).

Hemp protein powder is brownish-green in color and has a taste that can be described as earthy or grassy for some, or nutty for others. It can have a grittier texture than other plant-based protein powders. So, it is best consumed blended with other ingredients (Small and Marcus 2002). For example, hemp powder is added to shakes or smoothies to boost protein intake. The cold-pressed hemp proteins are digestible. Protein powders containing essential amino acids, fibers, unsaturated fats, minerals, e.g. magnesium and iron, and antioxidants are popular nutritional supplements used by vegan persons, athletes, and bodybuilders to increase muscle mass (Bouloc 2013; Cherney and Small 2016; Fike 2016; Chandra et al. 2017; Johnson 2018).

However, there is a debate in the literature on the exact amount of the essential amino acids present in hemp protein powder and on their real impact. In addition, while hemp protein powder is safe for most people, there can be potential side effects, e.g. it can cause digestive problems. It is not recommended for pregnant or lactating women, and people with anemia or with allergies (Cherney and Small 2016; Chandra et al. 2017).

From a fundamental research point of view, there are a considerable effort to improve both nutritional and functional properties of hemp proteins through the exploitation of innovative processing conditions (Malomo et al. 2014; Malomo and Aluko 2015a, 2015b; Korus et al. 2017; Pihlanto et al. 2017; Hadnađev et al. 2018; Dapčević-Hadnađev et al. 2019; Fathordoobady et al. 2019; King 2019; Wang and Xiong 2019; Zajac et al. 2019).

A technological challenge is the incorporation of hemp proteins, both as technological and biofunctional agents, into food products. Indeed, the utilization of these proteins in foods is limited because their behaviour highly depend on their structure and composition, environmental factors, e.g. pH, ionic strength, type of salt used, temperature, and isolation technique (Yin et al. 2008, 2009; Malomo and Aluko 2015a).

Various technologies have recently been proposed to improve the properties of hemp proteins, in particular gelling, foaming, and emulsifying properties, and antioxidant activity (Pihlanto et al. 2017; Hadnađev et al. 2018; Dapčević-Hadnađev et al. 2019; Zajac et al. (2019).

Zajac et al. (2019) recently reported that hemp products as valuable sources of nutrients such as fatty acids, proteins and minerals, could be used to create functional meat products. This topic is interesting since the research on meat products containing hemp ingredients is limited. The authors compared the quality of pork loaves produced with the addition of hemp seeds (5%), de-hulled hemp seeds (5%), hemp flour (5%), and hemp protein (5%).

The results showed that addition of hemp ingredients increased the products' hardness and the fibre content. Magnesium, manganese, iron and copper content was also higher in the products with hemp. Polyunsaturated fatty acids content increased in products with de-hulled and whole hemp seeds. Oxidation is decreased in products with hemp ingredients containing hemp shell. There was no change in the microbial growth after the addition of all the tested ingredients.

The overall acceptability was lower for the products with hemp ingredients, but the taste of meat loaf with de-hulled hemp seeds was comparable with the control product. Zajac et al. (2019) concluded that the most promising ingredients in terms of improving the products' nutritional value were hemp seed and de-hulled hemp seed.

It was possible to create functional meat products using hemp ingredients, with the suggestion to make an optimal mixture, to increase the content of nutritional components without decreasing the consumers' acceptance. However, to encourage the consumers to consume hemp enriched meat products, the information about the healthy ingredients should be provided (Zajac et al. 2019).

Although most research has focused on hemp seeds as a source of oils, proteins and essential fatty acids and minerals, recent studies also showed that that hemp inflorescences from fiber plants, i.e. hemp flowers and upper leaves, could be a good source of essential oils for flavoring in foods. However, the economic potential of these oils remains undefined (Fike 2016; Johnson 2018).

2.4.3 Nutraceutical Potential of Hemp Seeds and Sprouts

Hemp-based foods are gaining popularity and the nutraceutical domain is developing fast due to several factors like the change of lifestyle, interest in alternative diets, and the increasing wareness about sustainable production of food.

Hemp seeds and oil are particularly interesting the nutraceutical domain because they are also an excellent source of protein, minerals, dietary fibre, essential fatty acids, amino acids, and other bioactive substances such as polyphenols. Polyphenols as antioxidants can protect the organism against free radicals attack by reducing or inhibiting cell damages due to the oxidation of lipids or other biomolecules (Conrad 1997; Hartsel et al. 2016; Frassinetti et al. 2018). However, the nutraceutical domain is in its infancy and needs to progress. In addition, food-grade hemp seed requires detailed functional characterization of component proteins (Malomo and Aluko 2015a).

Frassinetti et al. (2018) recently reported that seeds and sprouts of *Cannabis sativa* were rich in phytochemical compounds, particularly polyphenols such as caffeoyltyramine and cannabisin A, B, and C, and also in amino-acids and saccharides. They possessed *in vitro* and *ex vivo* antioxidant activity, and also antimutagenic activity on yeast *Saccharomyces cerevisiae*. Due to the presence of bioactive compounds, seeds and sprouts can be used for nutraceutical and/or therapeutic purposes.

The et al. (2016) and Girgih et al. (2014a) previously reported the antioxidant and anti-hypertensive properties of hemp seed peptides and protein hydrolysate. Werz et al. (2014) also suggested the use of hemp sprouts as a novel anti-inflammatory hemp food product finding that germination and sprouting processes induced the production of anti-inflammatory compounds, prenylflavonoids cannflavins A and B, while cannabinoids were not present at sprout stage.

The byproducts of the oil production also contain secondary metabolites namely phenolic acids and flavonoids that have not been studied extensively. The and Birch (2014) showed that the application of ultrasound for optimization of yield of polyphenols from seed cakes contributed to industrial applications economically and environmentally since it reduced the usage of organic solvent and extraction time. The application of this technique was able to increase polyphenol extraction yields from cakes and aided in enhancing antioxidant capacity of the extracts.

The incorporation of heat during ultrasonic extraction resulted in higher polyphenol yields from the seed cakes compared to the ultrasonic treatment without heat. The polyphenols extracted from the seed cakes by ultrasonic treatment were a good source in product development for nutraceuticals and functional foods that could extend product shelf-life.

2.4.4 Hemp and Beverages

In the last decade, plant-based foods and beverages are gaining popularity and the market is developing fast (Chalupa-Krebzdak et al. 2018; Jeske et al. 2018; Jørgensen et al. 2019). For instance, there has been an expansion of milk alternative beverages originating from plant-based sources including soy, coconut, nuts, and hemp, e.g. Bjorg®, Evernat® and Pacific™ Foods. The reasons for the emergence of market for milk substitutes of plant origin could be attributed to several factors such as cow's milk intolerance, e.g. lactose intolerance, cow's milk allergy, e.g. milk protein allergies, cultural reasons or diet choice, e.g. veganism, flexitarian diet, etc.

Among the various products on the market, hemp seed milk is a popular vegetable alternative to cow's milk, e.g. it is beneficial for people who are lactose intolerant or who avoid dairy products, soy or gluten (Small and Marcus 2002). It is also a good choice for those on a vegan diet. Hemp milk can easily be created at home by mixing water with the seed. This milk is rich in high-quality plant protein, healthy fats and minerals. The hemp-based milk alternative has a protein content of 0.83 g 100 mL^{-1}. This hemp milk also contains alpha-linoleic acid, an essential omega-3 fatty acid, at 0.4 g 100 mL^{-1}, i.e. 25% of recommendend 1.6 g day^{-1} intake (Chalupa-Krebzdak et al. 2018). Some commercial varieties are also fortified with vitamins and minerals.

Compared to whole cow's milk, hemp milk has fewer calories, less protein and carbs but roughly the same amount of fat. It has an earthy, nutty flavor and a creamy consistency. It can be used in place of cow's milk in smoothies, coffee or cereal. Indeed, due to its creamy consistency and protein content, hemp milk is excellent for making lattes, cappuccinos and other coffed drinks. A major concern people have about hemp milk is that it may contain THC as ingredient which is not the case from a regulatory point of view (Bouloc 2013; Jonaitiene and Stuoge 2016; Chandra et al. 2017; Johnson 2018).

Hemp is steadily creeping into a wide range of beverage products (Fig. 2.9), e.g. protein shakes, infusions, hemp-infused beers, e.g. Turn®, Cannabia®, Mandrin®, Coors Light®, and Appenzeller Hanfblüte™, hemp-infused wines, hemp cocktails, e.g. Hempfy tonic gin, alcohols (hempseed used as a flavorant), e.g. Hempfy Martini, lemonades, tea, e.g. HempTea, and coffee nog. Hemp seeds can also be used in a smoothie or a yogurt. All these products have a niche market, based on natural food and beverages, and specialty food outlets.

2.5 Hemp for Cosmetics and Hygiene

All the advantages associated with hemp give hemp seed oils a high market value and make their use likely not only in human food and nutritional supplements, but also in cosmetology and skin care (Fig. 2.10), aromatherapy, and medicine (Conrad 1997; Small and Marcus 2002; Bertoli et al. 2010; Kolodziejczyk et al. 2012; Ionescu et al. 2015; Ligeza et al. 2016; Bonini et al. 2018).

Fig. 2.10 Hemp-based cosmetics. (Source: G. Crini, Chrono-environnement, Besançon, France)

Hemp oil is indeed a good alternative to the chemicals present in many petroleum-based lotions and cosmetics. It is widely used in wellness and body care stores, e.g. for skin hydratation, hand protector, hand sanitizer, body butter, body wash, etc. Its use is intented for people who are both sensitive to their well-being and to the protection of the environment. Consumers have a preference for natural ingredients with little or no impact on the environment.

In cosmetology, hemp is considered as a valuable resource for green cosmetics due to the high content of oil containing interesting substances for skin care with technological and therapeutic effects (Small and Marcus 2002; Vogl et al. 2004). In particular, hemp oils are interesting as natural ingredients/additives due to their high concentration of fatty acids, minerals and vitamins. Fatty acids are structural compounds of phospholipids in cell membranes that influence several cell membrane functions such as hormone activity.

Hemp oils and extracts have also captured a special attention due to their high antioxidant potential (Ionescu et al. 2015; Ligeza et al. 2016). Soothing and restructuring, natural cold pressed oils can be applied to the skin, on the face and on the body (Conrad 1997). Thanks to its natural emollient and moisturizing properties, hemp oil is a common ingredient in body care products, e.g. soaps, shampoos, creams (SATIVA™, BodyShop®), lotions, conditioners, and many other hair and beauty products, e.g. hair care, hair hydration and nourishment, scrubs, perfumes (Cannavis Santal Eau de Parfum), sunless bronzers (HEMPZ®), lipstick (HempOrganics™), etc. The commercial products are paraben-free,

THC-free, and 100% vegan. Recent studies claim that oil reduces wrinkles and keeps the skin a youthful appearance (Ionescu et al. 2015; Ligeza et al. 2016).

The inflorescences of industrial hemp, obtained during cultivation, represent a consistent byproduct that is underutilized. Its great availability make it a potential additional resource to exploit and valorize at industrial level to produce niche products not only for food industry flavoring in foods, but also for cosmetics and pharmacy industry.

Indeed, hemp flowers and upper leaves contain essential oils that they can be used as a scent in perfumes, soaps and candles (Bertoli et al. 2010). These essential oils also have interesting antimicrobial and insecticidal activities (Górski et al. 2009). They can also be used in medicinal formulations (Fernandez-Ruiz et al. 2013). However, research on these topics is still poor (Benelli et al. 2018).

Cosmetic, functional foods and nutraceutical markets appear to be currently driving the hemp revival. However, products remains relatively expensive and their consumption is likely to be restricted.

2.6 Hemp in Medicine

Just as hemp has been used for centuries for domestic and industrial purposes, hemp and cannabis/marijuana have been used for almost as long for medicinal purposes. Hemp seeds and its oil have been used to treat various disorders for thousands of years in traditional Asian medicine (Conrad 1997; de Padua et al. 1999; Leson 2013; Richard and Dejean 2013; Bonini et al. 2018). The medicinal properties of hemp have been exploited since 5000 BC in China and India, e.g. for a number of indications including fever, rheumatic pains, menstrual pains and constipation. The Ancient Greeks and Romans also used cannabis roots for medicinal purposes. For a historical review of this topic, see Russo (2007) and Hanus (2009).

As we have already said, hemp/industrial hemp and marijuana/medical cannabis are often confused (Holler et al. 2008; Sawler et al. 2015; VanDolah et al. 2019). However, hemp and marijuana are different varieties of the plant *Cannabis sativa* L., the first being grown for seeds, oil, fibre and therapeutic hemp, the second for its high content of the psychoactive substance tetrahydrocannabinol (cannabidiol oils).

Industrial hemp versus therapeutic hemp? Hemp is grown differently than marijuana, and the extract is obtained from specific parts of the plant which do not contain tetrahydrocannabinol. Most of the tetrahydrocannabinol content is found in the buds and flowers of the cannabis plant, but industrial hemp is not cultivated to produce buds, so this explains the different content of tetrahydrocannabinol in marijuana versus hemp.

What is therapeutic hemp? This term refers to strains of cannabis that do not contain enough tetrahydrocannabinol to render a psychoactive reaction when ingested by the user. Therapeutic hemp is used to make cannabis concentrate oils that are high in cannabidiol rather than tetrahydrocannabinol. The oil derived from therapeutic hemp is commonly referred to as cannabidiol hemp oil. Charlotte's Web Medical Hemp Act defines therapeutic hemp as "hemp that doesn't make you high".

VanDolah et al. (2019) recently summarized the current legal status of cannabidiol and hemp oils in the United States and provided a guide to identifying higher-quality products so that clinicians could advise their patients on the safest and most evidence-based formulations.

Hemp seeds are a good source of cannabidiol but special extraction techniques are required for its production. From the 1970s, the pharmaceutical industry developed various cannabinoid-based formulations, e.g. Cesamet©, a synthetic derivative of tetrahydrocannabinol used to treat vomiting and nausea and against neurological pain (Small and Marcus 2002), Marinol© also used in the treatment of nausea and vomiting and it can stimulate the appetite of cachectic AIDS patients (Richard and Dejean 2013).

Essential oils containing cannabinoids are approved and marketed, e.g. Sativex® Nabiximol as adjunctive treatment for symptomatic relief of spasticity in adult patients. These products are indicated for the treatment of multiple sclerosis, epilepsy and side effects of cancer chemotherapy. In 2018, the first cannabidiol-based drug, Epidiolex®, was approved by the US Food and Drug Administration for treatment of rare, severe epilepsy, further putting the spotlight on cannabidiol and hemp oils (VanDolah et al. 2019).

Currently, in China, hemp is widely used as a component not only for food but also for medicine: medicines that heal wounds, temper chronic pain, rheumatic pains, menstrual pains and nausea, reduces seizures in epileptics, to reduce acne breakouts and improve skin conditions, to treat dermatitis and eczema, etc.

There are other benefits attributed to compounds derived from various parts of the hemp plant, particularly seed due to its polyunsaturated fatty acids and cannabidiol content, such as anticancer, anti-inflammatory and anti-thrombosis properties, stimulation of general metabolism, neuroprotective effect, promotion of fat burning, hypertension and oxidative stress treatments, help to regulate the immune system, and inflammatory bowel disease (Rodriguez-Leyva and Pierce 2010; Cassano et al. 2013; Russo and Reggiani 2013; Girgih 2014a, b; Cherney and Small 2016; Hartsel et al. 2016; Parian and Limketkai 2016; Pathak et al. 2016; Bonini et al. 2018; Zhou et al. 2018; Devi and Khanam 2019a; Fathordoobady et al. 2019; Fiorini et al. 2019; Mamone et al. 2019).

Some research has suggested that hemp seeds could protect the brain, e.g. positive effects in Parkinson's disease, Alzeimer's disease, and multiple sclerosis (Sativex® is approved in Canada), boost heart health, e.g. reduce the risk of arrhythmias, reduce inflammation (type 2 diabetes), relieve rheumatoid arthritis, and improve skin conditions, e.g. reduce acne symptoms (Rodriguez-Leyva and Pierce 2010; Cassano et al. 2013; Russo and Reggiani 2013; Girgih 2014a, b; Cherney and Small 2016; Hartsel et al. 2016; Parian and Limketkai 2016; Pathak et al. 2016; Bonini et al. 2018; Zhou et al. 2018; Devi and Khanam 2019a; Fathordoobady et al. 2019; Fiorini et al. 2019; Mamone et al. 2019).

However, although the properties and virtues of hemp have been recorded for several thousand years, their medical applications are still not universally recognized and continue to stir much controversy (Fike 2016; Żuk-Gołaszewska and Gołaszewski 2018). Much work is needed to verify various claims about hemp's efficacy.

As reported by Fike (2016) "excitement over this potential needs to be cautious because data in the literature regarding hemp's nutritional and medicinal benefits are somewhat limited and variable, and much of the work has been conducted with animal models". Indeed, great care must be continually taken when reading information on therapeutic hemp and wellness.

2.7 Essential Oils from Hemp as an Effective Tool for Insect Pest Management

Hemp flowers and upper leaves, considered as low-cost byproducts, contain essential oils that they can be used as flavoring in foods or as a scent in cosmetics (Thoma et al. 2000; Small and Marcus 2002). Interestingly, these oils have also been shown to be toxic mosquitoes larves (Thoma et al. 2000). Other valuable properties often cited are their antimicrobial (Verma et al. 2014) and nematicidal (Mukhtar et al. 2013) properties. Thus, byproducts of industrial hemp could also represent an exploitable material to produce biopesticides for agrochemistry sector.

The use of hemp in agrochemistry seems to be a promising field as reported in various studies (Mukhtar et al. 2013; Isman 2015; Bedini et al. 2016; Pavela and Benelli 2016; Abé et al. 2018; Benelli et al. 2018; Fiorini et al. 2019). The following example is interesting.

The Asian tiger mosquito *Aedes albopictus* is acknowledged as the most invasive mosquito species worldwide. Because of its aggressive daytime human-biting behavior and its ability to transmit many pathogens and parasites, including dengue yellow fever and chikungunya, it represents a key threat for millions of people worldwide. The freshwater pan-pulmonate snail *Physella acuta* is another problematic invasive species that shares the same habitats of the *Aedes albopictus* larvae and it is considered a plague in rice fields. Nowadays, pests are largely controlled by synthetic pesticides. However, the continuous use of organophosphates and insect growth regulators has caused the rising of resistant mosquito strains. Besides, currently employed molluscicides are limited in number, expensive and also have negative effects on human health and the environment. Thus, there is a growing interest for alternative eco-friendly control tools for pest management.

The potential of essential oil from industrial hemp as an environmental-friendly botanical insecticide was studied by Bedini et al. (2016). These authors reported that this essential oil was effective against larvae of mosquito vectors and moth pests, as well as against flies and snails. They concluded that hemp essential oils represented new low-cost environmentally friendly insecticides and molluscicides. In addition, these oils are interesting because they are characterized by low toxicity towards other non-target organisms.

Another recent work has been published by the same group (Benelli et al. 2018). The authors used an essential oil obtained by fresh inflorescences of hemp, monoecious *cv.* Felina 32 by steam-distillation. Its composition, analyzed by gas chromatography and gas chromatography-mass spectrometry, was dominated by monoterpene and sesquiterpene hydrocarbons, with (*E*)-caryophyllene (45.4%), myrcene (25.0%) and α-pinene (17.9%) as the most abundant derivatives. The oil was tested against the filariasis vector *Culex quinquefasciatus*, the peach-potato aphid *Myzus persicae*, the housefly *Musca domestica* and the tobacco cutworm *Spodoptera littoralis*. To prove its harmlessness on non-target invertebrates, it was also tested on the multicolored Asian lady beetle, *Harmonia axyridis*, and *Eisenia fetida* earthworms, and compared with α-cypermethrin as the positive control.

Results from insecticidal tests showed that the essential oil from inflorescences of industrial hemp *cv.* Felina 32 was highly toxic to *Myzus persicae* aphids (LC50 of 3.5 mL L^{-1}) and *Musca domestica* flies (43.3 μg $adult^{-1}$), while toxicity was moderate towards *Spodoptera littoralis* larvae (152.3 μg $larva^{-1}$), and scarce against *Culex quinquefasciatus* larvae (LC50 of 252.5 mL L^{-1}) and adults (LC50 > 500 μg cm^{-2}). Contrary to α-cypermethrin, the hemp *cv.* Felina 32 essential oil was not toxic to non-target invertebrate species, including third instar larvae and adults of *Harmonia axyridis* ladybugs and adults of *Eisenia fetida* earthworms. The authors concluded that the essential oil from industrial hemp byproducts is an effective tool for insect pest management in organic crops, particularly to manage aphid and housefly populations (Benelli et al. 2018). The chemical characterization is a crucial step before any kind of biological assay.

Abé et al. (2018) also reported the insecticidal activity of *Cannabis sativa* L. leaf essential oil on the malaria vector *Anopheles gambiae* s.l. larvae. Their work showed high insecticidal effect of this oil against both larvae and adult after 24 h of exposition in controlled conditions. The authors concluded that *Cannabis sativa* L. leaf essential oil could be a serious alternative to insecticides. Further studies are required to assess the same effect in the environment.

2.8 New Uses for Hemp in Energy Production

Industrial hemp is an attractive biomass not only for bioplastic production but also for energy production. Indeed, biomass conversion to biofuels and bioproducts has generated in the last three decades a lot of interest due to the increasing demand for producing a sustainable energy supply that can be incorporated to the existing fuel system.

Traditionally, biofuels have been produced based on starches or sugars such as wheat, corn, sugar beets and sugarcane. New opportunities to use hemp biomass as solid fuel or feedstock in biogas and bioethanol production have been reported. A number of claims have been made that hemp could be used in energy production, as a fuel source with no sulfur emissions, either burnt directly or converted into liquid fuels such as bioethanol (Burczyk et al. 2008; Rice 2008; Li et al. 2010; Sipos et al.

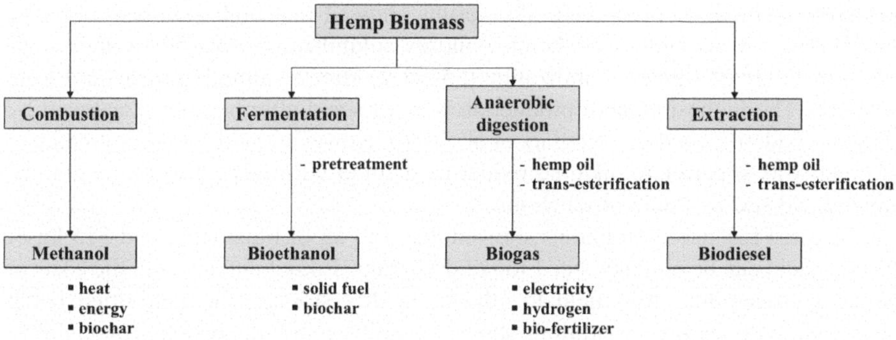

Fig. 2.11 Bioenergy pathways based on hemp biomass conversion. (Adapted from Rehman et al. 2013)

2010; Ahmad et al. 2011; Kreuger et al. 2011; Prade et al. 2011, 2012; Gomes 2012; Finnan and Styles 2013; Rehman et al. 2013; Kuglarz et al. 2014, 2016; Fernando et al. 2015; Das et al. 2017; Schluttenhofer and Yuan 2017).

Industrial hemp is valuable due to its high biomass and energy yields per hectare. For centuries, hemp oil was used as lamp oil. Today, hemp oil can be used to create biofuels to replace gasoline for diesel engines without any needed modifications. These biofuels are renewable and produce less of greenhouse gas carbon monoxide, potentially helping relieve global warming.

Figure 2.11 shows the possible pathway of bioenergy production from industrial hemp biomass (Rehman et al. 2013). This annual plant can be used to produce different products in a biorefinery concept which includes production of vehicle fuels, e.g. biogas from anaerobic digestion or bioethanol from fermentation, heat from briquettes or pellets, electricity from baled biomass, feed and biochemical such as succinic acid. Advantages over other energy crops are also found outside the energy balance, e.g. low pesticide requirements, good weed competition and in crop rotations (annual cultivation). The main competitors for hemp are maize and sugar beets for biogas production and the perennial crops willow, reed canary grass and miscanthus for solid biofuel production (Prade et al. 2012).

Hemp can provide two types of fuels/biofuels, biodiesel made from the oil of the pressed seed and bioethanol and methanol made from the fermented stalk. Biodiesel is considered as a clean and renewable energy alternative to petroleum-based diesel fuel. Bioethanol is also considered as one of the most promising biofuels as it can be easily incorporated into existing fuel systems and can partially substitute fossil fuels used in transportation. However, the most important problem is to understand whether industrial hemp can yield biofuel quantities comparable to the other biomass feedstocks. Another problem is the development of efficient pretreatment technologies to remove lignin and facilitate enzyme access to the cellulose for sugar release.

Angelidaki's group demonstrated that industrial hemp can be used for cellulosic bioethanol and succinic acid production in a biorefinery concept (Kuglarz et al. 2014, 2016). Two types of pretreatments, i.e. dilute-acid treatment and alkaline oxidative method, were studied. The results showed that high cellulose recovery (>95%) as well as significant hemicelluloses solubilization (49–59%) after acid-based method and lignin solubilization (35–41%) after alkaline H_2O_2 method were obtained. The highest ethanol production was achieved after hemp pretreatment by alkaline oxidative method (Kuglarz et al. 2016). However, acid-based pretreatment of hemp was superior to alkaline oxidative method with respect to the combined ethanol and succinic acid production.

The mass balance calculations showed that 149 kg of bioethanol and 115 kg of succinic acid can be obtained per 1 ton of dry hemp. Taking into account the costs of biomass processing, from field to ethanol facility storage, the field-dried hemp pretreated at the optimal conditions showed positive economic results. Angelidaki's group previously showed that the type of hemp cultivation, i.e. organic or conventional, did not influence significantly the effectiveness of the pretreatment as well as subsequent enzymatic hydrolysis and ethanol fermentation (Kuglarz et al. 2014).

Das et al. (2017), using a combined agronomic, experimental and economic analysis approach, reported that industrial hemp is a potential bioenergy crop in comparison with kenaf, switchgrass and biomass sorghum. For instance, the authors reported a predicted ethanol yield of 82 gallons/dry ton hemp stems which was comparable to the other three tested feedstocks. However, despite numerous studies related to the biofuels potential of hemp, its technical and economic feasibility still remains unclear (Fike 2016; Johnson 2018).

2.9 Environmental Applications

One of the most interesting applications for hemp is in cleaning up soil contamination through phytoremediation and phytoextraction processes. Hemp has been tested with favorable results as phytoextractor in areas where lands were contaminated with various pollutants such as metals, radioactive elements, organics including pesticides and fertilizers, oils and solvents (Linger et al. 2002; Small and Marcus 2002; Citterio et al. 2003; Vandenhove and Van Hees 2003; Gomes 2012; Gupta et al. 2013; Ahmad et al. 2016; Morin-Crini et al. 2018; Saxena et al. 2020).

Linger et al. (2002) previously reported that hemp was able to decontaminate metal polluted soils. All parts of plants, i.e. seeds, leaves, fibers and hurds, contained metals but the metal accumulation in these different parts was extremely different. The highest concentrations of Ni, Pb and Cd were accumulated in the leaves. In the field trial, hemp demonstrated a phytoextraction potential of 126 g Cd (ha vegetation period)$^{-1}$. The authors showed that the high quality of the fibers and hurds, which were not affected by the metal contamination, allowed them to be used in special products like combine material (Linger et al. 2002).

Hemp plants were shown to be effective in cleaning the soil around the site of Russia's Chernobyl nuclear disaster (Vandenhove and Van Hees 2003). They were also considered for use near Fukushima (Morin-Crini et al. 2018).

Another interesting environmental field is the removal of pollutants present in aqueous solution by hemp-based biosorbents. Indeed, with the increasing focus on renewable materials and sustainability issues, the development of non-conventional materials from natural resources and possessing complexing and chelating properties such as hemp is currently an area of extensive research due to their potential applications in pollutant removal, e.g. in water and wastewater treatment.

This is an interesting challenge because the majority of commercial organic resins are derived from petroleum-based raw materials using processing chemistry that is not always safe or environmental friendly. Today, there is growing interest in developing natural low-cost alternatives to synthetic resins and polymers. Hemp as biosorbent could be a promising alternative.

In the last decade, several research groups published numerous works on the ability of hemp to act as an effective biosorbent for the removal of metals from aqueous solutions or industrial effluents (Kostić et al. 2008, 2010; Pejić et al. 2008, 2009, 2011; Rosas et al. 2009; Tofan et al. 2009, 2010a, b, c, 2013, 2015, 2016a, b; Zou et al. 2012; Vukčević et al. 2012, 2014a, b, 2015; Yang et al. 2011, 2012; Rezić 2013; Sun et al. 2013; Balintova et al. 2014; Lupul et al. 2015a, b; Kyzas et al. 2015; Wang et al. 2015; Bugnet et al. 2017a, b; Loiacono et al. 2017a, b, c, 2018a, b; Morin-Crini et al. 2018, 2019).

Many materials containing hemp in raw, modified or carbon forms, have been proposed. All these results clearly demonstrated that hemp had a high affinity for metals such as Cd, Co, Cr, Cu, Mn, Ni, Pb and Zn. Hemp is able to remove metals from mono- and polycontaminated solutions and the performances are in general almost independent of pH between 4 and 6. Strong bonding of metal ions by carboxylic present in hemicelluloses, pectin and lignin, phenolic (lignin and extractives), carbonyl (lignin) and hydroxyl (polysaccharides) groups are responsible of the adsorption through chemisorption. However, the performance depends on the hemp form used, i.e. fibers, shives or felt, and also on the residual concentration of the metal in solution. In addition, all the works are in the stage of laboratory-scale study using often standard solutions containing one or a few metals. Future research needs to demonstrate the possibilities on an industrial scale using real polycontaminated wastewaters and discharge waters (Morin-Crini et al. 2018, 2019).

2.10 Miscellaneous Applications

Other domains for hemp uses include paints, e.g. Chanvre Mat, Milk Paint, varnishes, ink and lubricants, solvents, detergents, and industrial cleaners (Small and Marcus 2002; Callaway 2004; Bouloc 2013).

Innovative applications such as coatings, nanotechnology, e.g. nanomaterials with similar properties as graphene, supercapacitors (TitanHemp) and nanosheets, hemp plastic for 3D printing (Kanesis HBP®), biocomposites for airplanes or solar panels, in cleaning up air, acoustic domain and biotechnologies (Khattab and Dahman 2019; Rossi et al. 2020) opens new challenges. For instance, hemp-based supercapacitors offer an affordable next generation energy source to replace rechargeable batteries for applications such as electric cars, power tools and mobile devices (Sun et al. 2016). Several patents have been filed in recent years.

2.11 Conclusion

Industrial hemp is commonly used to refer to Cannabis strains cultivated for industrial use, i.e. for non-drug use. Although industrial hemp is a niche crop (Food and Agriculture Organization of the United Nations, www.faostat.fao.org), its production is currently experiencing a renaissance, particularly in Europe and North America. China is the largest producer and exporter of hemp in the world. Hemp is a valuable crop for the bio-based economy because of its unique properties and environmental benefits, and the high yield of natural products it provides.

Hemp is a multi-purpose crop delivering stalks, seeds and leaves, which find numerous applications. Its uses are indeed manifold: construction materials, textiles, paper, food and beverages, automotive sector, furniture, luxury market, cosmetics and personal care items, etc. There are an estimated 25,000 products derived from industrial hemp. The importance of functional foods and nutraceuticals containing hemp products is related to health promotion and diseases risk reduction. Neutraceutical products, cosmeceuticals, medical and therapeutic domains should be the next market in the development of industrial hemp. Other potential uses and innovative applications opens new challenges, e.g. phytoremediation, wastewater treatment, energy production, biofuels, bio-based plastics, and bio-pesticides.

The economics of the use of hemp products are a subject of ongoing debate, research and development, and trade analysis (Fike 2016). Although these available economic data are still limited, the hemp industry continues to evolve and invest (Source: BDS Analytics Newsletter 2019; https://bdsa.com/resources-summary/). Whether the hemp industry will grow depends on the political and economic framework in the European Union and other countries such as Canada and the U.S. The future development of hemp will also depend strongly on market demand for green products that are both beneficial to human health and have no impact on the environment.

References

Abd Rashid AF, Yusoff S (2015) A review of life cycle assessment method for building industry. Renew Sust Energ Rev 45:244–248

Abé H, Foko Dadji AF, Nkondjio CA, Awono-Ambene PH, Tamesse JL (2018) Insecticidal activity of *Cannabis sativa* L leaf essential oil on the malaria vector *Anopheles gambiae* sl (Giles). Int J Mosquito Res 5:65–74

Ahmad M, Ullah K, Khan MA, Zafar M, Tariq M, Ali S, Sultana S (2011) Physicochemical analysis of hemp oil biodiesel: a promising non edible new source for bioenergy. Energy Sources A 33:1365–1374. https://doi.org/10.1080/15567036.2010.499420

Ahmad R, Tehsin Z, Malik ST, Asad SA, Shahzad M, Bilal M, Shah MM, Khan SA (2016) Phytoremediation potential of hemp (*Cannabis sativa* L.): identification and characterization of heavy metals responsive genes. Clean Soil Air Water 44:195–201. https://doi.org/10.1002/clen.201500117

Aït Oumeziane Y, Moissette S, Bart M, Lanos C (2016) Influence of temperature on sorption process in hemp concrete. Constr Build Mater 106:600–607. https://doi.org/10.1016/j.conbuildmat.2015.12.117

Allegret S (2013) The history of hemp. In: Bouloc P, Allegret S, Arnaud L (eds) Hemp – industrial production and uses. CAB International, Wallingford

Amaducci S (2005) Hemp production in Italy. J Ind Hemp 10:109–115

Amaducci S, Gusovious HJ (2010) Chapter: Hemp – cultivation, extraction and processing. In: Müssig J (ed) Industrial applications of natural fibres: structure, properties and technical applications, vol 5. Wiley, Chichester, pp 109–135. https://doi.org/10.1002/9780470660324.ch5

Amaducci S, Scordia D, Liu FH, Zhang Q, Guo H, Testa G, Cosentino SL (2015) Key cultivation techniques for hemp in Europe and China. Ind Crop Prod 68:2–16. https://doi.org/10.1016/j.indcrop.2014.06.041

Amziane S, Arnaud L (2013) Bio-aggregate-based building materials: applications to hemp concretes. ISTE Ltd./Wiley, London. ISBN: 978-1848214040

Andre CM, Hausman JF, Guerriero G (2016) *Cannabis sativa*: the plant of the thousand and one molecules. Front Plant Sci 7:1–17. https://doi.org/10.3389/fpls.2016.00019

Ansell MP, Mwaikambo LY (2009) Chapter 2: The structure of cotton and other plant fibers. In: Eichhorn SJ, JWS H, Jaffe M, Kikutani T (eds) Handbook of textile fibre structure, Natural, regenerated, inorganic and specialist fibres, vol 2. Woodhead Publishing, Ltd, Cambridge, pp 62–94. https://doi.org/10.1533/9781845697310.1.62

Arizzi A, Viles H, Martin-Sanchez I, Cultrone G (2016) Predicting the long-term durability of hemp-lime renders in inland and coastal areas using Mediterranean, tropical and semi-arid climatic simulations. Sci Total Environ 542:757–770. https://doi.org/10.1016/j.scitotenv.2015.10.141

Arnaud L, Gourlay E (2012) Experimental study of parameters influencing mechanical properties of hemp concretes. Constr Build Mater 28:50–56. https://doi.org/10.1016/j.conbuildmat.2011.07.052

Bajpai P (2018) Biotechnology for pulp and paper processing. Springer, Singapore, pp 1–588. ISBN: 978-981107853-8

Balintova M, Holub M, Stevulova N, Cigasova J, Tesarcikova M (2014) Sorption in acidic environment – biosorbents in comparison with commercial adsorbents. Chem Eng Trans 39:625–630. https://doi.org/10.3303/CET1439105

Barberà L, Pèlach MA, Pérez I, Puig J, Mutjé P (2011) Upgrading of hemp core for papermaking purposes by means of organosolv process. Ind Crops Prod 34:865–872. https://doi.org/10.1016/j.indcrop.2011.02.005

Bedini S, Flamini G, Cosci F, Ascrizzi R, Benelli G, Conti B (2016) *Cannabis sativa* L. and *Humulus iupulus* essential oils as novel control tools against the invasive mosquito *Aedes albopictus* and fresh water snail *Physella acuta*. Ind Crop Prod 85:318–323. https://doi.org/10.1016/j.indcrop.2016.03.008

Benelli G, Pavela R, Petrelli R, Cappellacci L, Santini G, Fiorini D, Sut S, Dall'Acqua S, Canale A, Maggi F (2018) The essential oil from industrial hemp (*Cannabis sativa* L.) by-products as an effective tool for insect pest management in organic crops. Ind Crop Prod 122:308–315. https://doi.org/10.1016/j.indcrop.2018.05.032

Bertoli A, Tozzi S, Pistelli L, Angelini LG (2010) Fibre hemp inflorescences: from crop-residues to essential oil production. Ind Crop Prod 32:329–337. https://doi.org/10.1016/j.indcrop.2010.05.012

Bhavani K (2015) Performance evaluation of various natural agro fibres in carpet making and their costing. Asian J Home Sci 10:296–300. https://doi.org/10.15740/HAS/AJHS/10.2/296-300

Bismarck A, Mishra S, Lampke T (2005) Plant fibers as reinforcement for green composites. In: Mohanty AK, Misra M, Drzal LT (eds) Natural fibers, biopolymers, and biocomposites. CRC Press, Boca Raton, pp 37–108. ISBN 9780849317415

Bócsa L, Karus M (1998) The cultivation of hemp: botany, varieties, cultivation and harvesting. Hemptech, Sebastopol, 184 p. ISBN: 9781886874039

Bonini SA, Premoli M, Tambaro S, Kumar A, Maccarinelli G, Memo M, Mastinu A (2018) *Cannabis sativa*: a comprehensive ethnopharmacological review of a medicinal plant with a long history. J Ethnopharmacol 227:300–315. https://doi.org/10.1016/j.jep.2018.09.004

Bono P, le Duc A, Lozachmeur M, Day A (2015) Materials: new fields of research and development for the valorization of technical plant fibres (flax fiber and hemp). OCL oilseeds & fats crops and lipids. EDP. Sciences 22:1–11. https://doi.org/10.1051/ocl/2015041

Bouloc P (2013) Hemp: industrial production and uses. CABI, Oxfordshire, 312 p. ISBN: 9781845937935

Bran M, Olaru S, Dobre I (2017) Flax, hemp and cotton plants in Romania – a study for reconsideration of the textile industry. Ind Text 68:186–192

Bugnet J, Morin-Crini N, Cosentino C, Chanet G, Winterton P, Crini G (2017a) Hemp decontamination of poly-metallic aqueous solutions. Environ Eng Manag J 16:535–542

Bugnet J, Morin-Crini N, Chanet G, Cosentino C, Crini G (2017b) Du chanvre pour dépolluer des eaux polycontaminées en métaux. Morin-Crini G, Crini G (ed) PUFC: Besançon, pp. 323–340, Chapter XI

Burczyk H, Grabowska L, Kołodziej J, Strybe M (2008) Industrial hemp as a raw material for energy production. J Ind Hemp 13:37–48. https://doi.org/10.1080/15377880801898717

Callaway JC (2004) Hemp seed as a nutritional resource: an overview. Euphytica 140:65–72. https://doi.org/10.1007/s10681-004-4811-6

Cassano R, Trombino S, Ferrarelli T, Nicoletta FP, Mauro MV, Giraldi C, Picci N (2013) Hemp fiber (*Cannabis sativa* L.) derivatives with antibacterial and chelating properties. Cellulose 20:547–557. https://doi.org/10.1007/s10570-012-9804-3

Chalupa-Krebzdak S, Long CJ, Bohrer BM (2018) Nutrient density and nutritional value of milk and plant-based milk alternatives. Int Dairy J 87:84–92. https://doi.org/10.1016/jdairyl.2018.07.018

Chandra S, Lata H, Elsohly MA (eds) (2017) *Cannabis sativa* L. – botany and biotechnology. Springer, Cham. ISBN: 978-3-319-54563-9

Chang CW, Yen CC, Wu MT, Hsu MC, Wu YT (2017) Microwave-assisted extraction of cannabinoids in hemp nut using response surface methodology: optimization and comparative study. Molecules 22:1894. https://doi.org/10.3390/molecules22111894

Cherney JH, Small E (2016) Industrial hemp in North America: production, politics and potential. Agronomy 6:1–24. https://doi.org/10.3390/agronomy6040058

Chernova T, Mikshina P, Salnikov V, Ageeva M, Ibragimova N, Sautkina O, Gorshlova T (2018) Development of hemp fibers: the key components of hemp plastic composites. IntechOpen, chapter 3, pp. 41–56. https://doi.org/10.5772/intechopen.70976

Cigasova J, Stevulova N, Schwarzova I, Sicakova A, Junak J (2015) Application of hemp hurds in the preparation of biocomposites. IMST 2015 IOP Conf Ser Mater Sci Eng 96:012023

Citterio S, Santagostino A, Fumagalli P, Prato N, Ranalli P, Sgorbati S (2003) Heavy metal tolerance and accumulation of Cd, Cr and Ni by *Cannabis sativa* L. Plant Soil 256:243–252. https://doi.org/10.1023/A:1026113905129

Collet F, Chamoin J, Pretot S, Lanos C (2013) Comparison of the hygric behaviour of three hemp concretes. Energ Buildings 62:294–303. https://doi.org/10.1016/j.enbuild.2013.03.010

Conrad C (1997) Hemp for health: the medicinal and nutritional uses of *Cannabis sativa*. Healing Arts Press, Rochester. ISBN: 0-89281-539-6

Da Porto C, Voinovich D, Decorti D, Natolino A (2012) Response surface optimization of hemp seed (*Cannabis sativa* L.) oil yield and oxidation stability by supercritical carbon dioxide extraction. J Supercrit Fluids 68:45–51. https://doi.org/10.1016/j.supflu.2012.04.008

Danielewicz D, Surma-Slusarska B (2017) Properties and fibre characterization of bleached hemp, birch and pine pulps: a comparison. Cellulose 24:5173–5186. https://doi.org/10.1007/s10570-017-1476-6

Dapčević-Hadnađev T, Hadnađev M, Lazaridou A, Moschakis T, Biliaderis CG (2019) Hempseed meal protein isolates prepared by different isolation techniques. Part II. Gelation properties at different ionic strengths. Food Hydrocoll 81:481–489. https://doi.org/10.1016/j.foodhyd.2018.03.022

Das L, Liu E, Saeed A, Williams DW, Hu H, Li C, Ray AE, Shi J (2017) Industrial hemp as a potential bioenergy crop in comparison with kenaf, switchgrass and biomass sorghum. Bioresour Technol 244:641–649. https://doi.org/10.1016/j.biortech.2017.08.008

De Candolle A (1883) Origine des plantes cultivées. Librairie Germer Baillière et Cie, Paris. ISBN: 2734801426

de Padua LS, Bunyapraphatsara N, Lemmens RHMJ (eds) (1999) Plant resources of south-East Asia: medicinal and poisonous plants, vol 1. Backhuys Publishers, Leiden, pp 167–175

Devi V, Khanam S (2019a) Comparative study of different extraction processes for hemp (*Cannabis sativa*) seed oil considering physical, chemical and industrial-scale economic aspects. J Clean Prod 207:645–657. https://doi.org/10.1016/j.jclepro.2018.10.036

Devi V, Khanam S (2019b) Study of ω-6 linoleic and ω-3 α-linoleic acids of hemp (*Cannabis sativa*) seed oil extracted by supercritical CO_2 extraction: CCD optimization. J Environ Chem Eng 7:102818. https://doi.org/10.1016/j.jece.2018.102818

Dubois S, Evrard A, Lebeau F (2014) Modeling the hygrothermal behavior of biobased construction materials. J Build Phys 38:191–213. https://doi.org/10.1177/1744259113489810

Dunford NT (2015) Hemp and flaxseed oil: properties and applications for use in food. In: Talbot G (ed) Specialty oils and fats in food and nutrition: properties, processing and applications. Elsevier, Amsterdam, pp 39–63. https://doi.org/10.1016/B978-1-78242-376-8.00002-8

Elfordy S, Lucas F, Tancret F, Scudeller Y, Goudet L (2008) Mechanical and thermal properties of lime and hemp concrete (hempcrete) manufactured by a projection process. Constr Build Mater 22:2116–2123. https://doi.org/10.1016/j.conbuildmat.2007.07.016

Fangueiro R, Rana S (eds) (2016) Natural fibers: advances in science and technology toward industrial applications. From science to market. Springer, Dordrecht. https://doi.org/10.1007/978-94-017-7515-1

Faruk O, Bledzki A, Fink HP, Sain M (2012) Biocomposites reinforced with natural fibers: 2000–2010. Prog Polym Sci 37:1552–1596. https://doi.org/10.1016/j.progpolymsci.2012.04.003

Fathordoobady F, Singh A, Kitts DD, Singh AP (2019) Hemp (*Cannabis sativa L*) extract: anti-microbial properties, methods of extraction, and potential oral delivery. Food Rev Int. https://doi.org/10.1080/87559129.2019.1600539

Feng JY, Zhang JC (2015) Preparation and oil/air filtration properties of hemp paper. J Ind Text 45:3–32. https://doi.org/10.1177/1528083714521071

Fernandez-Ruiz J, Sagredo O, Pazos MR, Garcia C, Pertwee R, Mechoulam R, Martinez-Orgado J (2013) Cannabidiol for neurodegenerative disorders: important new clinical applications for this phytocannabinoid? Br J Clin Pharmacol 75:323–333. https://doi.org/10.1111/j.1365-2125.2012.04341.x

Fernando AL, Duarte MP, Vatsanidou A, Alexopoulou E (2015) Environmental aspects of fiber crops cultivation and use. Ind Crop Prod 68:105–115. https://doi.org/10.1016/j.indcrop.2014.10.003

Fernea R, Manea DL, Tamas-Gavrea DR, Miron OI (2017) Preliminary research on using industrial plants in constructions. Sci Pap Ser Manag Econ Eng Agric Rural Dev 17:203–208

Figueiredo JA, Ismael MI, Anjo CMS, Duarte AP (2010) Cellulose and derivatives from wood and fibers as renewable sources of raw materials. Top Curr Chem 294:117–128
Fike J (2016) Industrial hemp: renewed opportunities for an ancient crop. Crit Rev Plant Sci 35:406–424. https://doi.org/10.1080/07352689.2016.1257842
Finnan J, Styles D (2013) Hemp: a more sustainable annual energy crop for climate and energy policy. Energy Policy 58:152–162. https://doi.org/10.1016/j.enpol.2013.02.046
Fiorini D, Molle A, Nabissi M, Santini G, Benelli G, Maggi F (2019) Valorizing industrial hemp (*Cannabis sativa* L.) by-products: cannabidiol enrichment in the inflorescence essential oil optimizing sample pretreatment prior to distillation. Ind Crop Prod 128:581–589. https://doi.org/10.1016/j.indcrop.2018.10.045
Frassinetti S, Moccia E, Caltavuturo L, Gabriele M, Longo V, Bellani L, Giorgi G, Giorgetti L (2018) Nutraceutical potential of hemp (*Cannabis sativa* L.) seeds and sprouts. Food Chem 262:56–66. https://doi.org/10.1016/j.foodchem.2018.04.078
Gallos A, Paës G, Allais F, Beaugrand J (2017) Lignocellulosic fibers: a critical review of the extrusion process for enhancement of the properties of natural fiber composites. RSC Adv 7:34638. https://doi.org/10.1039/c7ra05240e
Gedik G, Avinc O (2018) Bleaching of hemp (*Cannabis sativa* L.) fibers with peracetic acid for textiles industry purposes. Fibers Polym 19:82–93. https://doi.org/10.1007/s12221-018-7165-0
George M, Chae M, Bressler DC (2016) Composite materials with bast fibres: structural, technical, and environmental properties. Prog Mater Sci 83:1–23. https://doi.org/10.1016/j.pmatsci.2016.04.002
Gibson K (2006) Hemp in the British Isles. J Ind Hemp 11:57–67
Girgih AT, Alashi A, He R, Malomo S, Aluko RE (2014a) Preventive and treatment effects of a hemp seed (*Cannabis sativa* L.) meal protein hydrolysate against high blood pressure in spontaneously hypertensive rats. Eur J Nutr 53:1237–1246
Girgih AT, Alashi AM, He R, Malomo SA, Raj P, Netticadan T, Aluko RE (2014b) A novel hemp seed meal protein hydrolysate reduces oxidative stress factors in spontaneously hypertensive rats. Nutrients 6:5652–5666
Girgih AT, He R, Malomo S, Offengenden M, Wu J, Aluko RE (2014c) Structural and functional characterization of hemp (*Cannabis sativa* L.) protein-derived antioxidant and antihypertensive peptides. J Funct Foods 6:384–394. https://doi.org/10.1016/j.jff.2013.11.005
Gomes HI (2012) Phytoremediation for bioenergy: challenges and opportunities. Environ Technol Rev 1:59–66. https://doi.org/10.1080/09593330.2012.696715
Gorchs G, Llovers J, Comas J (2000) Effect of hemp (*Cannabis sativa* L.) in a crop rotation hemp-wheat in the humid cool areas of north-eastern Spain. In: Proceedings of the conference on crop development for the cool and wet regions of Europe, Pordenone, Italy, pp 581–589
Górski R, Szklarz M, Kaniewski R (2009) Efficacy of hemp essential oil in the control of rosy apple aphid (*Dysaphis plantaginea* Pass.) occurring on apple tree. Prog Plant Prot 49:2013–2016
Gourlay E, Gle P, Marceau S, Foy C, Moscardelli S (2017) Effect of water content on the acoustical and thermal properties of hemp concretes. Constr Build Mater 139:513–523. https://doi.org/10.1016/j.conbuildmat.2016.11.018
Gupta AK, Verma SK, Khan K, Verma RK (2013) Phytoremediation using aromatic plants: a sustainable approach for remediation of heavy metal polluted sites. Environ Sci Technol 47:10115–10116. https://doi.org/10.1021/es403469c
Hadnađev M, Dapčević-Hadnađev T, Lazaridou A, Moschakis T, Michaelidou AM, Popović S, Biliaderis CG (2018) Hempseed meal protein isolates prepared by different isolation techniques. Part I. physicochemical properties. Food Hydrocoll 79:526–533. https://doi.org/10.1016/j.foodhyd.2017.12.015
Hanus LO (2009) Pharmacological and therapeutics secrets of plant and brain (endo) cannabinoids. Med Res Rev 29:213–271
Harris AT, Riddlestone S, Bell Z, Hatwell PR (2008) Towards zero emission pulp and paper production: the bioregional MiniMill. J Clean Prod 16:1971–1979. https://doi.org/10.1016/j.jclepro.2008.02.005

Hartsel JA, Eades J, Hickory B, Makriyannis A (2016) *Cannabis sativa* and hemp. In: Gupta RC (ed) Nutraceuticals – efficacy, safety and toxicity. Elsevier, Amsterdam, pp 735–754. https://doi.org/10.1016/B978-0-12-802147-7.00053-X

Hautala M, Pasila A, Pirilä J (2004) Use of hemp and flax in composite manufacture: a search for new production methods. Compos Part A 35:11–16. https://doi.org/10.1016/j.compositesa.2003.09.023

Holler JM, Bosy TZ, Dunkley CS, Levine B, Past MR, Jacobs A (2008) Δ9-Tetrahydrocannabinol content of commercially available hemp products. J Anal Toxicol 32:428–432

Ingrao C, Lo Giudice A, Bacenetti J, Tricase C, Dotelli G, Fiala M, Siracusa V, Mbohwa C (2015) Energy and environmental assessment of industrial hemp for building applications: a review. Renew Sust Energ Rev 51:29–42. https://doi.org/10.1016/j.rser.2015.06.002

Ionescu N, Popescu M, Bratu A, Istrati D, Ott C, Meghea A (2015) Valuable Romanian vegetable oils and extracts with high pharmaco-cosmetic potential. Rev Chim 66:1267–1272

Ip K, Miller A (2012) Life cycle greenhouse gas emissions of hemp-lime wall constructions in the UK. Resour Conserv Recycl 69:1–9. https://doi.org/10.1016/j.resconrec.2012.09.001

Isman MB (2015) A renaissance for botanical insecticides? Pest Manag Sci 71:1587–1590. https://doi.org/10.1002/ps.4088

Iucolano F, Boccarusso L, Langella A (2019) Hemp as eco-friendly substitute of glass fibres for gypsum reinforcement: impact and flexural behavior. Compos Part B Eng 175:107073. https://doi.org/10.1016/j.compositesb.2019.107073

Jami T, Rawtani D, Agrawal YK (2016) Hemp concrete: carbon-negative construction. Emerg Mater Res 5:1–7. https://doi.org/10.1680/jemmr.16.00122

Jami T, Karade SR, Singh LP (2019) A review of the properties of hemp concrete for green building applications. J Clean Prod 239:117852. https://doi.org/10.1016/j.jclepro.2019.117852

Jeske S, Zannini E, Arendt EK (2018) Past, present and future: the strength of plant-based dairy substitutes based on gluten-free raw materials. Food Res Int 110:42–51. https://doi.org/10.1016/j.foodres.2017.03.045

Jiang Y, Lawrence M, Ansell MP, Hussain A (2018) Cell wall microstructure, pore size distribution and absolute density of hemp shiv. Royal Soc Open Sci 5:171945. https://doi.org/10.1098/rsos.171945

Johnson P (1999) A critical review of claimed potentials for *Cannabis sativa*. TAPPI J 82:113–123

Johnson R (2018) Hemp as an agricultural commodity. Congressional Res Serv 7-5700:1–48. www.crs.gov

Jonaitiene JZ, Stuoge I (2016) Hemp cultivation opportunities and perspectives in Lithuania. In: Fangueiro R, Rana S (eds) Natural fibers: advances in science and technology toward industrial applications. From science to market. Springer, Dordrecht, pp 407–415. https://doi.org/10.1007/978-94-017-7515-1

Jørgensen CE, Abrahamsen RK, Rukke EO, Hoffmann TK, Johansen AG, Skeie SB (2019) Processing of high-protein yoghurt – a review. Int Dairy J 88:42–59. https://doi.org/10.1016/j.dairyl.2018.08.002

Kallakas H, Narep M, Narep A, Poltimae T, Kers J (2018) Mechanical and physical properties of industrial hemp-based insulation materials. Proc Est Acad Sci 67:183–192. https://doi.org/10.3176/proc.2018.2.10

Karaduman Y, Ozdemir H, Karaduman NS, Ozdemir G (2018) Interfacial modification of hemp fiber-reinforced composites. IntechOpen, Chapter 2:17–39. https://doi.org/10.5772/intechopen.70519

Karus M, Vogt D (2004) European hemp industry: cultivation, processing and product lines. Euphytica 140:7–12. https://doi.org/10.1007/s10681-004-4810-7

Khattab MM, Dahman Y (2019) Production and recovery of poly-3-hydrobutyrate bioplastics using agro-industrial residues of hemp hurd biomass. Bioprocess Biosyst Eng:1–13. https://doi.org/10.1007/s00449-019-02109-6

King JW (2019) The relationship between cannabis/hemp use in foods and processing methodology. Curr Opin Food Sci 28:32–40. https://doi.org/10.1016/j.cofs.2019.04.007

Kinnane O, Reilly A, Grimes J, Pavia S, Walker R (2016) Acoustic absorption of hemp-lime construction. Constr Build Mater 122:674–682. https://doi.org/10.1016/j.conbuildmat.2016.06.106

Kiruthika AV (2017) A review on physico-mechanical properties of bast fibre reinforced polymer composites. J Build Eng 9:91–99. https://doi.org/10.1016/j.jobe.2016.12.003

Kolodziejczyk P, Ozimek L, Kozlowska J (2012) The application of flax and hemp seeds in food, animal feed and cosmetics production. In: Kozlowski RM (ed) Handbook of natural fibres, vol 2. Elsevier, pp 329–366. https://doi.org/10.1016/B978-1-84569-698-6.50011-5

Korus J, Witczak M, Ziobro R, Juszczak L (2017) Hemp (*Cannabis sativa subsp. sativa*) flour and protein preparation as natural nutrients and structure forming agents in starch based gluten-free bread. Lebensm Wiss Technol 84:143–150. https://doi.org/10.1016/j.lwt.2017.05.046

Kostić MM, Skundric PD, Milosavljevic S (2003) Hemp is coming back (in Serbia). Tekstilna Industrija 51:9–16

Kostić MM, Pejić BM, Skundric PD (2008) Quality of chemically modified hemp fibres. Bioresour Technol 99:94–99. https://doi.org/10.1016/j.biortech.2006.11.050

Kostić MM, Pejić BM, Asanovic KA, Aleksic VM, Skundric PD (2010) Effect of hemicelluloses and lignin on the sorption and electric properties of hemp fibres. Ind Crop Prod 32:169–174. https://doi.org/10.1016/j.indcrop.2010.04.014

Kostić M, Vukčević M, Pejić B, Kalijadis A (2014) Hemp fibers: old fibers – new applications. In: Ibrahim MD, Mondal M (eds) Textiles: history, properties and performance and applications. Nova Science Publishers, Inc, New York, pp 399–446

Kostić M, Pejić B, Vukčević M (2018) Chapter 1: Waste hemp (*Cannabis sativa*) fibers as a biosorbent and a precursor for biocarbon sorbents: influence of their chemical composition on Pb(II) removal. In: Stevanovic T (ed) Chemistry of lignocellulosics: current trends. Taylor & Francis Group, CRC Press, Boca Raton, pp 3–21. https://doi.org/10.1201/b20936

Kozlowski R, Baraniecki P, Barriga-Bedoya J (2005) Chapter 2: Bast fibres (flax, hemp, jute, ramie, kenaf, abaca). In: Blackburn RS (ed) Biodegradable and sustainable fibres, Cambridge: Woodhead publishing in textiles. The textile institute. CRC Press/Woodhead Publishing Ltd, Cambridge, pp 36–88. https://doi.org/10.1533/9781845690991.36

Kreuger E, Sipos B, Zacchi G, Svensson SE, Björnsson L (2011) Bioconversion of industrial hemp to ethanol and methane: the benefits of steam pretreatment and co-production. Bioresour Technol 102:3457–3465. https://doi.org/10.1016/j.biortech.2010.10.126

Kuglarz M, Gunnarsson IB, Svensson SE, Prade T, Johansson E, Angelidaki I (2014) Ethanol production from industrial hemp: effect of combined dilute acid/steam pretreatment and economic aspects. Bioresour Technol 163:236–243. https://doi.org/10.1016/j.biortech.2014.04.049

Kuglarz M, Alvarado-Morales M, Karakashev D, Angelidaki I (2016) Integrated production of cellulosic bioethanol and succinic acid from industrial hemp in a biorefinery concept. Bioresour Technol 200:639–647. https://doi.org/10.1016/j.biortech.2015.10.081

Kyzas GZ, Terzopoulou Z, Nikolaidis V, Alexopoulou E, Bikiaris DN (2015) Low-cost hemp biomaterials for nickel ions removal from aqueous solutions. J Mol Liq 209:209–218. https://doi.org/10.1016/j.molliq.2015.05.060

Lamberti DD, Sarkar AK (2017) Hemp fiber for furnishing applications. 17th world textile conference AUTEX 2017. IOP Publ Mat Sci Eng 254:192009. https://doi.org/10.1088/1757-899X/254/19/192009

Latif E, Lawrence M, Shea A, Walker P (2015) Moisture buffer potential of experimental wall assemblies incorporating formulated hemp-lime. Build Environ 93:199–209. https://doi.org/10.1016/j.buildenv.2015.07.011

Leson G (2013) Hemp seeds for nutrition. In: Bouloc P, Allegret S, Arandu L (eds) Hemp – industrial production and uses. CABI, Wallingford, pp 229–238

Li SY, Stuart JD, Li Y, Parnas RS (2010) The feasibility of converting *Cannabis sativa* L. oil into biodiesel. Bioresour Technol 101:8457–8460. https://doi.org/10.1016/j.biortech.2010.05.064

Liberalato D (2003) Prospect of hemp utilization in the European textile industry. Agroindustria 2 (3):147–148

Ligeza M, Wygladacz D, Tobiasz A, Jaworecka K, Reich A (2016) Natural cold pressed oils as cosmetic products. Fam Med Prim Care Rev 18:443–447. https://doi.org/10.5114/fmpcr.2016.63699

Linger P, Müssig J, Fischer H, Kobert J (2002) Industrial hemp (*Cannabis sativa* L.) growing on heavy metal contaminated soil: fibre quality and phytoremediation potential. Ind Crops Prod 16:33–42

Liu M, Thygesen A, Summerscales J, Meyer AS (2017) Targeted pre-treatment of hemp bast fibres for optimal performance in biocomposites materials: a review. Ind Crop Prod 108:660–683. https://doi.org/10.1016/j.indcrop.2017.07.027

Liuzzi S, Sanarica S, Stefanizzi P (2017) Use of agro-wastes in building materials in the Mediterranean area: a review. Energy Procedia 126:242–249

Loiacono S, Morin-Crini N, Cosentino C, Torri G, Chanet G, Winterton P, Crini G (2017a) Simultaneous removal of Cd, Co, Cu, Mn, Ni and Zn from synthetic solutions on a hemp-based felt: experimental design. J Appl Polym Sci 134:1–11. https://doi.org/10.1002/app.44422

Loiacono S, Crini G, Martel B, Chanet G, Cosentino C, Raschetti M, Placet V, Torri G, Morin-Crini N (2017b) Simultaneous removal of Cd, Co, Cu, Mn, Ni and Zn from synthetic solutions on a hemp-based felt. 2: chemical modification. J Appl Polym Sci 134:1–16. https://doi.org/10.1002/app.45138

Loiacono S, Morin-Crini G, Cosentino C, Chanet G, Winterton P, Torri G, Crini G (2017c) La canapa: un material interessante per il trattamento delle acque contaminate da metalli. Chimica e Industria 2:8–12. https://doi.org/10.17374/CI.2017.99.2

Loiacono S, Morin-Crini N, Martel B, Chanet G, Bradu C, Torri G, Crini G (2018a) Zinc, copper, and manganese complexation by hemp: chemical abatement and ecotoxicological impact. Environ Risque Santé 17:240–252. https://doi.org/10.1684/ers.2018.1174

Loiacono S, Crini G, Chanet G, Raschetti M, Placet V, Morin-Crini N (2018b) Metals in aqueous solutions and real effluents: biosorption behavior of a hemp-based felt. J Chem Technol Biotechnol:1–10. https://doi.org/10.1002/jctb.5612

Lu X, Clarke RC (1995) The cultivation and use of hemp (*Cannabis sativa* L.) in ancient China. J Int Hemp Assoc 2:26–32. http://www.druglibrary.org/olsen/hemp/iha/iha02111.html

Lupul I, Yperman J, Carleer R, Gryglewicz G (2015a) Adsorption of atrazine on hemp stem-based activated carbons with different surface chemistry. Adsorption 21:489–498. https://doi.org/10.1007/s10450-015-9689-1

Lupul I, Yperman J, Carleer R, Gryglewicz G (2015b) Tailoring of porous texture of hemp stem-based activated carbon produced by phosphoric acid activation in steam atmosphere. J Porous Mat 22:283–289. https://doi.org/10.1007/s10934-014-9894-4

Malomo SA, Aluko RE (2015a) A comparative study of the structural and functional properties of isolated hemp seed (*Cannabis sativa* L.) albumin and globulin fractions. Food Hydrocoll 43:743–752. https://doi.org/10.1016/j.foodhyd.2014.08.001

Malomo SA, Aluko RE (2015b) Conversion of a low protein hemp seed meal into a functional protein concentrate through enzymatic digestion of fibre coupled with membrane ultrafiltration. Innov Food Sci Emerg Technol 31:151–159. https://doi.org/10.1016/j.ifset.2015.08.004

Malomo SA, He R, Aluko RE (2014) Structural and functional properties of hemp seed protein products. J Food Sci 79(8):C1512–C1521. https://doi.org/10.1111/1750-3841.12537

Mamone G, Picariello G, Ramondo A, Nicolai MA, Ferranti P (2019) Production, digestibility and allergenicity of hemp (*Cannabis sativa* L.). Food Res Int 115:562–571. https://doi.org/10.1016/j.foodres.2018.09.017

Mazhoud B, Collet F, Pretot S, Lanos C (2017) Mechanical properties of hemp-clay and hemp stabilized clay composites. Constr Build Mater 155:1126–1137. https://doi.org/10.1016/j.conbuildmat.2017.08.121

Miao C, Hui LF, Liu Z, Tang X (2014) Evaluation of hemp root bast as a new material for papermaking. Bioresources 9:132–142

Mikulcová V, Kašpárková V, Humpolíček P, Buňková L (2017) Formulation, characterization and properties of hemp seed oil and its emulsions. Molecules 22:700. https://doi.org/10.3390/molecules22050700

Mikulec A, Kowalski S, Sabat R, Skoczylas L, Tabaszewska M, Wywrocka-Gurgul A (2019) Hemp flour as a valuable component for enriching physicochemical and antioxidant properties of wheat bread. LWT Food Sci Technol 102:64–172. https://doi.org/10.1016/j.lwt.2018.12.028

Milanovic J, Kostić M, Milanovic P, Skundric P (2012) Influence of TEMPO-mediated oxidation on properties of hemp fibres. Ind Eng Chem Res 51:9750–9759. https://doi.org/10.1021/ie300713x

Miller SA (2018) Natural fiber textile reinforced bio-based composites: mechanical properties, creep, and environmental impacts. J Clean Prod 198:612–623. https://doi.org/10.1016/j.jclepro.2018.07.038

Mirski R, Boruszewski P, Trociński A, Dziurka D (2017) The possibility to use long fibres from fast growing hemp (*Cannavis sativa* L.) for the production of boards for the building and furniture industry. Bioresources 12:3521–3529. https://doi.org/10.1016/j.indcrop.2017.07.027

Morin-Crini N, Loiacono S, Placet V, Torri G, Bradu C, Kostić M, Cosentino C, Chanet G, Martel B, Lichtfouse E, Crini G (2018) Chapter 1: hemp-based materials for metal removal. In: Crini G, Lichtfouse E (eds) Green adsorbents for pollutant removal, Environmental Chemistry for a sustainable world, vol 19. Springer Nature Switzerland AG, Cham, pp 1–34. https://doi.org/10.1007/978-3-319-92162-4_1

Morin-Crini N, Loiacono S, Placet V, Torri G, Bradu C, Kostić M, Cosentino C, Chanet G, Martel B, Lichtfouse É, Crini G (2019) Hemp-based adsorbents for sequestration of metals: a review. Environ Chem Lett:1–16. https://doi.org/10.1007/s10311-018-0812-x

Moujalled B, Aït Ouméziane Y, Moissette S, Bart M, Lanos C, Samri D (2018) Experimental and numerical evaluation of the hygrothermal performance of a hemp lime concrete building: a long term case study. Build Environ 136:11–27. https://doi.org/10.1016/j.buildenv.2018.03.025

Mukhtar T, Kayani MZ, Hussain MA (2013) Nematicidal activities of *Cannabis sativa* L. and *Zanthoxylum alatum* Roxb. against *Meloidogyne incognita*. Ind Crop Prod 42:447–453. https://doi.org/10.1016/j.indcrop.2012.06.027

Münder F, Fürll C, Hempel H (2004) Advanced decortication technology for unretted bast fibres. J Nat Fibers 1:49–65. https://doi.org/10.1300/J395v01n01_04

Musio S, Müssig J, Amaducci S (2018) Optimizing hemp fiber production for high performance composite applications. Front Plant Sci 9:1702. https://doi.org/10.3389/fpls.2018.01702

Müssig J (2010) Industrial applications of natural fibres. Structure, properties and technical applications. Willey, Chichester. ISBN: 978-0-470-69501-1

Niyigena C, Amziane S, Chateauneuf A, Arnaud L, Bessette L, Collet F, Lanos C, Escadeillas G, Lawrence M, Magniont C, Marceau S, Pavia S, Peter U, Picandet V, Sonebi M, Walker P (2016) Variability of mechanical properties of hemp concrete. Mater Today Commun 7:122–133. https://doi.org/10.1016/j.mtcomm.2016.03.003

Nováková P (2018) Use of technical hemp in the construction industry. MATEC Web Conf 146:03011. https://doi.org/10.1051/matecconf/201814603011

Nunes L (2017) Nonwood bio-based materials. In: Jones D, Brischke C (eds) Performance of bio-based building materials. Elsevier, pp 97–186. https://doi.org/10.1016/B978-0-08-100982-6.00003-3

Nurazzi NM, Khalina A, Sapuan SM, Laila AHAMD, Rahmah M, Hanafee Z (2017) A review: fibres, polymer matrices and composites. Pertanika J Sci Technol 25:1085–1102

O'Brien C, Arathi HS (2019) Bee diversity and abundance on flowers of industrial hemp (*Cannabis sativa L.*). Biomass Bioenergy 122:331–335. https://doi.org/10.1016/j.biombioe.2019.01.015

Pacaphol K, Aht-Ong D (2017) Preparation of hemp nanofibers from agricultural waste by mechanical defibrillation in water. J Clean Prod 142:1283–1295. https://doi.org/10.1016/j.jclepro.2016.09.008

Parian A, Limketkai BN (2016) Dietary supplement therapies for inflammatory bowel disease: Crohn's disease and ulcerative colitis. Curr Pharm Des 22:180–188

Pathak T, Kaur J, Kumar R, Kuldeep K (2016) Development of electrochemical biosensor for detection of asparagine in leukemic samples. Int J Pharm Sci Res 7:783–788

Pavela R, Benelli G (2016) Essential oils as ecofriendly biopesticides? Challenges and constraints. Trends Plant Sci 21:1000–1007. https://doi.org/10.1016/j.tplants.2016.10.005

Pejić BM, Kostić MM, Skundric PD, Praskalo JZ (2008) The effects of hemicelluloses and lignin removal on water uptake behavior of hemp fibres. Bioresour Technol 99:7152–7159. https://doi.org/10.1016/j.biortech.2007.12.073

Pejić BM, Vukčević MM, Kostić MP, Skundric PD (2009) Biosorption of heavy metal ions from aqueous solutions by short hemp fibres: effect of chemical composition. J Hazard Mater 164:146–153. https://doi.org/10.1016/j.jhazmat.2008.07.139

Pejić BM, Vukčević MM, Pajić-Lijaković ID, Laušević MD, Kostić MM (2011) Mathematical modeling of heavy metal ions (Cd^{2+}, $Zn^{2+}$and Pb^{2+}) biosorption by chemically modified short hemp fibres. Chem Eng J 172:354–360. https://doi.org/10.1016/j.cej.2011.06.016

Pichardo PP, Martínez-Barrera G, Martínez-López M, Ureña-Núñez F, Ávila-Córdoba LI (2018) Waste and recycled textiles as reinforcements of building materials. IntechOpen, chapter 6:89–105. https://doi.org/10.5772/intechopen.70620

Pihlanto A, Mattila P, Mäkinen S, Pajari AM (2017) Bioactivities of alternative protein sources and their potential health benefits. Food Funct 8:3443–3458. https://doi.org/10.1039/C7FO00302A

Pil L, Bensadoun F, Pariset J, Verpoest I (2016) Why are designers fascinated by flax and hemp fibres composites? Compos Part A 83:193–205. https://doi.org/10.1016/j.compositesa.2015.11.004

Pittau F, Krause F, Lumia G, Habert G (2018) Fast-growing bio-based materials as an opportunity for storing carbon in exterior walls. Build Environ 129:117–129. https://doi.org/10.1016/j.buildenv.2017.12.006

Placet V, Meteau J, Froehly L, Salut R, Boubakar ML (2014) Investigation of the internal structure of hemp fibres using optical coherence tomography and focused ion beam transverse cutting. J Mater Sci 49:8317–8327. https://doi.org/10.1007/s10853-014-8540-5

Placet V, Day A, Beaugrand J (2017) The influence of unintended field retting on the physico-chemical and mechanical properties of industrial hemp bast fibres. J Mater Sci 52:5759–5777. https://doi.org/10.1007/s10853-017-0811-5

Prade T, Svensson SE, Andersson A, Mattsson JE (2011) Biomass and energy yield of industrial hemp grown for biogas and solid fuel. Biomass Bioenergy 35:3040–3049. https://doi.org/10.1016/j.biombioe.2011.04.006

Prade T, Svensson SE, Mattsson JE (2012) Energy balances for biogas and solid biofuel production from industrial hemp. Biomass Bioenergy 40:36–52. https://doi.org/10.1016/j.biombioe.2012.01.045

Pretot S, Collet F, Garnier C (2014) Life cycle assessment of a hemp concrete wall: impact of thickness and coating. Build Environ 72:223–231. https://doi.org/10.1016/j.buildenv.2013.11.010

Przybysz Buzala K, Kalinowska H, Przybysz P, Malachowska E (2017) Conversion of various types of lignocellulosic biomass to fermentable sugars using kraft pulping and enzymatic hydrolysis. Wood Sci Technol 51:873–885. https://doi.org/10.1007/s00226-017-0916-7

Ranalli P (1999) Advances in hemp research. Food Product Press, Binghamton. 272 p. ISBN 9781560228721

Ranalli P, Venturi G (2004) Hemp as a raw material for industrial applications. Euphytica 140:1–6. https://doi.org/10.1007/s10681-004-4749-8

Rehman MSU, Rashid N, Saif A, Mahmood T, Han JI (2013) Potential of bioenergy production from industrial hemp (*Cannabis sativa*): Pakistan perspective. Renew Sust Energ Rev 18:154–164. https://doi.org/10.1016/j.rser.2012.10.019

Rezić I (2013) Cellulosic fibres – Biosorptive materials and indicators of heavy metal pollution. Microchem J 107:63–69

Rice B (2008) Hemp as feedstock for biomass-to-energy conversion. J Ind Hemp 13:145–156. https://doi.org/10.1080/15377880802391274

Richard D, Dejean C (2013) Chapter 15: Medicinal uses of hemp. In: Bouloc P (ed) Hemp – industrial production and uses. CABI International, Wallingford, pp 222–228. ISBN: 978-1-84593-793-5

Rijavec T, Janjić S, Ačko DK (2017) Revitalization of industrial hemp *Cannabis sativa* L. var. *sativa* in Slovenia: a study of green hemp fibres. Tekstilec 60:36–48. https://doi.org/10.14502/Tekstilec2017.60.36-48

Robinson R (1996) The great book of hemp. The complete guide to the environmental, commercial and medicinal uses of the world's most extraordinary plant. Park Street Press, Rochester. ISBN: 978-089281541-8

Rodriguez-Leyva D, Pierce GN (2010) The cardiac and haemostatic effects of dietary hempseed. Nutr Metab 7:32–40

Rosas JM, Bedia J, Rodriguez-Mirasol J, Cordero T (2009) Hemp-derived activated carbon fibres by chemical activation with phosphoric acid. Fuel 88:19–26. https://doi.org/10.1016/j.fuel.2008.08.004

Rossi E, Di Nicolantonio M, Barcarolo P, Lagatta J (2020) Sustainable 3D printing: design opportunities and research perspectives. Adv Intell Syst Comput 975:3–15. https://doi.org/10.1007/978-3-030-20216-3_1

Roulac JW (1997) Hemp horizons: the comeback of the world's most promising plant, Series: the real goods solar living book, 1st edn. Chelsea Green Pub Co, White River Junction, 226 p. ISBN: 978-0930031930

Russo EB (2007) History of cannabis and its preparations in saga, science and sobriquet. Chem Biodivers 4:1614–1648

Russo R, Reggiani R (2013) Variability in antinutritional compounds in hempseed meal of Italian and French varieties. Plant 1:25–29. https://doi.org/10.11648/j.plant.20130102.13

Sadrmanesh V, Chen Y (2019) Bast fibres: structure, processing, properties and applications. Int Mater Rev 64:381–406. https://doi.org/10.1080/09506608.2018.1501171

Salentijn EMJ, Zhang Q, Amaducci S, Yang M, Trindade LM (2015) New developments in fiber hemp (*Cannabis sativa L.*) breeding. Ind Crop Prod 68:32–41. https://doi.org/10.1016/j.indcrop.2014.08.011

Salentijn EMJ, Petit J, Trindade LM (2019) The complex interactions between flowering behavior and fiber quality in hemp. Front Plant Sci 10:164. https://doi.org/10.3389/fpls.2019.00614

Sandler LN, Gibson KA (2019) A call for weed research in industrial hemp (*Cannabis sativa* L). Weed Res 59:255–259. https://doi.org/10.1111/wre.12368

Sarasini F, Fiore V (2018) A systematic literature review on less common natural fibres and their biocomposites. J Clean Prod 195:240–267. https://doi.org/10.1016/j.jclepro.2018.05.197

Sawler J, Stout JM, Gardner KM, Hudson D, Vidmar J, Butler L, Page JE, Myles S (2015) The genetic structure of marijuana and hemp. PlosOne 10:e133292. https://doi.org/10.1371/journal.pone.0133292

Saxena G, Purchase D, Mulla SI, Saratale GD, Bharagava RN (2020) Phytoremediation of heavy metal-contaminated sites: eco-environmental concerns, field studies, sustainability issues, and future prospects. Rev Environ Contam Toxicol 249:71–131. https://doi.org/10.1007/398_2019_24

Schluttenhofer C, Yuan L (2017) Challenges towards revitalizing hemp: a multifaced crop. Trends Plant Sci 22:917–929. https://doi.org/10.1016/j.tplants.2017.08.004

Schultes RE (1970) Random thoughts and queries on the botany of Cannabis. In: Joyce RB, Curry SH (eds) The botany and chemistry of Cannabis. J&A Churchill, London, pp 11–34. ISBN: 0700014799

Sepe R, Bollino F, Boccarusso L, Caputo F (2018) Influence of chemical treatments on mechanical properties of hemp fiber reinforced composites. Compos Part B 133:210–217. https://doi.org/10.1016/j.compositesb.2017.09.030

Shahzad A (2012) Hemp fiber and its composites – a review. J Compos Mater 46:973–986. https://doi.org/10.1177/002199831143623

Sipos B, Kreuger E, Svensson SE, Réczey K, Björnsoon L, Zacchi G (2010) Steam pretreatment of dry and ensiled industrial hemp for ethanol production. Biomass Bioenergy 34:1721–1731. https://doi.org/10.1016/j.biombioe.2010.07.003

Small E, Marcus D (2002) Hemp: a new crop with new uses for North America. In: Janick J, Whipkey A (eds) Trends in new crops and new uses. ASHS Press, Alexandria, pp 284–326

Spierling S, Knüpffer E, Behnsen H, Muderbasch M, Krieg H, Springer S, Albrecht S, Herrmann C, Endres HJ (2018) Bio-based plastics – a review of environmental, social and economic impact assessments. J Clean Prod 185:476–491. https://doi.org/10.1016/j.jclepro.2018.03.014

Sponner J, Toth L, Cziger S, Franck RE (2005) Chapter 4: Hemp. In: Franck RE (ed) Bast and other plant fibres, Woodhead publishing series in textiles. Elsevier, pp 176–206. https://doi.org/10.1533/9781845690618.176

Stevulova N, Cigasova J, Estokova A, Terpakova E, Geffert A, Kacik F, Singovszka E, Holub M (2014) Properties and characterization of chemically modified hemp hurds. Materials 7:8131–8150. https://doi.org/10.3390/ma7128131

Subratti A, Lalgee LJ, Jalsa NK (2019) Liquified dimethyl ether (DME): a green solvent for the extraction of hemp (*Cannabis sativa* L.) seed oil. Sustain Chem Pharm 12:100144. https://doi.org/10.1016/j.scp.2019.100144

Sun JJ, Gao JM, Hao XM, Cui XY, Ma T, Yang Y (2013) Effects of air-phosphoric acid activation on pore structure and adsorption property of mesoporous carbon from hemp stem. Book Ser Adv Mater Res 821-822:41–46. https://doi.org/10.4028/www.scientific.net/AMR.821-822.41

Sun W, Lipka SM, Swartz C, Williams D, Yang F (2016) Hemp-derived actrivated carbons for supercapacitors. Carbon 103:181–192. https://doi.org/10.1016/j.carbon.2016.02.090

The SS, Birch J (2014) Effect of ultrasonic treatment on the polyphenol content and antioxidant capacity of extract from defatted hemp, flax and canola seed cakes. Ultrason Sonochem 21:346–353. https://doi.org/10.1016/j.ultsonch.2013.08.002

The SS, Bekhit AED, Carne A, Birch J (2016) Antioxidant and ACE-inhibitory activities of hemp (*Cannabis sativa* L.) protein hydrolysates produced by the proteases AFP, HT, Pro-G, actinidin and zingibain. Food Chem 203:199–206. https://doi.org/10.1016/j.foodchem.2016.02.05

Thoma TG, Sharma SK, Prakash A, Sharma BR (2000) Insecticidal properties of essential oil of *Cannabis sativa* Linn. Against mosquito larvae. Entomon 25:21–24

Thomas S, Paul SA, Pothan LA, Deepa B (2011) Natural fibres: structure, properties and applications. In: Kalia S, Kaith BS, Kaur I (eds) Cellulose fibers: bio- and nano-polymer composites. Springer, Berlin, pp 3–42. https://doi.org/10.1007/978-3-642-17370-7_1

Tofan L, Păduraru C, Volf I (2009) Concentration of cadmium (II) trace amounts from large volumes of aqueous samples of chemically modified hemp fibres. Sci Pap J Agron Ser 52:506–511

Tofan L, Păduraru C, Volf I (2010a) Comparative study concerning the retention thermodynamics of some heavy metal ions on hemp fibres. Buletinul Institului Politechnic Iasi LVI(LX):137–145

Tofan L, Păduraru C, Volf I, Balan C (2010b) Kinetic and thermodynamic profile of Pb(II) sorption by untreated hemp fibres. Sci Pap J Agron Ser 53:146–150

Tofan L, Păduraru C, Volf I, Balan C (2010c) Removal of lead(II) from aqueous solution by sorption by on natural hemp fibres. Sci Pap J Agron Ser 53:150–153

Tofan L, Teodosiu C, Păduraru C, Wenkert R (2013) Cobalt(II) removal from aqueous solutions by natural hemp fibres: batch and fixed-bed columns studies. Appl Surf Sci 285:33–39. https://doi.org/10.1016/j.apsusc.2013.06.151

Tofan L, Păduraru C, Teodosiu C, Toma O (2015) Fixed bed columns study on the removal of chromium (III) ions from aqueous solutions by hemp fibres with improved sorption performance. Cellul Chem Technol 49:219–229

Tofan L, Păduraru C, Toma O (2016a) Zinc remediation of aqueous solutions by natural hemp fibres: batch desorption/regeneration study. Desalin Water Treat 57:12644–12652. https://doi.org/10.1080/19443994.2015.1052566

Tofan L, Wenkert R, Păduraru C (2016b) Natural and waste materials as green sorbents for Cd (II) removal from aqueous effluents. Environ Eng Manag J 15:1049–1058

Ummartyotin S, Pechyen C (2016) Strategies for development and implementation of bio-based materials as effective renewable resources of energy: a comprehensive review on adsorbent technology. Renew Sust Energ Rev 62:654–664. https://doi.org/10.1016/j.rser.2016.04.066

Usmani MA, Anas M (2018) Study of natural fibre reinforced composites. 1st Int Conf contemporary res mechanical Eng ICCRME-2018. Book series: IOP conference series. Mater Sci Eng 404:012048. https://doi.org/10.1088/1757-899X/404/1/012048

Väisänen T, Batello P, Lappalainen R, Tomppo L (2018) Modification of hemp fibers (*Cannabis sativa* L.) for composite applications. Ind Crop Prod 111:422–429. https://doi.org/10.1016/j.indcrop.2017.10.049

Vandenhove H, Van Hees M (2003) Fibre crops as alternative land use for radioactively contaminated arable land. J Environ Radioact 81:131–141. https://doi.org/10.1016/j.jenvrad.2005.01.002

VanDolah HJ, Bauer BA, Mauck KF (2019) Clinicians's guide to cannabidiol and hemp oils. Mayo Clin Proc 94:1840–1851. https://doi.org/10.1016/j.mayocp.2019.01.003

Vavilov NI (1992) Origin and geography of cultivated plants. Cambridge University press, Cambridge, 498 p. ISBN: 9780521404273

Verma RS, Padalia RC, Verma SK, Chauhan A, Darokar MP (2014) The essential oil of bhang (*Cannabis sativa* L.) for non-narcotic applications. Curr Sci 107:645–650

Vogl CR, Mölleken H, Lissek-Wolf G, Surböck A, Kobert J (2004) Hemp (*Cannabis sativa* L.) as a resource for green cosmetics: yield of seed and fatty acid compositions of 20 varieties under the growing conditions of organic farming in Austria. J Ind Hemp 9:51–68. https://doi.org/10.1300/J237v09n01_06

Vukčević M, Kalijadis A, Radisic M, Pejić B, Kostić M, Laušević Z, Laušević M (2012) Application of carbonized hemp fibres as a new solid-phase extraction sorbent for analysis of pesticides in water samples. Chem Eng J 211:224–232. https://doi.org/10.1016/j.cej.2012.09.059

Vukčević M, Pejić B, Laušević M, Pajić-Lijaković I, Kostić M (2014a) Influence of chemically modified short hemp fiber structure on biosorption process of Zn^{2+} ions from waste water. Fibres Polym 15:687–697. https://doi.org/10.1007/s12221-014-0687-9

Vukčević M, Pejić B, Kalijadis A, Pajić-Lijaković I, Kostić M, Laušević Z, Laušević M (2014b) Carbon materials from waste short hemp fibres as a sorbent for heavy metal ions – mathematical modeling of sorbent structure and ions transport. Chem Eng J 235:284–292. https://doi.org/10.1016/j.cej.2013.09.047

Vukčević M, Kalijadis AM, Vasiljevic TM, Babic BM, Laušević ZV, Laušević MD (2015) Production of activated carbon derived from waste hemp (*Cannabis sativa*) fibres and its performance in pesticide adsorption. Microporous Mesoporous Mater 214:156–165. https://doi.org/10.1016/j.micromeso.2015.05.012

Walker R, Pavía S (2014) Moisture transfer and thermal properties of hemp-lime concretes. Constr Build Mater 64:270–276. https://doi.org/10.1016/j.conbuildmat.2014.04.081

Walker R, Pavía S, Mitchell R (2014) Mechanical properties and durability of hemp-lime concretes. Constr Build Mater 30:340–348. https://doi.org/10.1016/j.conbuildmat.2014.02.065

Wang QL, Xiong YL (2019) Processing, nutrition, and functionality of hempseed protein: a review. Compr Rev Food Sci Food Saf 18:936–952. https://doi.org/10.1111/1541-4337.12450

Wang HM, Postle R, Kessler RW, Kessler W (2003) Removing pectin and lignin during chemical processing of hemp for textile applications. Text Res J 73:664–669. https://doi.org/10.1177/004051750307300802

Wang Y, Yang R, Li M, Zhao ZJ (2015) Hydrothermal preparation of highly porous carbon spheres from hemp (*Cannabis sativa* L.) stem hemicellulose for use in energy-related applications. Ind Crop Prod 65:216–226. https://doi.org/10.1016/j.indcrop.2014.12.008

Werz O, Seerges J, Shaible AM, Weinigel C, Barz D, Koeberle A, Allegrone G, Pollastro F, Zampieri L, Grassi G, Appendino G (2014) Cannaflavins from hemp sprouts, a novel cannabinoid-free hemp food product, target microsomal prostaglandin E2 synthase-1 and 5-lypoxygenase. Pharma Nutr 2:53–60. https://doi.org/10.1016/j.phanu.2014.05.001

Yang R, Liu GQ, Xu XH, Li M, Zhang JC, Hao XM (2011) Surface texture, chemistry and adsorption properties of acid blue 9 of hemp (*Cannabis sativa* L.) bast-based activated carbon fibres prepared by phosphoric acid activation. Biomass Bioenergy 35:437–445. https://doi.org/10.1016/j.biombioe.2010.08.061

Yang R, Liu GQ, Li M, Zhang JC, Hao XM (2012) Preparation and N_2, CO_2 and H_2 adsorption of super activated carbon derived from biomass source hemp (*Cannabis sativa* L.) stem. Microporous Mesoporous Mater 158:108–116. https://doi.org/10.1016/j.micromeso.2012.03.004

Yin SW, Tang CH, Cao JS, Hu EK, Wen QB, Yang XQ (2008) Effects of limited enzymatic hydrolysis with trypsin on the functional properties of hemp (*Cannabis sativa* L.) protein isolate. Food Chem 106:1004–1013. https://doi.org/10.1016/j.foodchem.2007.07.030

Yin SW, Tang CH, Wen QB, Yang XQ (2009) Functional and structural properties and in vitro digestibility of acylated hemp (*Cannabis sativa* L.) protein isolates. Int J Food Sci Technol 44:2653–2661. https://doi.org/10.1111/j.1365-2621.2009.02098.x

Yu Y, Zhan JB, Wan H, Zhang Y, Li LW, Yu J, Yu TT, Xie J, Yue BS (2017) Manufacturing technology and application of hemp cigarette paper with dense ash integration. 3rd Int Conf Energy Mat Environ Eng Book series: IOP conference series earth and environmental science 61:UNSP 012078. https://doi.org/10.1088/1755-1315/61/1/012078

Zabalza Bribián I, Valero Capilla A, Aranda Usón A (2011) Life cycle assessment of building materials: comparative analysis of energy and environmental impacts and evaluation of the eco-efficiency improvement potential. Build Environ 46:1133–1140. https://doi.org/10.1016/J.BUILDENV.2010.12.002

Zajac M, Guzik P, Kulawik P, Tkaczewska J, Florkiewicz A, Migdal W (2019) The quality of pork loaves with the addition of hemp seeds, de-hulled hemp seeds, hemp protein and hemp flour. LWT Food Sci Technol 105:190–199. https://doi.org/10.1016/j.lwt.2019.02.013

Zhang H, Zhong Z, Feng L (2016) Advances in the performance and application of hemp fiber. Int J Simul Syst Sci Technol 17:18.1–18.5. https://doi.org/10.5013/IJSSST.a.17.09.18

Zhou Y, Wang S, Lou H, Fan P (2018) Chemical constituents of hemp (*Cannabis sativa* L.) seed with potential anti-neuroinflammatory activity. Phytochem Lett 23:57–61. https://doi.org/10.1016/j.phytol.2017.11.013

Zou XQ, El Fallah J, Goupil JM, Zhu GS, Valtchev V, Mintova S (2012) Green removal of aromatic organic pollutants from aqueous solutions with a zeolite-hemp composite. RSC Adv 2:3115–3122. https://doi.org/10.1039/c2ra01176j

Żuk-Gołaszewska K, Gołaszewski J (2018) *Cannabis sativa* L. – cultivation and quality or raw material. J Elem 23:971–984. https://doi.org/10.5601/jelem.2017.22.3.1500

Chapter 3
Industrial Hemp in the USA: A Brief Synopsis

John H. Fike, Heather Darby, Burton L. Johnson, Larry Smart, and David W. Williams

Abstract Industrial hemp has a long history of use in the U.S.A. Historically the crop was grown for fiber for home use as well as for ship cordage, rigging and sails, and this made the crop of strategic importance. The end of slavery, changing technologies, and other competing fibers all played roles in hemp's decline as an important commodity in the U.S.A., but prohibition through various federal legislation essentially created a 70-year hiatus for the crop. Changes in U.S. federal legislation have made hemp once again legal, and the crop is returning to production fields. However, this time around the majority of effort with the crop is for flowers production. Most regions of the country have seen rapid expansion of acres in hemp flower production, largely due to the potential returns significantly greater than possible with other commodity crops. Opportunities for other hemp products – i.e., grain and fiber – remain more restricted as processing capacity, particularly for fibers, is limited. However, regions such as the Mid-West and Great Plains are

Author David W. Williams deceased at the time of publication.

J. H. Fike (✉)
School of Plant and Environmental Sciences, Virginia Tech, Blacksburg, VA, USA
e-mail: jfike@vt.edu

H. Darby
Department of Plant and Soil Science, University of Vermont, Burlington, VT, USA
e-mail: Heather.Darby@uvm.edu

B. L. Johnson
Department of Plant Sciences, North Dakota State University, Fargo, ND, USA
e-mail: burton.johnson@ndsu.edu

L. Smart
School of Integrative Plant Science, Horticulture Section, Cornell University, Geneva, NY, USA
e-mail: Lbs33@cornell.edu

G. Crini, E. Lichtfouse (eds.), *Sustainable Agriculture Reviews 42*,
Sustainable Agriculture Reviews 42, https://doi.org/10.1007/978-3-030-41384-2_3

likely to be able to grow hemp grain in rotation with other crops given that this part of the U.S.A. has grain suitable infrastructure capable of handling hemp. Both long-term farmers and agricultural neophytes are engaging in efforts to grow hemp flowers, and it is likely that the market for cannabidiol (CBD) hemp growers will remain volatile over the next few years. Still-evolving Federal guidelines provides further uncertainty to these markets. Current production models rely on labor-intensive production, harvest and processing systems, and the application of technologies such as mechanization is likely to introduce significant changes in costs, value, and opportunity for this new industry. To date, research has largely focused on plant varieties and management. Efforts to address issues such as low seed vigor and seed shatter would likely have value across all hemp production systems. Research work also is being conducted on disease resistance and management, weed management and herbicide susceptibility. After a long interruption, U.S.A. efforts to improve industrial hemp are in the beginning stages of an exciting new chapter for this crop.

Keywords Hemp · Markets · Fiber · Grain · Cannabinoids

Abbreviations

CBD Cannabidiol
THC Δ9 Tetrahydrocannabinol

3.1 Introduction

Renaissance of an industrial hemp industry has been slower in the U.S.A. than in Europe and Canada, perhaps because of historic drug policies and the consequent cultural stigma associating all things *Cannabis* with recreational marijuana. Interestingly, people on opposite ends of the political spectrum found common ground in their efforts to legalize industrial hemp production in the U.S.A. At a state level, Colorado and Kentucky – which historically have had somewhat different attitudes towards recreational *Cannabis* – made some of the first efforts to push hemp forward. Former Kentucky Commissioner of Agriculture, James Comer, has told how his campaign platform to legalize hemp production allowed him to garner support both from rural conservatives as well as urban liberal voters during the early days of the hemp revival. The public's changing of opinions about industrial hemp may reflect more relaxed attitudes toward marijuana use for medicinal and recreational purposes, as well as a growing perception that hemp has potential as a renewable fiber, food, fuel, and pharmacological crop. Many claims surrounding the utility and sustainability aspects of hemp have yet to be borne out – it's really too early to tell. Much of the early (and continued) boosterism surrounding hemp was based on the idea that the crop may be grown multiple purposes and multiple products. However, whether management for multiple outputs (i.e., for any combination of feed/food, fibers, and flowers) makes economic sense remains to be seen. Any multi-product system is likely to involve some level of tradeoffs and will require optimizing the product outputs on the basis of their value in the market. Anecdotal evidence also suggests that hype may be an important factor behind corporate production and market decisions. For example, recent U.S.A. research suggests kenaf (*Hibiscus cannabinus*) may be just as useful and more productive than hemp as a fiber crop. However, we know of a fiber processor capable of processing both fiber species that works with growers producing hemp – not because hemp produces the highest yields or quality, but because of the market's particular interest in (and demand for) the crop.

In the U.S.A., hemp is at an early stage as a crop in the marketplace, having only been legalized for production (outside of state research programs) with passage of the 2018 Farm Bill. As of this writing (spring 2019), only a few states remain holdouts with respect to legalizing hemp production. But what does "hemp production" really mean? There are significant regional differences as well as similarities in approach to hemp production across the U.S.A., and our purpose here is to provide some historical context and a brief synopses of the hemp systems we see developing on a regional basis as this "new" crop gains traction.

3.2 A Brief Historical Context for Hemp in North America

Hemp has a long history in North America, with the British colonists growing the crop along the Atlantic seaboard from Canada and New England to Georgia, southernmost of the 13 U.S. colonies. The crop largely was used for fiber for "homespun" cloth as well as for rope, but it was not particularly competitive with the European hemp grown and processed for high quality canvas given differences in production techniques and trade laws that prevented the shipping of finished products to Europe (Duvall 2014). Hemp production in the colonies was further constrained by the greater value of food and cash crops, particularly tobacco. Hemp traveled into the heartland as colonists moved beyond the Appalachian Mountains, and it would become a primary crop for setters in Kentucky and parts west. Ironically, one fiber industry supported another, as hemp largely was grown for rope and cordage to bundle and bale the cotton (*Gossypum hisrutum*) that was grown in Deep South states. During the U.S. Civil War, Midwestern states lost their market for hemp fibers, and following the war, freed men and women had little desire to return the labor of their enslavement. By the turn of the twentieth century, development of new technologies such as metal binding for cotton bales, the rise of steam- and fossil-fueled ships, and the availability of cheaper imported fibers were reducing the value and markets for hemp. Despite these changes, hemp was still considered a crop or strategic interest, and USDA research on the mechanization of hemp production began in a similar timeframe and lasted about two decades.

Early U.S. researchers identified the basic production practices to aid growers in successful production. Dewey (1902) reported low hemp production on poorly drained, infertile, and drought prone soils. In established stands, Dewey noted hemp's extensive taproot enabled soil water extraction from depths up to 3 m into the soil profile. Non-uniform and reduced stands were observed on heavy clay soils, which caused plants to branch and produce larger stems, resulting in harvest difficulties and reduction of fiber quality. In 1919, Haney reported seeding hemp between May 10 and May 20 in North Dakota, depending on spring conditions with emphasis on soils having sufficiently warmed to promote rapid emergence and stand establishment. Haney (1919) also noted hemp in thick stands could be competitive with weeds, including perennial grasses and thistles, depending on establishment timing. Haney (1919) stated, "Trials of hemp have proven satisfactory and with the machinery that has been developed recently makes this an important and valuable crop."

Although significant strides were made in plant breeding and mechanization during this time (Wright 1918; Dewey 1928), such progress was insufficient to keep the crop within the good graces of those politicians who were bent on outlawing the hemp as a psychoactive plant material. Despite opposition from the American Medical Association, the U.S. government passed the Marihuana Tax Act (MTA) in 1937, placing cultivation of all *Cannabis* under the U.S. Treasury Department's control (USDA 2000). This measure effectively constrained hemp production in the U.S.A., but concerns over lost access to cheap foreign fibers during World War II led the U.S. government to support a "Hemp for Victory" campaign in which several thousand farmers were recruited to grow the crop (Johnson 1999). Following the war, hemp production rapidly declined in the face of cheaper imported materials and the development of synthetic fibers. By the time the Controlled Substances Act made all *Cannabis* a Schedule I drug, opportunity to grow the crop was already out of reach for farmers, and it would remain so for over 30 more years, when the Agriculture Act of 2014 opened the possibility of research with hemp in the U.S.A.

The Farm Act of 2014 authorized U.S. states to define pilot research programs for the cultivation of industrial hemp, and several embraced the opportunity to initiate research into hemp cultivation and passed legislation authorizing industrial hemp pilot programs. Although this poised those states to become producers of industrial hemp – providing healthy hempseed oil and fiber for textiles, insulation and other uses – there was no foundational knowledge of hemp cultivation or infrastructure in place to initiate a hemp industry. However, based on historic cultivation of hemp, there was and is great potential to incorporate hemp grain, fiber, and CBD crops into rotation with field crops and vegetables into U.S. agriculture, allowing farmers to diversify their production and improve sustainability.

3.3 Renaissance of Hemp in the U.S.A – Regional Variations on a Theme

Following what was essentially a 70-plus-year absence in production, industrial hemp (*Cannabis sativa* L.) is again being grown across the U.S.A. Previously published research regarding agronomic guidelines for industrial hemp production in the United States are severely dated and limited – and new uses for the crop are changing the research needs. Initial university research trials and efforts from state-pilot programs have thus begun the process of defining the basic agronomic guidelines for hemp production that will aid in successful crop commercialization. While we will highlight efforts from various regions of the U.S.A., we start with the specific case of Kentucky, both because it was an early leader in the country's hemp production efforts and because that state's experiences have been common for many states that have followed.

3.3.1 Hemp in Kentucky and the Southeast

In the spring of 2014, the Kentucky Department of Agriculture (KDA), then led by Commissioner James Comer, began work towards the establishment of a pilot research program with industrial hemp under the provisions of the newly passed Farm Bill. Being a brand-new initiative on all levels, there were several bumps in the road as those efforts moved forward. However, through the efforts of KDA and several universities, hemp was planted across the state in 2014. While the quality of the scientific data derived from these initial efforts was limited, the experiences gained at all levels, *i.e.* administrative and agronomic, were extremely valuable, enabling the program to move forward efficiently and effectively.

In 2015, much more work (both in numbers of farms and acres) was conducted on private lands than by university researchers. The KDA administrators of the pilot research program considered and deduced that under the language of the Farm Bill, research efforts involving marketing must inherently also involve both the ability of farmers to sell to processors, and for processors to sell goods to the public. Hence, a renewed hemp industry in Kentucky was underway in earnest. At that point in time, there was one very striking difference in industry efforts relative to subsequent and current efforts. A slight majority of total interest focused on fiber and grain, with a slight minority of the efforts focused on cannabinoid production (Fig. 3.1).

However, from 2016 to today, the majority of efforts have focused on "flower" production for cannabinoid-based products and relatively very small efforts towards fiber and/or grain. By 2019, there were 1075 applicants to the KDA hemp program; 1009 of which applied to focus on cannabinoid production (93% of all applicants).

KDA Industrial Hemp Research Pilot Program
Annual Overview

Production Year	# University Projects	Approved Processors	Approved Growers	KY Counties with Hemp	Approved Acres	Planted Acres	Harvested Acres	% Grain or Seeds	% Fiber	% CBD	% Grain & CBD	% Grain & Fiber
2014	7	9	20	14	-	33	-	47%	32%	21%		
2015	8	29	99	41	1,742	922	500	47%	6%	47%		
2016	17	45	137	60	4,600	2,300	2,000	34%	6%	60%		
2017	17	49	204	71	12,800	3,200	2,300	36%	5%	27%	32%	
2018 (Sept)	14	72	210	73	16,100	6,700	TBD	18%	4%	61.5%	14%	2.5%

Fig. 3.1 Data from the Kentucky Department of Agriculture industrial hemp pilot research program 2014–2018

Similar responses have occurred in other states across the U.S.A., with larger numbers of people getting into the business of hemp production, whether they know anything of crop production or not. Interest is driven by the perceived potential profit from cannabinoid production, which has attracted a very large and diverse group of program participants. Data in Fig. 3.2 indicates that in Kentucky in 2017, average yields and average prices for the crop clearly favored cannabinoid production relative to fiber and grain by highly significant amounts. This potential has created a real interest in cannabinoid agriculture by life-long, highly successful farmers alongside those who literally have never cultured a crop of any kind before. The diversity of interest is staggering, as is the likely potential range of success and failure.

Although a lone southern state (Mississippi) currently holds out against hemp production, other states in the Southeast that have engaged in hemp production have seen similar patterns of interest and exponential increases in production efforts as occurred in Kentucky. E.g., Virginia went from a handful of producers growing hemp in 2018 for university research to over 750 registered growers with over 7500 registered acres in 2019 – almost all for flowers. Interestingly, those states that historically had hemp industries (particularly in the Southeast and mid-Atlantic regions of the U.S.A.) were more aggressive in developing hemp research (and then full-scale production) programs. Many of these same states also have had a long tradition of tobacco production. Until very recent times, tobacco was produced on very small family farms as well as large, corporate operations. Once it was established that tobacco's consumption and use could contribute to negative health outcomes, the federal government slowly removed support for the crop, which

Production Report Data*

2017	Grain	Fiber	Floral
Growers	20	14	112
Acres Planted	1,227	159	1,883
Acres Harvested	826	120	1,371
Avg. Yield lb/acre**	518	2,556	1,024
Avg. Price per lb	$1.70	$0.34	$7.20

*Preliminary; *analysis of 2017 production report data is not complete.*
***Average Yield was calculated only from plots with reported yields above zero. (Does not include dual purpose Grain & Floral)*

Kentucky Proud. Kentucky Department of Agriculture

Fig. 3.2 Average yield and price data provided by participants in the 2017 Kentucky industrial hemp pilot research program. (Source: Kentucky Department of Agriculture)

inevitably led to very large reductions in production. As tobacco acres across the region have declined, many growers and politicians have come to view hemp as a means to revitalize their flagging agricultural economies and rural communities. We will return to this tobacco connection in a subsequent consideration of flower production models.

3.3.2 Hemp in the Mid-Atlantic and New England States

As in other colonies, hemp production in the Mid-Atlantic and New England historically and primarily was for household use and generally less important than flax as a household fiber (Dewey 1914). New England in particular has a history tied to hemp through its shipping industries which manufactured cordage for ropes – but much of the raw material likely was imported. Despite the demand for naval stores, government incentives, experimentation with hemp in early settlements, and the desire of New England farmers to have an agricultural staple, hemp never became a major crop in New England, and local production never met the demands of New England shipyards (Bidwell and Falconer 1941). The failure of hemp to achieve staple crop status in New England has been attributed to the crop's being outcompeted by grain crops for a limited supply of fertile land in the region (Bidwell and Falconer 1941).

New England's land constraint has relevance today, and in a modern context, hemp production in the region is almost entirely driven to CBD production. Although currently it is not legal in all states. Industrial hemp may be grown for commercial and research purposes in Vermont, Maine, Massachusetts, and Rhode Island. As of 1 February 2019, hemp cannot be grown commercially in New Hampshire and Connecticut, although feasibility studies have been enacted in both states and special permissions have been given to research universities (National Conference of State Legislatures 2019). No official production statistics of hemp in recent years have been published by the USDA. A publication by the Congressional Research Service, *Hemp as an Agricultural Commodity*, reports national statistics for industrial hemp in the US and Canada in the 1990s and 2000s. While there is data for Canada and US imports, only 2016–2017 acreage data by state is presented. From 2016 to 2017, the number of acres in production in Maine increased from 1 to 30, and in Vermont from 180 to 575 (Johnson 2013). Official state data of registered growers and acreage in Vermont indicate 427 registrations and 2711 acres in production in the 2018 growing season (Clithero 2018). The State of Rhode Island's Department of Business Regulation records four state-approved industrial hemp licenses on their website as of 24 May 2019 (State of Rhode Island 2019). For 2019, all states have reported a significant increase in registrants and potential acres of hemp to be grown. In Vermont as of April 2019, there were currently 300 farmers registered to plant 4500 acres of hemp (Vermont Agency of Agriculture 2019). The Vermont Agency of Agriculture reports that over 90% of acreage will be planted with intended use for CBD. As of May 2019, Massachusetts reported issuing 77 licenses for growing and processing hemp. It is projected that 430 acres of hemp will be grown primarily for flower production for the CBD market. In Maine, there were 143 licensed growers planning to produce 2150 acres of hemp in the 2019 growing season. An informal regional survey from 2019 indicates about 90% of farmers are primarily interested in CBD production, with a majority growing 10 or fewer acres. A significant majority (58%) of these producers were in their first year in the hemp industry.

In the Mid-Atlantic, New York has perhaps been one of the most aggressive of states at promoting industrial hemp. Following initial authorization of hemp research (2014) New York modified and greatly broadened the scope of its pilot program in 2016. The first licenses under this program were issued to several institutions of higher education, although a few pioneering growers also obtained their licenses. The first hemp crop, totaling approximately 30 acres, was planted in 2016. New applications for hemp farming were encouraged in 2017 and the state took a unique approach by granting funds to Cornell University to engage farmers in hemp cultivation. About 1700 ac of the dual-purpose grain/fiber hemp cultivar 'Anka' was grown across 24 different farms. This project was aimed at demonstrating full field-scale cultivation of industrial hemp on farms across the state to attract pioneer producers and de-risk grower and processor participation in hemp enterprises. Through one season of hemp cultivation on a relatively large scale, these farmers gained the most basic knowledge about the logistics, agronomics, and economics of industrial hemp, while stimulating interest among farmers to expand the hemp enterprise. New York's approach also was novel in that farmers were compensated $350 per planted acre and they were not required to harvest the fields. Due to the lateness of initiating this project and delays in importing seed from Canada, the earliest fields were planted on July 10, 2017 and very few of the fields even achieved maturity. By late 2017, the limits on hemp growers' licenses were removed and licenses were issued to just over 100 applicants, most of whom were interested in growing CBD-rich cultivars. In early 2018, the application procedure for growing CBD-rich cultivars was closed until early December 2018, when it was reopened for a month to allow more than 200 new applications. At the start of the 2019 growing season, there are close to 300 licensed hemp growers and the application process was reopened indefinitely.

The first hemp production in New York was for grain to be used for food products and hempseed oil for personal care products. Early food products that were marketed include culinary hempseed oil, hemp-based pasta and flour mixes, and hemp baby green salad mixes. The 2018 growing season was dominated by production of CBD-rich hemp cultivars in a horticultural system in fields of 5–20 acres with a total of approximately 2200 acres across the state. The 2019 growing season will also be predominantly represented by production of CBD-rich cultivars, and the acreage is likely to expand significantly across the state and in larger fields on average.

3.3.3 Hemp in the Mid-West and Great Plains

Hemp production on the Great Plains (and the rest of the U.S.A.) may benefit from historical research. Although many early hemp boosters proclaimed hemp capable of growing with few nutrients on difficult soils and with no pest or weed challenges, these advocates would do well to review existing literature from the region. During the previous century, Wilsie et al. (1944) showed that heavy Canada thistle (*Cirsium arvense* L. Scop.) and quackgrass [*Agropyron repens* (L.) Beauv] infestations in Iowa occurred when establishment was slow or in hemp stands that were poor. Soil moisture and temperature, fertility, and seed quality also were shown to be important factors influencing stand establishment. Seeding rate did not affect yield, but was influenced fiber percentage with higher fiber percentage from the higher seeding rate (Wilsie et al. 1944). Seeding date recommendations were between 5 and 20 May for hemp produced in Iowa. Robinson (1943) recommended planting hemp after small grains and before corn (*Zea mays* L.). Better hemp performance was noted when the previous crop was a legume rather than non-legume grass crops such as oat (*Avena sativa* L.) and sorghum (*Sorghum bicolor* L.), primarily due to the legume's nitrogen contribution to the subsequent hemp crop. Excessive nitrogen reduced fiber quality because of high interplant competition and self-thinning of stands to low levels that resulted in short plants with thick stems (Howard et al. 1946). Howard et al. (1946) noted greater hemp yield at higher nitrogen levels, but a reduced stem breaking

Fig. 3.3 Varietal evaluation for grain and fiber crops is a critical component of research programs aimed at re-establishing industrial hemp as a viable row crop. (Source: Burton Johnson)

strength at soil nitrogen levels of 80 lbs./acre. Howard stated nitrogen fertility should maximize yield and fiber quality.

Current field research in the U.S.A. and Canada is engaged in redefining basic production guidelines due to changes in technology, cultivars, regulations, and markets. Hemp cultivars in Canada and the U.S.A. must not contain more than 0.3% THC in plant tissue. Hemp THC concentrations largely are determined by genotype, but climatic conditions during the growing season can also influence THC level (Small et al. 2003). This presents challenges for developing varieties that both are suitably productive over large regions and consistently below 0.3% THC.

As should be apparent, much of the work from Plains states is viewed through the lens of traditional row-crop production systems and early work has focused varietal evaluation (Fig. 3.3). Hemp grain and dual-purpose varieties used in the region are primarily from Canadian seed companies Hemp Genetics International Inc. and Parkland Industrial Hemp Growers, given the similarity in climatic and latitudinal conditions, although a "homegrown" company, New West Genetics, is in the process of releasing U.S.-bred varieties developed in Colorado. Grain yields in North Dakota research trials primarily with Canadian varieties have averaged about 1270 lb/acre, with better yields at lower plant densities.

In North Dakota's initial hemp pilot-program with the Department of Agriculture, five producers planted a total of 70 acres of hemp. An in-state point of sale was available for producers to contract grain at \$0.90/lb, and producer yields ranged from 400 to 1100 lbs/acre. This generated net returns higher than mainstream commodity crops such as wheat (*Triticum aestivum*), corn (*Zea mays*), and soybean (*Glycine max*). Producer interest increased to 35 growers and slightly more than 3000 acres in 2017 with grain prices down from 2016 to the \$0.65/lb. range. However, the crop was still more profitable, considering higher input costs and declining grain prices for traditional crops. Lower grain prices and higher seed costs in 2018 discouraged what would have been record-breaking production near 10,000 acres by 55 producers. Instead, 28 growers grew approximately 2800 acres, with a grain value of \$0.45/lb. Market prices, market size, and market type are in developmental stages and have not outpaced progress when compared with advances on the agronomy side of hemp production. Successful commercialization requires both processes to evolve together with market demand directing the scale of agronomic production. We consider this in a brief look at current market conditions in Sect. 3.4.

3.3.4 Hemp in the Western States

Western states have been early leaders in the *Cannabis* industry, largely from a willingness to buck Federal regulations surrounding the production of recreational marijuana. Many of the current hemp varieties used for CBD production have been bred from marijuana strains grown in states such as Oregon and Colorado, which were leaders in legalizing recreational use. (We use the industry term "strain" here rather than "cultivar," given that the more exacting protocols for cultivar development found in the commercial crop industry are rarely applied to hemp.) Interestingly, tensions between hemp and marijuana communities have arisen in states such as Oregon, because marijuana growers fear the effects of cross-pollination from hemp fields. Pollination would lower the value of the marijuana crop if fertilized, and the subsequent genetics of the new seed would likely have lower expression of psychoactive THC. This is leading many marijuana growers to move their production indoors.

Like other states, large increases in grower numbers and have accompanied the enthusiastic rush to get into the hemp business. However, experience from these states may also serve as a precautionary tale, as marijuana growers in Oregon have seen significant drop in prices as production in the state has far outpaced demand. Many of these growers are now moving to the production of CBD-rich hemp strains.

3.4 Markets

At the time the 2014 Farm Bill was signed, market demand for hemp-based feedstocks along with market-ready products already existed in the U.S.A. At that point, a number of companies were purchasing imported hemp materials in order to manufacture and market hemp-based goods. Of course, at that time there were no hemp farmers or primary processors from whom to purchase U.S.-produced materials. The products included fiber-based packaging materials, food products from grain and hempseed oil, and personal care products incorporating hempseed oil. Interest remains in domestically producing both grain and fiber for such uses, although some regions may be better prepared to meet consumer need. In the Great Plains, grower interest has focused on grain production because the infrastructure and a local market already exists. The 2018 U.S. Farm Bill, which allowed interstate movement of hemp grain further supported this by opening market opportunities beyond the confines of state borders. Grain markets currently are more limited in volume, however, and opportunities to scale up will likely require consumer education about the healthful qualities of hemp seed in order to increase demand. The Great Plains also would be well-suited to fiber production, but few fiber processors currently exist and are typically some distance from growers. Those growers interested in collecting hemp straw following grain production also would face the added penalty of grain crops having less stem yield, with fibers of lower quality. These factors, coupled with the fact that fiber value is relatively low, make for a difficult positive net return. As markets for this fiber do not readily exist, producers often end up burning the hemp stalks since they are a nuisance in the field in terms of residue management. If fiber value was slightly greater than costs associated with baling stalks, producers might be inclined to bale instead of burn. This would have many positive effects on soil health, ecosystem services, and reducing carbon dioxide emissions.

Production of actual fiber cultivars will also be well-suited to the Mid-West and portions of the Great Plains, but processing remains a limitation. Where fiber processors are in operation and have developed a local grower base, fiber processing is more competitive with traditional row crops. Long-term, questions about global competitiveness also must be addressed for both grain and fiber production if the U.S.A. is to be competitive with as countries such as China, that have more mature production and processing capacity.

3.4.1 Hemp Flowers – What Goes Up Must Come Down?

Given the limitations in the grain market and the lack of processors for fibers, it should not be surprising that as hemp comes on line, nearly all the interest and efforts for its development have been with flower production systems. Of course, as noted above this is not just a system interest, but a function of the existing economic reality (see Fig. 3.2). Indeed, many hemp enthusiasts have been known to express that their initial (and long-term) interests in hemp were and are for its potential environmental benefits as a fiber and food crop for all manner of sustainable bioproducts. . .but that they grow flowers to pay today's bills.

One of the oddest aspects of the phenomenal growth in the flowers market is that there are so many hemp program participants who expect this market to continue as it exists today, perhaps even in perpetuity. Many participants simply have no concept that a crop which provides such huge profits per acre will soon be produced at such high levels that supply of the product will far exceed any demand, thus resulting in a significant drop in the value of the crop. In other words, there are many participants that have no actual experience in production agriculture or familiarity with the basic concepts of agricultural economics and/or commodity agriculture. Or course, even our observation here assumes that an agricultural production model of some form will win out. However, efforts are afoot to develop laboratory methods to synthesize cannabinoids, and should they prove economically viable, labor-intensive flowers production would likely be limited to sales into specialty markets.

Complicating this entire scenario is continued ambiguity regarding the ultimate regulatory framework under which cannabinoids will be managed by the U.S. Food and Drug Administration (FDA). Passage of the 2018 Farm Bill provided several very positive changes for the continued evolution of a U.S. hemp industry. Hemp as defined in the Bill was removed from the definition of marijuana within the Controlled Substances Act, thus making it legal within the purview of an approved hemp program. The new Farm Bill also provided that extracts from legal hemp were also removed from the definition of marijuana. Lastly and very importantly, the Bill contained language that provided full oversight and regulatory authority of the cannabinoids to the FDA. The Bill was signed into law on 20 December 2018, and on that same day, the Commissioner of the FDA released a statement acknowledging the new legal status of hemp, but also called for science-based, clinical research in support of utilizing cannabinoids in essentially any type of product available to the public. Despite the as-of-yet unknown regulatory status of the cannabinoids, individuals and entities continue to make multi-million-dollar investments in infrastructure to grow, process, extract, formulate, and sell cannabinoid products both in brick and mortar stores and across the internet. This has created a gold-rush mentality to meet a perceived public demand for cannabinoid products, such that it is literally controlling the evolution of the entire hemp industry today. Efforts in fiber and/or grain production are miniscule compared to efforts in cannabinoid production; rarely even considered at any significant level. Additionally, capital investments in infrastructure to process fiber and/or grain crops are a mere fraction of investments towards the cannabinoid market. This is true all across the U.S.A. at present.

How long will this evolution continue? Logic would dictate until the price of cannabinoids is affected by supply exceeding demand. At that time, it could be most likely that the price of the molecules from that point forward will fit more closely within a typical, commodity-based economic model. However, it is still possible that the FDA could decide to regulate cannabinoids such that broad-acre production in outdoor systems might not be feasible. If the ultimate regulatory framework continues to tightly control cannabinoids at any level as is the case today (e.g., Epidiolex®), and if that classification is enforced by federal agencies, then production would almost surely be similar to other horticultural crops where quality control and predictability of yields would be greatly increased by indoor production models; such parameters are unachievable by outdoor production models. Other examples would include recreational and medical forms of *Cannabis*, the values of which easily support high-input, indoor production systems. We further explore these aspects of production in the next section.

3.5 Flower Production Models

One very important and as-of-yet unanswered question in hemp farming is which production model will be most cost-efficient? When considering hemp grain and/or fiber, we can immediately know that on an industrial scale, these will be standard row crops cultured almost entirely by mechanical means. There are several reasons for this simple conclusion. First and foremost, the value of hemp grain and fiber today will not support production by higher-input models. We know that hemp grain and/or fiber can be profit-competitive with normal commodity crops like corn and soybeans. Any increases in input costs that might reduce profit potential from hemp would push farmers to grow a more profitable crop in support of their business. When considering 100s or 1000s of acres, just a small difference in input costs and/or income per acre will make a huge difference in gross farm income. Farmers will generally make wholly business-based or profit-based decisions when choosing crops. We have definitely noted exceptions to this premise as hemp has become more widely distributed, but when push comes to shove, farmers will almost certainly rely on their business skills to estimate profits from potential crops, evaluate the agronomic benefits/costs of their decision, and choose the most appropriate crop accordingly. We should also note that consumer demands for artisan, craft, and locally produced hemp grain and fiber-products will be a factor. The value of these products will be much higher than for industrial products, thus justifying potentially higher-input production models in support of the final product (e.g., produced organically with higher human inputs).

Hemp grown for cannabinoids today is nearly always produced by much higher input production models than the standard, highly mechanized, broad-acre row crops. Early adopters of cultivation of CBD-rich cultivars of hemp in most states have been biased toward specialty crop growers, since they have expertise and equipment for cultivation in raised beds with plastic mulch. Surprisingly, these growers often have a leg up on former marijuana growers who could not apply these techniques to illicit (marijuana) grow systems.

In the "plasticulture" model, plants generally are started from seed or clonally propagated from cuttings. Seeds may be considered more robust than clones and generally are cheaper, but they have the disadvantage of being less uniform and perhaps having males – which need to be eliminated. In either model, the plants are typically started in greenhouses or controlled environments then transplanted to the field, typically at planting spacing of 1200–2000 plants per acre. This horticultural approach to hemp cultivation had heavy labor requirements for harvest, drying and post-harvest processing of plant material (Fig. 3.4). Examples of existing models

Fig. 3.4 Current cannabinoid production models typically involve growing hemp under intensive management, often using drip irrigation under plastic film which is used both to conserve moisture and provide weed control. A relatively new planting (left) on wide intra- and inter-row spacing fills in over the course of the growing season. Photo on right shows the plants at mid-season. (Source: Jabari Byrd)

include either tobacco or tomato production systems, both of which require extremely high inputs relative to row crops.

Interestingly, when considering hemp, there's no science behind the usage of these high-input models for cannabinoid production. Rather, growing single female plants while maintaining an unfertilized state (not pollinated) is purely anecdotal from marijuana production systems. The scientific literature is still very poor regarding cannabinoid production (e.g., the effects of pollination alone on CBD yields). New research will address these questions post-haste, but until then, the CBD world is depending on the anecdotal production systems of old. It should be noted here that the value of the CBD molecule today more than validates just about any production model one could possibly dream up, but as mentioned above, this must be a temporary economic condition. Once an equilibrium of supply and demand are met or supply grossly exceeds demand, the price will adjust accordingly which will certainly impact the desire or need for increasing cost efficiencies.

In addition to production, harvest systems are likely to undergo significant transformation in the next few years. The current model of labor-intensive hand harvest (Fig. 3.5) is already being challenged by mechanization more typical of grain or forage systems. In grain-type systems, combines have been modified (Fig. 3.6) to collect both seed and chaff, which may have value for extraction Equipment is being used to bale or chop the hemp plants, and in some cases the material is harvested wet then wrapped in plastic to ensile. While these production methods are likely to greatly lower labor inputs, it remains to be seen how the effective these models will be, given the potential increase in processing and extraction costs if the harvested material is of low cannabinoid concentration and quality.

Fig. 3.5 Current hemp production systems involve labor-intensive hand harvest (top). Many producers currently hang plants in barns to dry (bottom), which adds to the labor and cost of production. (Source: John Fike)

Fig. 3.6 This combine has been modified by placing a bin underneath to collect hemp chaff which comes from around the seed grain and could potentially be suitable for extraction. (Source: John Fike)

3.6 Research Needs

Research will be central to the advancement of hemp production in the U.S.A., and indeed, the development of state-based industrial hemp research programs was central to the first (2014) Farm Bill. Most of the initial work conducted by universities involved cultivar appraisal and agronomics of grain or fiber production. While these issues are still of interest, they likely will be a lower priority for most regions of the U.S.A. under the current market climate of high flower demand. Regardless of production system, variety development and evaluation and agronomic management practices are some of the first issues that come to mind for growers. As well, fertility, suitable herbicides and pest control often loom large on their minds.

Given the semi-domesticated nature of the plant, improvement in a few agronomic traits could go a long way in improving the productivity and sustainability of hemp production systems. For example, hemp seed is prone to shatter and also subject to geminating on the stalk (Fig. 3.7). The shatter issue further complicates harvest by forcing producers to gather in hemp seed at elevated moisture and before they are all ripe. Reducing shatter to support greater seed harvest at maturity will be important for increasing grain yield and also for lowering the energy required to dry the crop.

Fig. 3.7 Hemp germinating on the stalk following a period of high humidity and wet weather. (Source: Josh Ellinger)

Across hemp types, poor seedling vigor can be a problem. This may be less of an issue in fiber production systems that rely on high seeding rates, because when growing in concert, the many seeds push together to get out of the soil. Lower seeding rates used in grain production systems may be inadequate to do this, particularly if seeding depth has not been well controlled or if the soils have a texture that is prone to crusting. This may have environmental consequence as well, as the general recommendation is to plant hemp in finely tilled fields. Thus, improving seed vigor to improve standability – and to increase the success of no-till establishment – is an important area of effort.

There is limited information about hemp susceptibility to insect pressures, although some preliminary research suggests that arthropods cause minimal damage to plants. This understanding may be premature, however, given that for many states, statewide and regional assessments of disease and arthropods are in their early stages. Certainly, empirical observations indicate species such as corn earworm (*Helicoverpa zea*) and several armyworm (*Spodoptera*) species consume hemp seedheads (Fig. 3.8), and there is some indication that injury from these assaults to

Fig. 3.8 Corn earworm (*Helicoverpa zea*) consuming industrial hemp grown for flowers (left) and fall armyworm (*Spodoptera frugiperda*) consuming hemp grown for grain (right). (Source: John Fike)

Fig. 3.9 Bud rot in the flowers of a CBD-rich hemp plant. (Source: John Fike)

the plant promotes the onset of fungal diseases such as bud rot (*Botrytis* spp.; Fig. 3.9). Efforts to compile information about the most common pests for industrial hemp that growers face and what damage they may cause the crop is critical to formulating best management practices for mitigating pests and to disseminate that information to farmers.

Improving plant disease resistance and determining best management practices to prevent disease outbreaks will be important for hemp (and particularly for cannabinoid) production systems. Several fungal diseases have been identified on hemp, and under poor establishment conditions (particularly cool moist soils) "damping off" – in which fungi attack new seedlings – can cause stand loss. While other fungal species appear to have only limited impact on plant productivity, widespread reintroduction of hemp to U.S.A. production systems is likely to increase the chance for a significant disease outbreak. This may have greater implications in flower production, given that mycotoxins may in turn be present and could potentially end up in food products or health aids.

This list of needs discussed here largely focuses on the agronomic realm. Along with these efforts, much opportunity exists to improve hemp through breeding and greater understanding. Of how to control specific constituents such as cannabinoids and terpenes, seed fatty acid profiles and concentrations, and fiber quality. Along with these efforts, work will be needed to develop and refine scalable extraction and processing, to understand the value of hemp feed and flower products for human and animal nutrition and health, and to determine the economic value of hemp for these purposes and its many other uses.

Hemp production also faces a set of issues common to any other new crop. There are no labeled pesticides for weed, insect or pathogen management for use with industrial hemp in the U.S.A. Growers also face a greater degree of risk with hemp at present, as Federal crop insurance programs common to other crops remain to be implemented.

3.7 Conclusion

Hemp has been an important crop in the U.S.A., but restrictions on the crop prevented its production for almost 70 years. Changes in the law have once again allowed production in the U.S.A., and the hemp industry is gearing up for a rapid expansion across the country. Much of this effort is directed to flower production for CBD, but there are particular questions about the strength and longevity of markets for CBD and other hemp products. Current flower production models are labor-intensive and likely to change; how these systems develop over time and their long-term economic demand remain to be determined. Researchers across the country are working to address a number of questions about this plant, its management and uses. Perhaps no other crop has ever been the subject of such enthusiasm – and perhaps misunderstanding – but the U.S.A. finally appears fully vested in determining how it might best be managed and used.

References

Bidwell PW, Falconer JI (1941) History of agriculture in the northern United States, 1620–1860, Carnegie Institute of Washington. Publication; no. 358. P. Smith Publisher, New York

Clithero G (2018) Registered hemp growers and acreage in production. Vermont center for geographic information and Vermont agency of agriculture, food, and markets

Dewey LH (1902) The hemp industry in the United States. Yearbook of the Unites States Department of Agriculture – 1901. Government Printing Office, Washington, pp 541–554

Dewey LH (1914) Hemp. Yearbook of the United States Department of Agriculture – 1913. Government Printing Office, Washington, DC, pp 283–346

Dewey LH (1928) Hemp varieties of improved type are result of selection. In: What's new in agriculture. Yearbook of the United States Department of Agriculture – 1927. Government Printing Office, Washington, pp 357–360

Duvall C (2014) Cannabis. Reaktion Books/University of Chicago Press, London, p 264

Haney JG (1919) American hemp -a new crop and weed fighter. In: Six years' results on IHC demonstration farms. International Harvester Co, Racine, pp 11–12

Howard HV, Lang AL, Enfield GH (1946) Effects of fertilizers on yields and breaking strengths of American hemp *Cannabis sativa*. J Am Soc Agron 38:551–563

Johnson P (1999) Industrial hemp: a critical review of claimed potentials for *Cannabis sativa*. TAPPI J 82:113–123

Johnson R (2013) Hemp as an agricultural commodity. CRS Report for Congressional Research Service, Washington, DC

National Conference of State Legislatures (2019) State industrial hemp statutes. Webpage. Available at http://www.ncsl.org/research/agriculture-and-rural-development/state-industrial-hemp-statutes.aspx. Updated 1 February 2019

Robinson BB (1943) Hemp. United States Department of Agriculture Farmer's Bulletin no. 1935

Small E, Pocock T, Cavers PB (2003) The biology of Canadian weeds. 119. *Cannabis sativa* L. Can J Plant Sci 83:217–237. https://doi.org/10.4141/P02-021

State of Rhode Island (2019) Approved industrial hemp licenses. Department of Business Regulation. Retrieved from: http://www.dbr.ri.gov/divisions/medicalmarijuana/hemp_approvals.php

USDA (2000) Industrial hemp in the United States: status and market potential. In: AgEcon search, research in agricultural & applied economics. Economic Research Service, p 43. https://ageconsearch.umn.edu/record/278840

Vermont Agency of Agriculture (2019). https://agriculture.vermont.gov/event/2019-industrial-hemp-conference

Wilsie CP, Black CA, Aandahl AR (1944) Hemp production experiments, cultural practices, and soil requirements, Iowa agricultural experiment station bulletin. Iowa State College, Ames, p 63

Wright AH (1918) Wisconsin's hemp industry. In: Agricultural experiment station of the University of Wisconsin, Madison, May 1948, p 47

Chapter 4
Hemp Fibers in Serbia: Cultivation, Processing and Applications

Biljana Pejić, Marija Vukčević, and Mirjana Kostić

Abstract Hemp (*Cannabis Sativa* L.), as one of the oldest cultivated plants, represents a renewable and sustainable source of fibers, mainly for textile production. However, owing to their specific chemical composition, structure and properties, such as high specific strength, nontoxic, biocompatible and biodegradable nature, hemp fibers become ideal candidates for a wide range of applications. Being lignocellulosic, they can be used alone, or combined with different kinds of polymers to provide a wide range of useful composites in textiles, construction, automotive industry, soil conservation. Also, hemp fibers have proven to be efficient biosorbents and suitable precursors for the production of low-cost carbon materials for different applications. This chapter deals with the cultivation, processing and applications of hemp fibers in Serbia, starting with the cultivation of the hemp plant, followed by the production and modification of fibers in order to attain the appropriate structure and chemical composition of hemp fibers that meet the requirements of specific application. Current and future applications of hemp fibers and hemp fiber-based products have also been discussed.

Keywords Hemp fibers · Cultivation · Modification · Carbonization · Adsorption · Water purification

4.1 Introduction: Hemp in Serbia

Hemp (*Cannabis sativa* L.) is one of the oldest cultivated crops in the world, its use ranging from the production of ropes, textiles, and paper, to oil and medicines. The oldest relics of human manufacture are bits of hemp fabric found in tombs dating back to 8000 BC. Hemp fiber, the so called fiber of a hundred uses, is also used for paper. The oldest piece of paper, over 2000 years old, found in China, was made

B. Pejić · M. Vukčević · M. Kostić (✉)
Faculty of Technology and Metallurgy, University of Belgrade, Belgrade, Serbia
e-mail: biljanap@tmf.bg.ac.rs; marijab@tmf.bg.ac.rs; kostic@tmf.bg.ac.rs

G. Crini, E. Lichtfouse (eds.), *Sustainable Agriculture Reviews 42*, Sustainable Agriculture Reviews 42, https://doi.org/10.1007/978-3-030-41384-2_4

from hemp and other fibers. The first bible, following Gutenberg's invention of the printing press, and just about every other printed thing up to the nineteenth century was printed on paper mainly containing hemp fibers. Hemp seed oils were used for a wide range of applications, from cooking to cosmetics, while extracts of hemp were used to treat a wide range of diseases (Lu and Clarke 1995; Kozlowski et al. 2005; Amaducci and Gusovius 2010).

Hemp was cultivated by the Chinese 8500 years ago; its cultivation is documented as early as 2700 BC in ancient Chinese writings. Cultivation then spread across Asia and Europe, arriving in Europe about 2000–2200 years ago (Lu and Clarke 1995; Milosavljevic et al. 2004).

Hemp arrived in the area of present-day Serbia very early; evidence of its cultivation exists on numerous frescoes, drawings and notes of monks from the time of the Nemanjić dynasty (twelfth century). Hemp production became a traditional part of agriculture, there was even a tax on hemp in Serbia in the fifteenth and sixteenth century, imposed by the Turks. The production and processing of hemp was something the whole country was familiar with, although it was primarily concentrated in the Province of Vojvodina. In 1765, following the recommendation of the Austro-Hungarian empress Maria Theresia that only high-quality hemp should be sown, the Imperial Chamber brought to the Bačka region new colonists from Northern Italy, in order to transfer the skill of growing and processing hemp. In these small colonial settlements, hemp growing was recognized as the main opportunity to prosper, which lead to small manufactories for hemp processing being built, eventually growing into larger factories for hemp fabric production. Thanks to the natural conditions (fertile soil rich in nutrients, river Mostonga very convenient for hemp retting), a small colonial settlement at the town of Odžaci became one of the most important industrial centers in the Bačka region. As the production grew, Odžaci became one of the biggest centers for hemp production in Europe. In 1779, a hemp fair was held for the first time, eventually gaining significance. In 1813, Odžaci were granted the privilege of holding a weekly fair and two yearly trade fairs, gaining the status of a stock marketplace. In 1907, a factory for cultivation of hemp and production of ropes was opened by the members of the wealthy Ertl family (Franc and Johan). However, the first organized production of hemp products (rope) in Serbia developed in Stojkovac, near Leskovac, as far back as 1884 (Milosavljevic et al. 2004; Stojanović 2016).

The impact of Italian varieties on Serbian hemp culture was very significant. At the very beginning of hemp production, landraces of Italian origin were grown in Serbia and named after the region of their intensive cultivation, like "Futoška", "Titelska", "Apatinska", "Leskovačka", etc. During the 1960s, certified seed of the Italian cultivars "Bologna" and "Carmagnola" was imported and even multiplied to some extent in Serbia. The monoecious French "Fibrimon" was also tried, though with limited success (Clarke 1996).

In order to enhance hemp research and production, the Institute of Field and Vegetable Crops in Novi Sad (Serbia's most complex and largest research institute in the field of agriculture), established its Research Station for Hemp, Hops and Sorghums in 1952 in Bački Petrovac, near Novi Sad. The first domestic cultivar,

"Novosadska konoplja", selected from the same Italian hemp used for selecting Hungarian cultivars the "F-hemp" and the "Kompolti", was officially registered in 1967 (Clarke 1996; Sponner et al. 2005). Due to a decreased interest in hemp production, the program on hemp almost came to a complete halt during the 1960s, but in 1992 it was renewed, and has been active ever since. In the meantime, three varieties have been developed within the hemp breeding program: a dioecious variety "Marina", a monoecious variety "Helena" and a hybrid variety "Diana", all three registered in 2002 (Berenji et al. 2008).

In spite of a long and well-documented history of hemp fiber and seed production in Serbia, due to the reduction in hemp production and consumption in the second half of the twentieth century, the lack of knowledge and experience in this area was noticeable. There was almost no background knowledge available in Serbia, since bast fiber research stopped by the end of the 1960s. However, in the middle of the 1990s, due to ecological trends and considerable exhaustion of raw material sources, the attention of the industrial and scientific community turned towards renewable and biodegradable raw materials, prompting the beginning of an unexpected, world-wide return to the almost forgotten bast fibers, among them hemp (Milosavljevic et al. 2004; Kozlowski et al. 2005).

This chapter gives an overview of scientific research on the cultivation, processing and applications of hemp fibers in Serbia. Special attention has been paid to fiber modification, performed in order to attain the appropriate structure and chemical composition of hemp fibers that meet the requirements of specific application.

4.2 Cultivation of Hemp and Fiber Production

Fast growing and not very demanding as to climate, soil quality, and nutrients, hemp was farmed all over the world until its ban in the 1930s by most Western countries due to its hallucinogenic properties. Furthermore, cheap and plentiful imported cotton and jute, as well as relatively cheap, simple and efficient chemical fibers production technology, made hemp uncompetitive. However, chemical fibers, created to replace natural fibers, after quite a long period of intensive application, are today considered inferior to natural ones, especially in respect to physiological, hygienic, and health properties, as well as comfort and ecological properties. Furthermore, due to the exhaustion of major organic chemicals resources (coal, gas, oil) necessary for the production of chemical fibers, in the 1990s the focus was shifted to the renewable and biodegradable raw materials. Since hemp is one of the most efficient plants, known for its ability to utilize sunlight and capture large quantities of CO_2 to photosynthesize, with an annual growth up to 5 m in height, production of hemp fiber is experiencing a renaissance. Hemp is an annual plant that grows in temperate climates, without agrochemicals and with optimum fiber yield of more than 2000 kg/ha. Currently, hemp is the subject of a European Union subsidy for

Table 4.1 World production of hemp fiber for the period between 1925 and 2017 and the leading producers

	Hemp fiber and tow production (in 1000 tons)[a]						
	1925	1965	1994	2006	2015	2016	2017
World	740.0	340.8	51.2	114.8	79.5	71.1	59.8
Union of Soviet Socialist Republics	448.9	105.0	–	–	–	–	–
Russian Federation	–	–	3.4	1.5	1.5	1.5	1.5
Italy	123.9	9.9	–	1.2	3.1	5.1	3.4
Yugoslavia	43.3	47.4	–	–	–	–	–
Serbia	–	–	0.8	–	–	–	–
Poland	26.9	19.1	0.4	0.1	0.1	0.1	0.1
Romania	13.1	13.9	1.1	2.4	3.5	3.4	3.5
Hungary	13.2	29.2	0.7	0.5	0.3	0.3	0.3
Netherland	–	–	–	1.5	14.6	17.4	9.5
France	5.6	3.0	0.2	2.5	1.1	1.1	1.2
China	18.2	75.0	22.0	82.9	31.4	18.5	16.6
Korea	20.8	–	–	–	–	–	–
Democratic People's Republic of Korea	–	2.2	12.0	13.0	14.6	14.6	14.8
Republic of Korea	–	5.6	0.4	0.1	–	–	–
Turkey	–	10.0	2.8	0.1	–	–	–
Chile	–	3.8	4.0	4.4	4.3	4.2	4.2

[a]Source: Milosavljevic et al. 2004; Food and Agriculture Organization 2019

non-food agriculture, and a considerable initiative for its further development in Europe is underway (Ranalli and Venturi 2004; Faruk et al. 2012; Kostic et al. 2014b).

The world production of hemp fiber for certain years between 1925 and 2017, along with the leading producers, are given in Table 4.1. Between the two world wars (in 1925), the production of hemp fiber recorded 740,000 tons, which accounted for about 12% of the world production of all fibers. More than 90% of it was produced in Europe, and about 5% in Asia. The world's largest producer of hemp was former Union of Soviet Socialist Republics, which was producing more than 60% of the world output. Italy ranked second with about 17%, and former Yugoslavia was third with 6% of the world production. Due to the previously mentioned reasons, hemp production decreased dramatically over the next 40 years. In 1965, the production of hemp fiber was 340,821 tons, which accounted for about 2% of the world production of all fibers (18,887 million tons). About 70% of it was produced in Europe, and about 28% in Asia. Major producers were the Union of Soviet Socialist Republics, China, Yugoslavia, Hungary and Poland (Milosavljevic et al. 2004; Oerlikon Textile 2010; Food and Agriculture Organization 2019).

In the next three decades, the production of hemp fiber had shown a steady decline. In 1994, the production of hemp fiber was about 51,000 tons, but about 73%

of it was produced in Asia, and only 20% produced in Europe. Major producers were China, the Democratic People's Republic of Korea, Chile and the Russian Federation. Hemp accounted for about 0.12% of the world production of all fibers (42,074 million tons), and less than 0.3% of the world production of natural fibers. After a period of stagnation, thanks to the renewed interest, the production of hemp fiber has shown a steady increase, i.e. the world production of hemp fiber grew from about 50,000 tons in 1994 to almost 115,000 tons in 2006, with more than 70% of it produced in China and only 13% in Europe. Major producers are China, Democratic People's Republic of Korea and Chile, while smaller production occurs in Austria, France, Romania, Netherlands and the Russian Federation. In the past 10 years, the production of hemp fiber has shown a slight decline, with scattering between 48,000 and 80,000 t. In 2017, the production of hemp fiber shows a decrease of about 11,500 tons compared to the previous year, with China as the leading producer (in spite of a production decrease of 11.5%), followed by the Democratic People's Republic of Korea (a production increase of about 1%), Netherlands (production decreased for 45%), Chile, Romania and Italy (Oerlikon Textile 2010; Food and Agriculture Organization 2019).

In Serbia, the whole country was, by and large, familiar with the production and processing of hemp. Serbia was, in fact, the biggest hemp producer in former Yugoslavia. Between the two world wars, there were seven hemp processing factories, organized as a family business or as an association of two families. After the Second World War, there was a drastic social change in Serbia (former Yugoslavia), and this change was reflected in the economic situation, since the industry was nationalized. In 1946, the Ministry of Industry of the People's Republic of Serbia founded the State Administration for hemp and flax, based in Novi Sad, responsible for operational planning and direct management of all affairs related to the hemp industry in Serbia, which included 201 companies. In that period, Odžaci once again become an official center of hemp production since the "Mostonga" company was established there, bringing together 13 hemp production factories in neighboring and more distant settlements: Bač, Bački Brestovac, Bački Gračac, Bačko Novo Selo, Bogojevo, Bođani, Vajska, Doroslovo, Ratkovo, Srpski Miletić, Pivnice, Plavna and Lalić. The peak production was reached in 1949, with the recorded maximum hemp area of as much as 108,215 ha. Even relatively recently, in the 1960s, in the region with the most intensive hemp production, the Province of Vojvodina, industrial hemp was cultivated on more than 20,000 ha, larger than the entire area dedicated to hemp cultivation in Europe in 2014. The first considerable reduction, by 80%, occurred in 1968. In 1988, a further reduction of the growing area by another 80% was observed. As a consequence, hemp is only a minor crop in Serbia today. Ambitious plans exist for its revitalization, with special emphasis on classical (threads, twines, ropes, cordage, rough canvas) as well as modern textile products and paper, seed oil, etc. Based on the rich and long-lasting history of hemp in this region, it is expected that hemp will easily become a major crop in Serbia once again (Clarke 1996; Stojanović 2016).

4.3 Fiber Separation: Primary Processing of Hemp Stalks

The purpose of primary processing is to separate the flexible fibers from the surrounding tissue as gently as possible or, when necessary, using more powerful biological and mechanical processes. The application of these processes is determined by how developed and mature the plant is, and economic reasons.

The first stage in hemp processing is harvesting. The optimal time for harvesting hemp grown for fiber is the early yellow maturity of the stem, when the stem, bark and fiber yield has reached its maximum. Hemp grown for seeds should be harvested at yellow maturity, when the stems are completely yellow, the bolls are yellow and the seeds are fully developed and beginning to turn brown. The harvesting of hemp is done by pulling the crops up, with roots, by hand or by mechanical pullers. Crops are pulled rather than cut in order to ensure that the entire length of the fibers, which run from the root to the top of the stalk, is harvested (Akin 2010; Amaducci and Gusovius 2010).

Further processing (the degree and type of processing) is determined by the destination of the crop. End users, including paper manufacturers, building product suppliers, the automotive industry and textile mills each require a supply of fiber crops in different forms, ranging from raw stalks or husks to fibers-only or hurds-only. For example, harvesting for chopped stalk essentially eliminates processing costs, but does not capture revenue for the raw fiber. Concerning bast fiber crops, there are three approaches to exploiting them commercially: separation of bast fibers from hurds in order to use them separately, using the whole stalk (in fiberboard production, for example) and using seeds for oil, protein meal, etc. (Kozlowski et al. 2005; Kostic et al. 2014b).

At advanced stages of raw crop processing, fibers are more marketable and less costly to transport than unprocessed crop. The extraction/separation of fibers, which are held together in stems, leaves or husks by woody matter and cellular tissue, may be done by using biological, mechanical or chemical methods or, more usually, a combination of retting, breaking and scutching.

Retting is the process of loosening the bond between the fiber bundles and surrounding tissue (decomposing the natural adhesive pectin). There are several retting methods: dew, water, chemical and enzyme retting; however, only dew and water retting are commonly used.

Dew retting (field retting) is the oldest but also the most common technique today. In dew retting the stalks are laid in swathes on the ground as they are harvested. The combined action of dew and showers of rain provide the necessary conditions for the development of the micro-organisms on the stalks. Dew retting tends to yield dark-colored fibers. In spite of lower quality of produced fibers compared to water retted hemp, lower labor costs and higher fiber yields make dew retting attractive and sustainable (Akin 2010).

Water retting method produces the highest-quality fibers. The retting process depends on the colonization and fermentation of the stalks by anaerobic bacteria, e.g. *Clostridium felsinium*, to degrade pectins and other matrix substances and free bast fibers. In days of old, the harvested stalks were bundled and submerged in natural running or still water (e.g. lakes, rivers, dams) for 5–7 days and then dried in the field for 1–2 weeks. Later on, the stalks were retted in industrial conditions in so-called retting mills, i.e. in special basins (tanks) in warm or cold water. Due to high costs of the process in terms of energy (required for heating water up to 30 °C) and waste water (20 tons of water per one ton of stalk and 10 tons for washing and rinsing) it is a considerable threat to the environment and it is not used any more in many countries (Kozlowski et al. 2005; Akin 2010).

Chemical retting was developed as a replacement for dew retting by using chemicals. It is carried out by treating the stalk with chemical solutions of oxalic acid, sulfuric acid, sodium hydroxide, acid and sodium carbonate, oxygenated water, sodium sulfite, etc. Chemical retting methods have not always proved satisfactory in terms of fiber quality, cost of production and fiber yield. On the other hand, hemp can be easily separated from the stem by a mechanical process, the so called green retting or mechanical retting, and "green hemp" is now produced commercially in this way. Enzyme retting was developed in order to overcome the previously mentioned drawbacks of dew, water and chemical retting. Successful enzyme retting could provide advantages, such as: high and consistent quality of fiber, tailored properties for specific applications and broadened geographic regions for the production of lignocellulosic fibers. A major research effort to develop enzyme retting took place in Europe in the 1980s. The strategy was to submerge pulled stems in an enzyme mixture containing pectinases, hemicellulases and cellulases and simulate water retting by replacing bacteria with cell-free enzymes (Akin 2010).

In primary processing of fiber crops, proper retting is a major problem. Plant development and maturity, as well as weather conditions in the case of dew and water retting, influence the quality of retting, which in turn determines both fiber yield and quality. Underretting, i.e. incomplete degradation of the matrix components (i.e. pectin and hemicelluloses), leaves the woody core cells and cuticularised epidermis still associated with fiber, reducing processing efficiency and fiber quality (fibers that are too coarse and contaminated with stiff shive particles). Contrary, overretting occurs from excessive microbial degradation, where fiber strength is reduced owing to excessive thinning of bundles and/or degradation of fiber cellulose, making such fibers not suitable for use in some applications, such as textiles or composites.

After retting, mechanical fiber separation, or decortication, follows to separate the flexible fibers from the stiff and more brittle ligneous woody parts (called hurds) of the stalks. This is achieved by carrying out a series of breaking, shaking and scutching operations repeated several times until the fiber is satisfactorily extracted and separated from the shive.

Breaking is typically done by breakers, which are a set of smooth and ribbed pairs of cylinders. The stalks are inserted between them and crushed. The weight and pressure of the smooth surface rolls splits the stalks lengthways and the ribbed rollers of gradually reduced pitches then transversally and progressively break the hurds into smaller and smaller pieces. Most of these separate from the fiber and fall through or out at the end of the machine before the start of the next operation. The broken stalks are then subjected to the scutching, i.e. operation which separates the unwanted woody matter from the fibers. The operation is carried out in a scutching machine consisting of two pairs of scutching turbines and associated mechanisms which separate and remove shorter fibers, hurds, dust and other waste matter from the long fibers. This is done by gripping the broken stalks and beating first the top portion and then the lower portion with blunt wooden or metal blades of the scutching turbines. As the long fiber bundles are beaten, short fiber bundles, called tow, are sorted out along with contaminants and cleaned separately. After scutching, the fibers are combed or hackled by drawing them through sets of pins, each successive set being finer than the previous one. This operation removes smaller contaminants, disentangles and aligns the long fiber bundles and separates the bundles without destroying length. A short fiber bundle fraction, termed hackling tow, is obtained as a byproduct of the long fibers. The technologies described above serve to process retted stalk. The raw stalk needs much more intense operation (decortication) to loosen the bonds between hurds and fiber, and yields fiber that requires extensive further processing to make it spinnable into yarns materials (Kozlowski et al. 2005).

4.4 Structure and Chemical Composition of Hemp Fibers

The cultivation of hemp in Serbia decreased rapidly since the 1960's. Consequently, all research in Serbia related to the production and processing of hemp, as well as the characterization of the fibers obtained from the stems of this plant, was stopped. However, owing to the high utilization of almost all parts of the hemp plant for various applications, and to the exceptional properties of hemp fibers, since the end of the twentieth century the production of hemp fibers in Serbia is experiencing a renaissance. The growing popularity of hemp has brought on an increase in scientific research in the field of hemp production and usage. At first, researchers in Serbia met with a lack of experience and standardized methods for hemp fiber examination. The necessity for research in this field was recognized by the Ministry of Science and Environmental Protection of the Republic of Serbia, who, for the first time in 2002, provided funds for it through the project "Development of agrocellulosic fibers and fibrous materials based on domestic naturally available bio-renewable resources (hemp) for the needs of the textile and high-quality paper industry". The successful implementation of this project was followed by a series of other projects, financed by the relevant Ministry, and related to the production, modification and characterization of hemp fibers and their applications in conventional and new ways. Broadening the scope of use of hemp fibers is primarily conditioned by the chemical composition and the structure of these fibers. For that reason, first hemp fibers studies in Serbia were directed towards determining their chemical composition and its influence on the structure and properties of hemp fibers. Methods and procedures for determining the chemical composition of hemp fibers, already developed elsewhere, did not provide accurate data on the content of the components. According to some data, upon determining the chemical composition of hemp fibers, the total content of all components was over 100% (Sedelnik 1999). This discrepancy could be related to the heterogeneous chemical composition of the hemp fiber and the specific position of the components in the fiber structure.

It is well known that hemp fibers belong to the class of technical fibers and that they are composed of elementary fiber bundles, obtained by primary processing of hemp (*Cannabis Sativa*) plant stalks. These fibers are characterized by a heterogeneous chemical composition and a fairly complicated structure. The main constituents are cellulose, hemicelluloses and lignin. The content of pectin, fats and waxes is significantly lower, but their role in the construction of the plant and fibers is very important.

The main component of the hemp fiber structure is cellulose, a natural polymer consisting of D-anhydro-glucoside repeating units held together by β-(1,4)-glycosidic linkages, with the degree of polymerization around 10,000 (Kostic et al. 2007; John and Anandjiwala 2008). Hydroxyl groups on C2, C3 and C6 of each repeating unit have the ability to form hydrogen bonds within the same cellulose chain and the surrounding cellulose chains. In that manner, parallel cellulose chains with regions of high order form a crystalline supermolecular structure, while regions of low order form an amorphous structure. The bundles of linear cellulose chains in the longitudinal direction form a microfibril which is oriented in the cell wall structure (Hashem et al. 2007).

Hemicelluloses comprise a group of polysaccharides composed of different monosaccharide units, such as the combination of 5- and 6-carbon ring sugars, with the degree of polymerization around 50–300, which is 10–100 times lower than that of native cellulose. Hemicelluloses are derived mainly from chains of pentose sugars, and act as a supportive matrix for cellulose microfibrils (Demirbas 2008).

Unlike cellulose and hemicelluloses, lignin is a hydrocarbon polymer with a complex three-dimensional structure and very high molecular weight, consisting of both aliphatic and aromatic constituents. Lignin is totally amorphous and hydrophobic in nature and it contains a variety of functional groups, such as hydroxyl, methoxyl and carbonyl, which impart a high polarity to the lignin macromolecule (Demirbas and Kucuk 1993; Hashem et al. 2007; Kostic et al. 2007).

Pectins are complex polysaccharides consisting mainly of esterified D-galacturonic acid resides, and they represent an adhesive matrix in the hemp fibers structure (Pérez et al. 2003).

The structure of elementary fiber is divided in interlamellar layer (i.e. middle lamella –ML), primary (P) wall and secondary (S) wall sub-divided in three layers (S1, S2 and S3) surrounding the central lumen (L), as it is shown in Fig. 4.1. The primary wall contains randomly oriented microfibrils, while the secondary wall consists of helically-wound microfibrils. Actually, each of the three sub-layers has a different microfibrilar orientation. Since the S2 layer is the thickest, the S2 microfibril angle (MFA) controls mechanical properties along the fiber axis. Plant fibers are more ductile if the microfibrils have a spiral orientation to the fiber axis. Fibers are inflexible, rigid, and have a high tensile strength if the microfibrils are

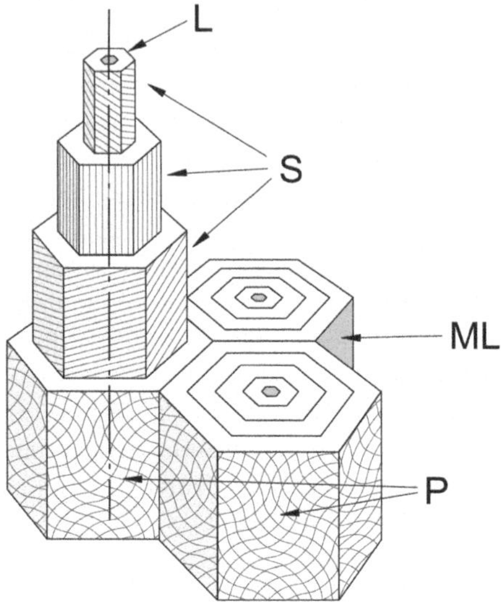

Fig. 4.1 The structure of plant fiber. The structure of elementary fiber is divided in interlamellar layer (i.e. middle lamella –ML), primary (P) wall and secondary (S) wall sub-divided in three layers surrounding the central lumen (L)

oriented parallel to the fiber axis, i.e. smaller value for MFA results in a higher fiber modulus (Bismarck et al. 2005; Ansell and Mwaikambo 2009; Thomas et al. 2011).

The chemical composition of hemp fibers may differ due to a number of reasons: hemp fibers may have been obtained from different varieties of the same plant species; the tests may have been carried out at different stages of maturity of the plants and by using different methods of analysis; the plants from which the fibers were extracted may have grown in different soils and under different conditions. Therefore, it is of great importance to find an appropriate method for determining the chemical composition of hemp fibers. As it was already mentioned, hemp fibers have a heterogeneous chemical composition and a specific position of each component within the fiber structure. For that reason it was difficult to find an adequate method for determining the chemical composition of hemp fibers. The team of researchers working within the mentioned project adopted a procedure for chemical composition analysis based on sequential removal of individual components from hemp fibers' structure, which was in accordance with the Soutar and Bryden research (Garner 1967).

The scheme for determining the chemical composition of hemp fibersis given in Fig. 4.2.

The proposed procedure was applied on several hemp fiber samples that have been obtained from different varieties of the hemp plant species, for determination of chemical composition. All of the hemp plants and fibers used were obtained from the Institute of Field and Vegetable Crops of Novi Sad and ITES Odžaci, both from Serbia. Samples obtained from ITES Odžaci are marked as O, while samples obtained from the Institute of Field and Vegetable Crops of Novi Sad are marked as I samples. The obtained results for fibers' chemical composition are given in Table 4.2. The average content of hemp fibers' components (A), found in literature (Demirbas 2008; Pejic et al. 2009), is also included in Table 4.2.

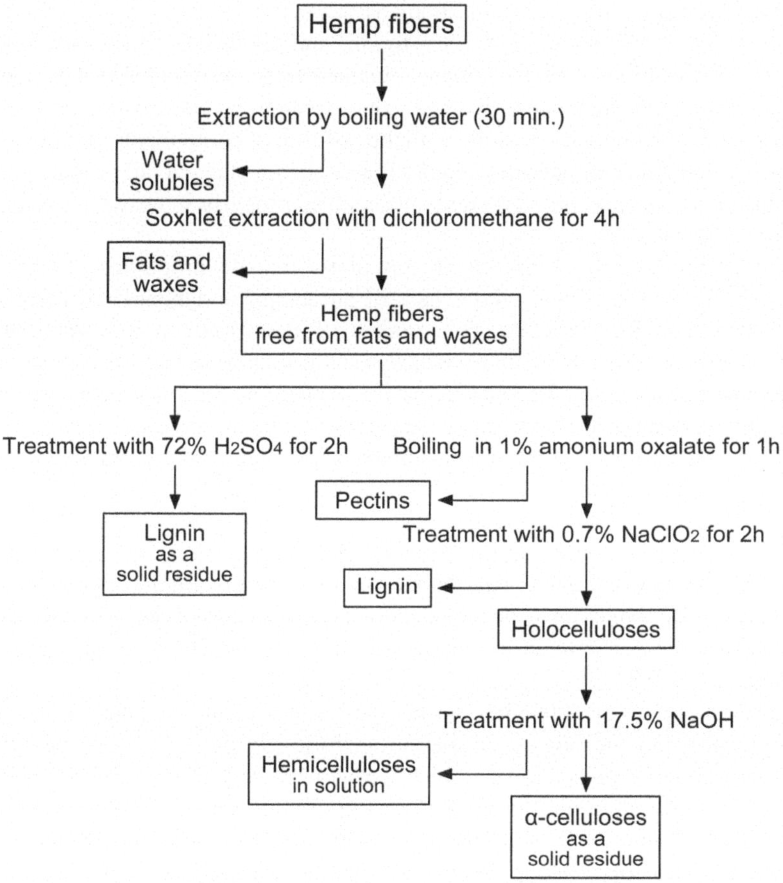

Fig. 4.2 Procedure for determining the chemical composition of hemp fibers. According to this procedure, chemical composition of hemp fibers can be determined through the sequential removal of individual components from hemp fibers' structure

Table 4.2 Chemical composition of hemp fiber samples

Hemp fibers sample	Component, %					
	Water solubles	Fats and waxes	Pectines	Hemicelluloses	Lignin	α-celluloses
O1	1.70	1.59	1.55	12.28	5.65	76.12
O2	1.51	1.26	1.42	10.41	6.13	79.10
O3	1.50	0.69	1.39	11.09	4.68	78.15
I1	1.65	0.23	11.95	11.55	5.51	69.11
I2	1.82	0.47	12.81	11.97	5.83	67.10
A	–	0.8	2.9–3.3	5.5–16.1	3.7–5.7	67.0–78.3

Generally, hemp fibers contain about 67.0–78.3% cellulose, 5.5–16.1% hemicelluloses, 0.8–2.5% pectin, 2.9–3.3% lignin, and some fats and waxes, and the chemical composition of tested samples obtained from Serbia (Table 4.2) does not deviate significantly from these average values (Demirbas 2008; Pejic et al. 2009).

The chemical composition, along with the position of components within the fiber structure, are the most important variables that determine the overall properties of the plant fibers, especially physical and mechanical properties, as well as sorption properties (Kostic et al. 2014b).

In an effort to characterize the hemp fibers' tensile properties, the team of researchers from Serbia met with a lack of an adequate standardized method for determining breaking strength. The process of developing a new method was accompanied by some difficulties, which were primarily related to the statistical measurement uncertainty. Namely, in order to obtain precise information regarding the breaking strength of fibers, many aspects of the hemp fiber structure should be considered. Hemp fibers, as natural bast fibers, show great variation in fiber diameter and relative roughness and unevenness, with small fibrillar ends pointing away from the fibers' surface, which is directly related to the heterogeneous chemical composition, as well as the nature and position of the components in the fiber structure. These irregularities in the fiber structure led to a very high variation of data obtained by measuring the breaking strength of individual technical fibers, which became a multidimensional problem in the characterization of the fibers' tensile properties (Kostic et al. 2008).

Measurement of the breaking strength of single technical hemp fibers gives precise results, but the process of measurement is laborious and time-consuming. In order to obtain representative results, it was necessary to measure the breaking force of a very large number of fibers, because of wide variations in the tenacity of single fibers. Namely, the coefficient of variation was around 30% even after 200 measurements of breaking force on single fibers (Kostic et al. 2008).

In the next step researchers were measuring the tenacity of fiber bundles, which gave average values. However, this method had certain disadvantages due to the fact that some fibers in the bundle may have structural irregularities and very low strength. These fibers will break first during the determination of the breaking strength, thereby reducing the bundle's cross-section size, and affecting the reduction of tenacity values.

This problem was overcome by measuring the breaking strength on hemp fiber bundles of different fineness (bundles containing different numbers of fibers) and lengths. The results of these tests showed that bundle tenacities decrease with the reduction of bundle fineness and with an increase in bundle length. The decrease in bundle tenacities by reducing its fineness is connected with an increased number of weak places inside the bundle. Likewise, with increasing bundle length, the number of weak places also increases along the bundle's axis, as is the case with small fibrillar ends pointing away from the fibers' surface. As the result of investigating the

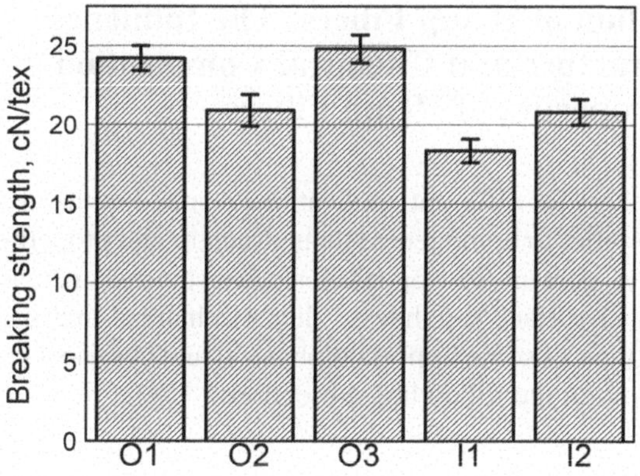

Fig. 4.3 Breaking strength for samples obtained from ITES Odžaci, marked as O1, O2 and O3, and samples obtained from the Institute of Field and Vegetable Crops of Novi Sad, marked as I1 and I2. It is notable that hemp fibers obtained from ITES Odžaci show higher breaking strength compared to the samples from Novi Sad. These results were obtained by using method for breaking strength measurements developed by Kostic et al. (2008)

influence of measuring conditions on the tensile properties of hemp fibers, an original method, based on the use of tenacity of 500 tex fiber bundle extrapolated to zero-test length, was developed. This method for measuring the breaking strength of hemp fibers requires much less time and skill, and at the same time the values obtained for flat bundle tenacity show a high correlation with values measured on single fibers (Kostic et al. 2008; Kostic et al. 2014b).

The results obtained for different hemp fiber samples (from ITES and Novi Sad) using the developed method for breaking strength measurements are given in Fig. 4.3. Tested samples have shown breaking strength in the range from 18.36 to 24.8 cN/tex, which is in agreement with the results obtained for hemp fibers and other lignocellulosic fibers (Bongarde and Shinde 2014; Faruk et al. 2014; Roy et al. 2014).

Based on the results presented in Table 4.2 and Fig. 4.3, it can be observed that breaking strength varies due to differences in the chemical composition of the fibers. However, direct dependence between the amount of individual components of the hemp fibers' chemical composition and the values of breaking strength was not observed. The obtained results indicated that the breaking strength does not depend solely on the chemical composition, but also on the position of the structural components within the fiber. These observations have indicated that further research into hemp should focus on changing the chemical composition and structure of hemp fibers by applying different methods of modification, as well as examining the impact of these treatments on physico-mechanical, chemical and sorption properties.

4.5 Modification of Hemp Fibers: The Influence of the Structure and Chemical Composition on the Properties of Hemp Fibers

The use of hemp fibers for different applications resulted in the necessity for further fiber processing in order to homogenize their structure and properties, clean the fiber surface, reduce the content of noncellulosic substances, increase the degree of elementarization and surface roughness, alter mechanical and sorption properties, etc. A brief description of some important physical and chemical treatments for fiber modification is given in the following subsections.

4.5.1 Physical Treatments

Physical treatments alter the structural and surface properties of the fibers without the use of chemical agents. According to the effects they produce, they are divided into treatments that promote separation of fiber bundles into more homogenous structures – elementary fibers, such as stretching, calendering, steam explosion, thermomechanical and ultrasound treatments; and treatments for fiber surface modification, such as corona, dielectric barrier and plasma discharges, and, more recently, laser and UV (ultraviolet) irradiation (Bledzki and Gassan 1999; Belgacem and Gandini 2005; Mukhopadhay and Fangueiro 2009; Faruk et al. 2012).

The steam explosion process separates the lignocellulosic fibers into their main components: cellulose fibers, amorphous lignin, and hemicelluloses, without the disintegration of single cells (Mukhopadhay and Fangueiro 2009). The combination of chemical and physical treatments, i.e. the impregnation of fiber bundles with NaOH (0.1–0.5%) before the steam explosion treatment, produces an effective loosening of the rigid fiber bundle structure and its splitting into elementary fibers (Kessler et al. 1998). Thermomechanical processes separate fiber bundles into individual fibers, just like the steam explosion process, providing clean fibers with rough surfaces and increased crystallinity. During thermal treatments, around the glass transition temperature of lignin, the lignin, and to some extent the hemicelluloses, are depolymerized into the lower molecule of aldehyde- and phenolic-functionalities, leaving single fibers which will have greater strength and stiffness than the fiber bundles (Li et al. 2000; Mukhopadhay and Fangueiro 2009). Exposure of cellulose fibers to ultrasound action may also cause morphological and structural changes in fibers due to cavitations produced by ultrasound, which act as strongly localized micro-reactors generating a temperature of several thousand degrees and a pressure in excess of one thousand atmospheres, causing strong hydrodynamic shear and/or extensional forces (Mason 1997). Laine and Goring reported an increase in the fiber wall porosity and a slight increase in the carbonyl group content of cellulose fibers after ultrasound treatments, while Surina and Andarassy reported changes in sorption properties of flax fibers with a simultaneous increase in the resistance to the action of tensile forces (Laine and Goring 1977; Surina and Andrassy 2013).

Plasma-based treatments (corona, low- and atmospheric pressure plasma) deserve special attention due to some unique properties and growing demands for the environmental friendliness of processes for surface modification and coating of textiles. They are based on the physico-chemical changes of the material surface without any change in bulk properties. The effect of the corona treatment on mechanical and/or sorption properties has been reported for main bast fibers, i.e. jute, hemp and flax. Among many different plasma configurations available, operating at low- or atmospheric pressure, atmospheric pressure plasma has many advantages, such as robustness, cost effectiveness, shorter processing time, etc. Furthermore, by controlling the plasma variables, such as the nature of gas, discharge power, pressure and exposure time, a great variety of surface effects can be achieved (reactive free radicals and new functional groups increase or decrease surface energy, cross-linking, etc.). The plasma treatment of hemp and other bast fibers has been studied extensively (Gassan and Gutowski 2000; Wong et al. 2001; Yuan et al. 2004; Herbert 2007; Skundric et al. 2007; Stegmaier et al. 2007; Pizzi et al. 2009; Ragoubi et al. 2010; Faruk et al. 2012; Kafi et al. 2012; Shen et al. 2015; Zhang et al. 2015).

There is not nearly as much work devoted to the modification of the lignocellulose fibers by laser treatment and UV irradiation (Laine and Goring 1977; Belgacem and Gandini 2005; Mukhopadhay and Fangueiro 2009).

4.5.2 Chemical Treatments

Simple chemical treatments can be applied to the fibers to selectively remove noncellulosic compounds, thereby getting cellulose-rich fibers with improved fiber strength and fineness, and/or to change surface tension and polarity through modification of the fiber surface. Hemp fibers, bearing hydroxyl groups from cellulose, hemicelluloses and lignin, can be easily functionalized by chemical modification since these groups may be involved in different reactions such as esterification and etherifications, oxidation, acetalation, chemical coupling. Other chemical treatments are alkalization, de-waxing, de-lignifications, fiber impregnation, etc. Chemical modifications may activate hydroxyl groups or can introduce new moieties that can be either used directly because of their special properties, e.g. hydrophilization or hydrophobization, or they are used as reactive "chemical hooks" for further chemical modification (Bledzki and Gassan 1999; Belgacem and Gandini 2005; Kostic et al. 2007; Kalia et al. 2009; Figueiredo et al. 2010; Faruk et al. 2012).

The most important cellulose derivatives, from the perspective of application, are esters and ethers. Esterification of cellulose fiber, especially the acetylation, is a well-known method for fiber modification originally applied to wood cellulose to improve dimensional stability, and stabilize the cell walls against moisture and environmental degradation. The main idea of acetylation is reacting the hydroxyl groups of fibers, responsible for their hydrophilic character, with molecules that have a more hydrophobic nature. Etherification has proven to be a far more effective and general strategy for regioselective substitution than esterification, due to the bulky reagents used for etherification and steric demands. Etherification of cellulose through methylation, carboxymethylation, cynaoethylation, hydroxypropylation, silylation, single or mixed, is one of the most important routes of cellulose functionalization (Li et al. 2000; Khalil et al. 2001; Mwaikambo and Ansell 2002; Kalia et al. 2009; Carter Fox et al. 2011; Varshney and Naithani 2011; Faruk et al. 2012; Haque et al. 2015).

Oxidation and alkalization are the most widely used chemical treatments for plant fibers. Traditionally, they have been used to remove noncellulosic substances and improve plant fiber quality. The alkali treatment of plant fibers, so called mercerization, leads to fibrillation which causes the breaking down of the composite fiber bundle into smaller fibers. Mercerization reduces fiber diameter, increases the aspect ratio, surface roughness and mechanical properties. Moreover, mercerization changes the fine structure of the native cellulose I to cellulose II and increases the number of possible reactive sites. The oxidation of cellulose fibers is used to oxidize hydroxyl groups into the corresponding carbonyl (aldehyde and keto) or carboxyl groups with simultaneous removal of noncellulosic substances. The extensive modifications which accompany this reaction, give rise to products whose chemical and physical properties depend upon the oxidizing reagent used, resulting in selective or non-selective oxidation patterns. The oxidation of cellulose with periodates and the catalytic oxidation using nitroxyl radicals of the TEMPO-type (TEMPO – 2,2,6,6-tetramethylpiperidine 1-oxyl radical) are highly selective reactions. The periodate oxidation of cellulose effects bond cleavage between C-2 and C-3 of the anhydroglucose units, with the concomitant introduction of aldehyde functionalities, while the second route introduces carboxyl functionalities at C-6 of the anhydroglucose units via the intermediate aldehyde stage. Both oxidations yield well defined products, but they are currently still too expensive for large scale applications. Non-selective oxidation, such as permanganate, peroxide, yields less defined products, but it is more suited for large scale applications (Bledzki and Gassan 1999; Mwaikambo and Ansell 2002; Princi et al. 2004; Saito and Isogai 2004; Potthast et al. 2007; Kostic et al. 2008; Kalia et al. 2009; Praskalo et al. 2009; Figueiredo et al. 2010; Faruk et al. 2012; Kostic et al. 2014a).

Chemical coupling methods using molecular or macromolecular agents provide a potential route for significant altering of cellulose fiber properties. Highly functionalized cellulose fibers take advantage of the intrinsic properties of the cellulose in combination with different coupling reagents (Bledzki and Gassan 1999; Belgacem and Gandini 2005).

An alternative to physical and chemical methods is represented by the rapidly expanding use of biological agents such as fungi and enzymes. Biological modifications offer several advantages over physical and chemical methods, such as the selective removal of hydrophilic pectic and hemicellulosic materials, lower energy input and the recycling of enzymatic systems after each use (Faruk et al. 2012; George et al. 2014).

To achieve the balance between environmental acceptability and cost efficiency, along with the improvement of the hemp fibers' properties, researchers in Serbia mainly used alkali and plasma treatment, as well as oxidation for the modification of hemp fibers.

4.5.3 Influence of Applied Treatments on Hemp Fibers' Properties

Treatment of the surface of hemp fibers by low-temperature plasma at atmospheric pressure represents a very promising economic and environmentally desirable technology by which it is possible to obtain similar, equal, or better effects in the fiber structure and properties modification than by applying traditional chemical or physico-chemical finishing processes. The team of researchers from Serbia used the dielectric-barrier discharge (DBD) in air and atmospheric pressure for the modification of hemp fibers. The source of DBD was developed in the Laboratory of Quantum Optics of the Faculty of Physics in Belgrade (Serbia) as a device prototype with plane-parallel geometry, for the continuous plasma-modification or treatment of textile materials. A relatively short DBD treatment leads to changes of the surface without changes in the bulk properties. The plasma oxidation increased the content of various hydrophilic functional groups on the fiber surface and thereby improved sorption properties (moisture content, capillarity and wetting time). In this way, fibers of high hydrophilicity, improved chemisorptions, biocompatibility and adhesive properties can be obtained (Skundric et al. 2007).

An alkali treatment of hemp fibers with sodium hydroxide solutions of different concentrations (5% and 17.5% w/v), at room and boiling temperature, for different periods of time, and both under tension and slack, selectively removed hemicelluloses, with the content of hemicelluloses decreasing up to 70% compared to unmodified fibers, while lignin showed low reactivity, mainly because of strong carbon–carbon linkages and aromatic groups and rings, which were very resistant to chemical attack. Changes in the chemical composition occurred during the alkaline treatment affect the structure of hemp fibers, leading to the disruption of the hydrogen bond network, conversion of cellulose I into cellulose II, and an increase in the amorphous cellulose content at the expense of crystalline cellulose. Additionally, the alkaline treatment increases the amount of cellulose exposed on the fiber surface and thereby increases the surface roughness. This type of modification was carried out in order to separate the fiber bundles and improve the sorption properties of hemp fibers. The quality of hemp fibers was characterized by determining their chemical composition, fineness, physico-mechanical and sorption properties. The alkali treated hemp fibers have become finer and more flexible, due to the lower content of hemicelluloses and lignin. Hemicelluloses removal leads to fiber liberation, with the fiber fineness being reduced from 21.5 tex for unmodified to 1.8 tex for the sample modified at 45 min. Additionally, the alkali treatment at both temperatures under slack conditions resulted in textured fibers. Also, in some cases (modification under tension) tensile properties of modified hemp fibers were improved (Wang et al. 2003; Kostic et al. 2008, 2010, 2014b; Pejic et al. 2008, 2009).

Modification of hemp fibers by sodium-chlorite (0.7% aqua solution), carried out at boiling temperature for different periods of time, could be used to selectively remove lignin. Treatment of hemp fibers with 0.7% $NaClO_2$ progressively removed lignin (about 50% in relation to unmodified fibers) while the content of hemicelluloses in modified hemp fibers decreased for about 17%. During oxidation, as well as during the alkali treatment, the content of noncellulosic components in modified fibers decreased in relation to unmodified hemp fibers, proportionally to the increase of modification time. This trend has been confirmed by the increase in weight loss (indicator of treatment severity) in both cases. Progressive removal of lignin also leads to fiber liberation and fiber fineness. Fineness of oxidized fibers was reduced 2–2.5 times in comparison to the unmodified hemp fibers, which is lower than in the case of alkali treatment. (Pejic et al. 2008; Kostic et al. 2014b; Vukcevic et al. 2014b).

Besides mechanical properties, the utilization of an appropriate modification method can improve the sorption properties of hemp fibers. After alkali and oxidative treatments, the degree of swelling increases in relation to the unmodified hemp fibers. The increase of the degree of swelling of hemp fibers modified with NaOH is most likely the consequence of removing the hemicelluloses from interfibrillar regions, and a decrease in lignin, fats and waxes content.

The degree of swelling of alkali modified hemp fibers samples is higher from 77% to 130% in relation to the unmodified fibers, and increases with the time of modification. It can be explained by the removal of hemicelluloses and hydrophobic components during the alkaline treatment. Once the hemicellulosic components have been progressively removed, interfibrillar regions become less dense and less rigid, which, along with the greater content of amorphous regions, enables easier penetration of larger quantity of water molecules into the hemp fiber structure (Pejic et al. 2009).

The degree of swelling of hemp fiber samples treated with $NaClO_2$ for 5 and 60 min, is higher for about 68% and 78%, respectively, in relation to the unmodified fibers. During oxidative treatment, progressive removal of lignin occurred mostly in the middle lamella. The decrease of lignin and hemicelluloses content brought the changes in hemp fiber structure that influenced an increase of degree of swelling in relation to the unmodified fibers. It can be noted that the degree of swelling of oxidized hemp fibers is lower in relation to the alkali treated fibers, since a greater part of hemicelluloses remained in the interfibrillar regions and their densities have not been reduced after oxidation (Pejic et al. 2009; Kostic et al. 2014b).

The changes in hemp fibers' surface and structure, induced by chemical modification, lead to the changes in total water holding capacity, which can be estimated by determining water retention values. All water absorbing and holding surfaces, cracks, and cavities are included in the water retention measurement. Alkali treated hemp fibers show almost unchanged water retention values compared to the value of unmodified fibers, while the prolongation of alkali treatment decreased these values further still. The decrease of water retention value of hemp fibers modified with NaOH with an increase of modification time is a consequence of structural changes, i.e. changes in the size and number of pores and microcracks in fibers during their modification, as well as reduction in the content of hydrophilic components, hemicelluloses and pectins. On the other hand, oxidative treatment significantly increases water retention value (for 20%) due to the fact that lignin removal increased the roughness of hemp fiber surfaces and induced new capillary spaces in inter-surfacial layers between completely or partially separated fibers within the modified technical hemp fiber (Wang et al. 2003; Kostic et al. 2008). These structural changes induced by progressive removal of hemicelluloses or lignin (alkali and oxidative treatments) also affect the capillarity of the hemp fibers through the increase in the capillary diffusion coefficient and in the coefficient related to the hydrodynamic radius of pores (Pejic et al. 2008).

It was shown that application of simple and inexpensive alkaline and oxidative treatments can increase sorption capacity of short hemp fibers up to 33% for Cd^{2+} and up to 85% for Zn^{2+} (Pejic et al. 2009; Vukcevic et al. 2014b). The observed increase in sorption properties was ascribed not only to the decrease of lignin or hemicelluloses content, but also to the location of these components in the hemp fiber, since lignin removal results in a more homogenous middle lamella due to the gradual elimination of micro-pores and the less rigid cell wall, while the removal of hemicelluloses makes inter-fibrillar regions of fibers less dense and rigid, and fibrils more capable of rearrangement (Buschle-Diller et al. 1999; Wang et al. 2003; Pejic et al. 2008; Kostic et al. 2010). Furthermore, the removal of both hemicelluloses and lignin from lignocellulosic fibers is followed by fiber fibrillation, surface peeling and an increase in roughness, which affects the fiber specific surface area. Besides changes in fiber surface and structure, the applied chemical treatments affect the amount and accessibility of functional groups incorporated in the fiber structure. Both of these chemical treatments remove the accompanying components from the fiber surfaces, leading to the liberation of the functional groups, and an increase in their amount. Since functional groups act as active sites for adsorption, chemical modification efficiently improves the sorption properties of lignocellulosic biosorbents (Pejic et al. 2009; Vukcevic et al. 2014b).

4.6 Carbonization of Hemp Fibers

The usage of lignocellulosic waste for production of carbon materials is very attractive from the point of decreasing the waste disposal costs and improving environment protection through waste recycling and producing useful products. In that way, different carbon sorbents produced from lignocellulosic waste were used for water purification by removal of specific pollutants, like dyes, heavy metals, pesticides and phenols (Budinova et al. 2009; Giraldo and Moreno-Pirajan 2008;

Hameed et al. 2008, 2009; Aber et al. 2009; El-Hendawy 2009; Valente Nabais et al. 2009; Salman and Hameed 2010; Chowdhury et al. 2011; de Lima et al. 2011; Hernández-Montoya et al. 2011; Salman et al. 2011; Sugumaran et al. 2012; Rahman et al. 2014; Vukcevic et al. 2014a, 2015).

Carbon materials can be obtained by controlled thermal decomposition of lignocellulosic materials, in an inert atmosphere, which undergo pyrolysis and carbonization. The pyrolysis process involves an initial softening of lignocellulosic material and the release of volatile matter which can lead to significant weight loss. The pyrolysis is controlled mainly by two predominant reactions, dehydration and depolymerization. Physical desorption of water is one of the first processes during pyrolysis, followed by dehydration of the cellulosic units. The dehydration reaction stabilizes the cellulose structure; during the dehydration, elimination of the hydroxyl groups results in double bonds, conjugated double bonds, and subsequently, in the formation of an aromatic structure. The polymeric structure is basically retained through dehydration and weight loss is mainly limited to the evaporation of water. Dehydration is followed by degradation of native cellulose fibers under inert atmosphere. Decomposition of cellulose and hemicelluloses takes place in the narrow temperature range of about 200–400 °C, while lignin decomposition starts at low temperatures and continues at a low rate until the temperature surpasses 900 °C (Mohamed et al. 2010). This low reactivity of lignin, compared to other biomass constituents, can be the consequence of its highly cross-linked nature (Sharma et al. 2004). It is confirmed that lignin is the main source of char, while cellulose and hemicelluloses are responsible for formation of volatile substances, leading to the major mass loss of the solid residue (Mohamed et al. 2010). Therefore, the lignocellulosic materials with higher lignin content produce a higher amount of char and a higher carbon material yield. Carbonization is the second step of controlled thermal decomposition of lignocelluloses and represents the conversion of a depolymerized structure into graphite-like layers through the re-polymerization. In the temperature range of 400–900 °C, the carbonaceous residue is converted into a more ordered carbon structure. Further heating, above 900 °C, initiates graphitization, and generally amorphous carbonaceous structures convert to a turbostratic carbon structure containing graphene layers. During carbonization, the carbon content is increased over 90%. A remarkable feature is that the carbon structure formed via pyrolysis retains some memory of the starting structure through the entire process (Dumanh and Windle 2012).

Activated carbons with well-developed internal pore structures and a large specific surface area are usually produced by physical or chemical activation. Physical activation is a two steps process that includes carbonization of a raw lignocellulosic material and activation of the carbonized material using water steam, carbon dioxide, air, or their mixture as activating agents. Chemical activation can be conducted in a single step, by thermal treatment of the activating agent and carbon precursor mixture, under inert atmosphere. In this way, the carbonization and activation are carried out simultaneously. Inclusion of pre-carbonization in the chemical activation process contributes to the formation of activated carbon with a higher specific surface area and microporosity than those without pre-carbonization

(Yun et al. 2001; Basta et al. 2009). They found that the so-called 2-steps activation, using KOH as an activating agent, is much more advantageous than the single-step activation for obtaining carbon materials with a high specific surface area. The activation process mechanism, along with the activation parameters used, plays an important role in determining the resulting adsorptive properties of activated carbons. Real time monitoring of the activation process using carbon material derived from lignocellulosic precursor can provide an insight into the mechanism of activation, and establish the connection between activation parameters and specific surface area of activated material (Lozana-Castelló et al. 2007). Monitoring of evolved gaseous products of activation indicated that KOH activation occurs gradually through the three distinct phases of H_2 evolution. Porosity development correlates with both H_2 and CO evolution, and the temperatures at which the major H_2 and CO evolution occurs during the activation process are shifted to higher values by increasing KOH/carbonized material ratio. It was found that the increase in the carbonization and activation temperature leads to a higher extent of activation and values of the specific surface area (Vukcevic et al. 2015).

Hemp fibers represent a good precursor for carbon production since they contain cellulose, hemicelluloses and lignin, which are rich in carbon. It was suggested that the complex structure and heterogeneous chemical composition of hemp fibers are one of the crucial factors that affect the specific surface area, amount of surface oxygen groups and morphology of the resulting carbon sorbent (Vukcevic et al. 2012, 2017). Additional factors that define the final properties of carbon materials are related to the conditions of production: carbonization and activation temperature, heating rate, type of activating agent and activating agent/carbon material ratio. Therefore, in order to obtain carbon materials with different characteristics, the Serbian team varied the temperature of carbonization, temperature of activation and amount of activating agent, and as a result the carbon sorbents with high specific surface area (up to 2200 m^2/g) and good sorption characteristics (i.e. high sorption capacity for different heavy metal ions and pesticides) were obtained (Vukcevic et al. 2012, 2014a, 2015, 2017; Mijailovic et al. 2017).

As it was already mentioned, the amount of lignin, hemicelluloses and cellulose in the carbon precursor affects the specific surface area of carbonized materials. Lignin has been found to be effective in creating pores (Kennedy et al. 2004). Furthermore, the specific surface area was found to be the highest for the carbon materials obtained from carbon precursors with the highest lignin content (Vukcevic et al. 2012). On the contrary, Khezami and associates have suggested that the microporosity of carbonized material is due to cellulose and not due to lignin and hemicelluloses. Additionally, it is suggested that polymorphic transformation of cellulose I to more reactive cellulose II primarily affects the porosity of hemp fibers-based carbon materials (Vukcevic et al. 2012). Taking into consideration the influence of the precursor's chemical composition, along with the carbonization and activation conditions, on textural characteristics of activated carbon, it is possible to select appropriate conditions of production and a specific structure of hemp fibers that would promote the formation of required surface porosity.

4.7 Application of Hemp Fibers and Hemp Fiber-Based Products

Nowadays, the mobilization of pollutants in the environment becomes a serious concern due to the increased industrial activities and discharge of effluents into the environment. Most of these effluents contain different organic and inorganic toxic substances, whose presence in the environment is of major concern due to their toxicity and threat to humans and other forms of life. Therefore, the pursuit of the optimal method for purification of contaminated water represents one of the major ongoing topics. A number of methods have been developed for water purification: photocatalytic degradation, combined photo-Fenton and biological oxidation, aerobic degradation, ozonation, chemical precipitation, chemical oxidation or reduction, electrochemical treatment, evaporative recovery, filtration, reverse osmosis, ion exchange and membrane technologies. Among them, adsorption on activated carbon has proved to be an efficient technology for the purification of drinking water and wastewater, although its large-scâle application is limited by the high cost of the carbon adsorbent. While adsorption may be a costly method, the use of inexpensive and renewable materials as an alternative to conventional activated carbon could make the adsorption process cost effective. Therefore, in recent years, a lot of attention has been devoted to the development of adsorbents from different kinds of agricultural and industrial waste, the so called biosorbents. The application of biosorbents for the purification of waste and drinking water is a very cost-effective method. It is the equivalent of using one waste for cleaning up another (Volesky 2007). On the other hand, following the general trend of finding low-cost adsorbents, different kind of wastes, especially lignocellulosic, were utilized as precursors for the production of carbon adsorbents (Mohamed et al. 2010; Nor et al. 2013).

4.7.1 Hemp Fibers for Sorbent Applications

Heavy metals are discharged into the environment from a number of industries (petroleum, petrochemical, dyeing and metal processing). The toxic effect of heavy metals on the environment is multifaceted due to their bio-accumulating tendency, persistence and possibility to transform into even more toxic forms under certain conditions. Heavy metals represent permanent pollutants because they cannot be subjected to any degradation process. As a result of this, their concentration in the soil, waterways and sediments often exceeds the permissible levels. In that way, heavy metals accumulate in the environment and in food chains where they profoundly disrupt biological processes. The influence of heavy metals on human health is reflected in a number of diseases such as anemia, severe diarrhea, neurological and renal disturbance, lung insufficiency, etc. The toxicology of heavy metals, however, will not be discussed here.

Conventional technologies used for the removal of heavy metals, besides a number of advantages, show several disadvantages, such as the ineffectiveness for lower concentrations of metal ions in aqueous solutions, and high cost. These disadvantages can be overcome by the implementation of biosorption as an efficient and low-cost alternative to the conventional methods for heavy metals removal. Biosorbents represent sustainable and renewable cheap filter materials, often with high affinity and capacity, and therefore the usage of biosorbents becomes an alternative to conventional water purification methods. Biosorption has been defined as the property of certain biomolecules, dead biomass as well as living plants and organisms, to bind and concentrate metal ions from aqueous solutions. Due to the complexity of biosorbent structure, the process of metal ion removal from aqueous solutions can occur through various biosorption mechanisms. Biosorption mechanisms can be based on the active metabolic transport of heavy metals through the cell wall of living organisms (bioaccumulation) or on the affinity between the dead biomass and heavy metals (biosorption). Application of microorganisms (specifically bacteria, algae, yeasts and fungi) as biosorbents for heavy metal removal has been received with growing interest due to high surface to volume ratio, large availability, rapid kinetics of adsorption, and desorption and low cost. However, there are some limitations to the usage of living organisms as sorbents, e.g. they cannot function at a low pH level, or at toxic levels of metal ions; as proteinaceous materials they may putrefy under moist conditions, while dead biomass is chemically and physically more robust (Bailey et al. 1999; Volesky 2007).

Therefore, the short and entangled hemp fibers obtained as waste from the Serbian textile industry were used as environmentally and economically friendly, natural materials for the removal of heavy metals. Hemp fibers that our research team used showed biosorption capacities of 3.86 mg/g for zinc ions, 6.63 mg/g for cadmium ions, and 15.54 mg/g for lead ions removal, which is in agreement with the results for hemp fibers presented by Demirbas and Tofan (Demirbas 2008; Tofan et al. 2010). A notable increase in fiber capacities (7.07 mg/g for zinc ions, 8.77 mg/g for cadmium ions and 16.16 mg/g for lead ions removal) after modification of hemp fibers is a consequence of differences in the chemical composition (content of cellulose, hemicelluloses, lignin, pectin and extractives), fiber structure (different thickness of cell walls and lumen, degree of crystallinity and fibrilar orientation) and specific surface area, i.e. the presence of micropores and microcracks (Milosavljevic et al. 2004; Pejic et al. 2009, 2011; Kostic et al. 2014b, 2018; Vukcevic et al. 2014b).

The process of heavy metals' biosorption onto hemp fibers implies metal ions bonding to carboxylic (primarily present in hemicelluloses, pectin and lignin), phenolic (lignin and extractives) and, to some extent, hydroxylic (cellulose, hemicelluloses, lignin, extractives and pectin) and carbonyl groups (lignin) through the mechanism of complexation and ion exchange. It was suggested that the ion exchange is at the root of the biosorption metal uptake, which implies that biosorption is determined by the amount, availability and ionization state of the functional groups involved in metal binding (Volesky 2007). Biosorbent capacity is affected by the acid-base behavior of lignocellulosic fiber, i.e. the point of zero charge pH (pH_{PZC}) and the initial pH value of the biosorption medium. The point of zero charge pH is a pH of the solution at which the overall observed charge on the lignocellulosic fiber surface is zero. When the sorbent is kept in a solution whose pH is less than pH_{PZC}, the protonation of functional groups occurs, and the sorbent behaves as positively charged. At this point, functional groups repel the positive metal ions. An increase in pH above pH_{PZC} makes the functional groups deprotonated, they act as negative species and attract and bind positive metal ions (Vukcevic et al. 2014b).

However, biosorption is a physico-chemical process that is not restricted only to complexation and ion exchange, but comprises several mechanisms such as sorption by physical forces and ion entrapment in inter- and intrafibrillar capillaries and spaces of fiber structure. In order to determine the biosorption mechanism and its potential rate-controlling step, pseudo-first and pseudo-second order kinetic models have been exploited to test the biosorption experimental data. In addition, information on the kinetics of metal uptake is required to select the optimum conditions for full-scale batch metal removal processes. In most biosorption experiments, pseudo-second order kinetic model is considered appropriate to represent the kinetic data in biosorption systems. This tendency comes as an indication that the rate-limiting step in biosorption of heavy metals are chemisorption, involving valence forces through the sharing or exchange of electrons between sorbent and sorbate, complexation, coordination and/or chelation. However, a good fitting model does not necessarily illustrate the real nature of the rate-limiting step. In some biosorption processes, diffusion, as opposed to the chemical reaction, can also be the rate-limiting step (Febrianto et al. 2009; Pejic et al. 2011). In order to investigate the diffusion mechanism during the biosorption process, experimental data was usually tested using the intraparticle diffusion model. When the intraparticle diffusion plot passes through the origin, intraparticle diffusion is the only rate-controlling step in the overall adsorption process. However, for a majority of biosorption systems, the obtained intraparticle diffusion plots are multi-linear, indicating that three consecutive steps occur in the adsorption process: the first linear part could be assigned to a surface adsorption; the second to the intraparticle diffusion stage, and the third linear part represents the final equilibrium stage (Pejic et al. 2011; Vukcevic et al. 2014b).

The research team from Serbia has shown that the influence of metal ion diffusion on biosorbent capacity and kinetics can be described by empirical equations with constants determined in an experimental manner. The effective diffusion coefficient of metal ions through the fibers, the profile of heavy metal ion concentration in the fibers and the effectiveness of sorption, determined by a mathematical model based on the second Fick's low, describes not only the mechanism of adsorption and ion transport, but also the structural characteristics of biosorbent. Mathematical models, generally, represent an efficient way to find the adequate biosorbent with appropriate sorption capacity toward certain metal ions, and optimal parameters of biosorption processes (Pejic et al. 2011; Vukcevic et al. 2014b).

Another study important for determining biosorbent capacity, along with the proper analysis and design of biosorption separation processes, is the examination of biosorption equilibrium data. In equilibrium, a certain relationship prevails between the solute concentration in solution and adsorbed state (i.e., the amount of solute adsorbed per unit mass of adsorbent). Their equilibrium concentrations are a function of temperature. Therefore, the adsorption equilibrium relationship at a given temperature is referred to as adsorption isotherm. The biosorption process equilibrium is most commonly studied by Langmuir and Freundlich isotherm models. The Langmuir model assumes monolayer adsorption, which can only occur at a finite number of sites that are identical and equivalent, with no lateral interaction and steric hindrance between the adsorbed species. This model refers to homogeneous adsorption which implies that all active sites for adsorption have equal affinity for the adsorbate. Maximum monolayer capacity of a biosorbent, along with the information related to sorption energy, can be obtained by treating the experimental data with the Langmuir isotherm model. On the other hand, the Freundlich isotherm is not restricted to the formation of the monolayer and can be applied to multilayer adsorption with uneven distribution of adsorption energy and affinity over the heterogeneous surface. The Freundlich isotherm constants, related to sorption capacity and surface heterogeneity of adsorbent, can be obtained by fitting adsorption equilibrium data. This isotherm is particularly suitable for fitting data from highly heterogeneous sorbent systems and, therefore, can adequately be used for modeling of most biosorption systems (Vukcevic et al. 2014b).

4.7.2 Hemp Fibers as Precursor for Carbon Sorbent Production

Activated carbons are known to be more efficient in adsorbing a greater amount of pollutants. Actually, adsorption on activated carbon has been proven to be an efficient technology for purification of drinking water and wastewater; however, its large-scale application is limited by the high cost of the carbon adsorbent. Therefore, our team utilized short and entangled hemp fibers as inexpensive and renewable materials for carbon production that could make the adsorption process cost effective (Vukcevic et al. 2012, 2014a, 2015, 2017). Modified and unmodified hemp fibers were carbonized up to 1000 °C with the carbonization rate of 5 °C/min, under nitrogen flow, as shown in Fig. 4.4.

It was shown that the distribution of celluloses, lignin and hemicelluloses in the hemp fiber structure, as well as the changes in their amount induced by chemical modification, affect the surface properties, morphology and, consequently, adsorption properties of the resulting hemp fibers-based carbon materials (some data given in Table 4.3). Progressive removal of both hemicelluloses and lignin from the precursor structure lead to liberation of elementary fibers, and that fibrous structure is preserved after carbonization.

Figure 4.5 shows the scheme of activated carbon production starting with waste hemp fibers as raw material. The activation of hemp fibers carbonized at 700 °C and 1000 °C was performed at 700 °C and 900 °C, in the presence of different amount of KOH, with the activation rate of 5 °C/min. In this way, following optimal production

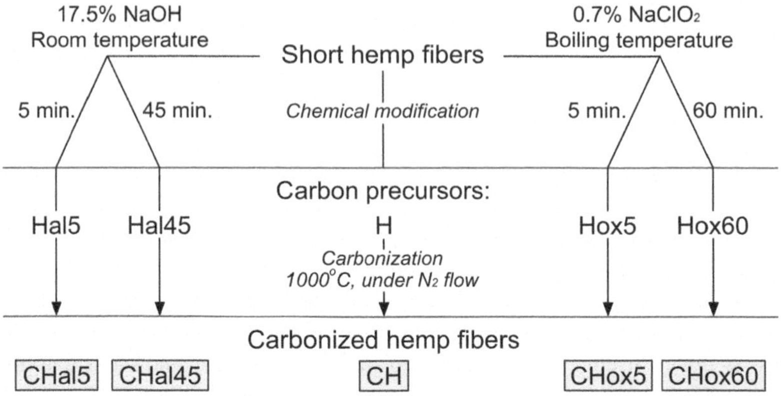

Fig. 4.4 Scheme of hemp fibers carbonization. Unmodified (H) and hemp fibers alkali treated for 5 (Hal5) and 45 (Hal45) minutes, as well as hemp fibers oxidized for 5 (Hox5) and 60 (Hox60) minutes were used as precursors for carbonization. Carbonized hemp fibers: CH, CHal5, CHal45, CHox5 and CHox60 were obtained after carbonization of H, Hal5, Hal45, Hox5 and Hox60, respectively, at 1000 °C, under N_2 flow

Table 4.3 Surface and adsorption properties of the hemp fibers-based carbon materials (for the sample description see Fig. 4.4)

	Samples of carbonized hemp fibers				
Surface and adsorption properties	CH	CHal5	CHal45	CHox5	CHox60
Amount of surface oxygen groups, mmol/g	0.810	1.045	1.060	0.854	1.459
Specific surface area, m^2/g	519	426	573	429	389
pH_{PZC}	10.58	10.72	10.95	11.05	10.87
q_e (Pb^{2+}), mg/g	28.37	43.28	42.03	38.89	46.60

parameters, microporous hemp fibers based activated carbons with a high specific surface area (up to 2192 m^2/g) and amount of surface oxygen groups were obtained.

The specific surface area and pore size distribution are dominant factors of influence in the case of physical adsorption of nonpolar organic molecules. However, the adsorption of inorganic and polar organic compounds is also influenced by the surface chemistry of the carbon sorbent. Generally, a developed surface porosity has a positive impact on the carbon sorbent efficiency. However, the specific surface area is not the only factor that influences the sorbent capacities of carbon materials derived from hemp fibers as lignocellulosic waste. Other factors that influence the adsorption capacity are related to adsorption conditions, material structure and amount of surface oxygen groups that may act as active sites for adsorption.

The structure of activated carbon can be characterized by the presence of heteroatoms (O, Cl, N, S) which may originate from the carbon precursor structure, or as a consequence of non-ideal carbonization and activation conditions. Oxygen is found to be predominant and exists in the form of different functional groups, such as carboxyl, carbonyl, phenols, lactone, carboxyl anhydride and others. Surface oxygen

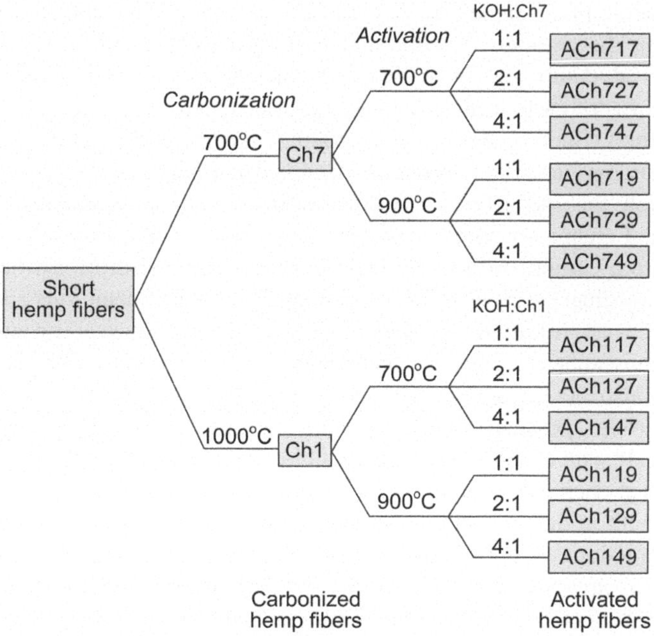

Fig. 4.5 Scheme of hemp fibers carbonization and activation. Hemp fibers carbonized at 700 °C (Ch7) and 1000 °C (Ch1) were activated using different amount of KOH at temperature of 700 °C and 900 °C. Activated hemp fibers (ACh) were marked in the way that first digit represents carbonization temperature (7 for 700 °C and 1 for 1000 °C), second digit represents the amount of KOH (1 for KOH:Ch = 1:1, 2 for KOH:Ch = 2:1 and 4 for KOH:Ch = 4:1), while third digit represents activation temperature (7 for 700 °C and 9 for 900 °C)

groups on carbon materials are usually determined by the temperature programmed desorption (TPD), X-ray photoelectron spectroscopy (XPS), Fourier transform infrared spectroscopy (FTIR) and Boehm titration method (Zhou et al. 2007; Kalijadis et al. 2011, 2015). TPD provides quantitative information on the total number of surface oxygen groups. Surface oxygen complexes on carbon materials decompose upon heating by releasing CO and CO_2, thus, the TPD peaks of CO and CO_2 at different temperatures correspond to specific oxygen groups. For example, CO_2 is released by decomposition of carboxylic groups at 100–400 °C, or lactone groups at 200–650 °C. Both CO and CO_2 peaks originate from the decomposition of carboxylic anhydrides in the temperature range of 350–650 °C. Phenols, ethers, carbonyls and quinones give rise to CO at 700–970 °C. The quantities of CO and CO_2 released during the TPD experiments correspond to the total amount of surface oxygen groups. The decomposition temperature is related to the bound strength of specific oxygen-containing groups. Thus, the position of the peak maximum at a defined temperature corresponds to a specific oxygen complex at the surface (Vukcevic et al. 2008; Kalijadis et al. 2011). These oxygen containing groups may be of acidic, basic or neutral nature. The total amount of acidic and basic groups, along with the amount of carboxyl, lactone and phenol groups can be determined by using the Boehm titration method (Vasiljević et al. 2004; Vukcevic et al. 2013). The ratio between the amount of acidic and basic groups, present at the surface of a carbon material, determines its acid/base character, i.e. the point of zero charge pH (pH_{PZC}). pH_{PZC} of carbon materials can be determined by a fast and simple mass titration method (Baćić-Vukcevic et al. 2006; Vukcevic et al. 2008).

Adsorption onto a carbon sorbent surface is usually a complex process that may occur through the mechanism of physisorption, i.e. ion-exchange, electrostatic attraction, and chemisorption, i.e. surface complexation. For example, metal ions' adsorption on a carbon surface can be considered as physisorption when it occurs through the mechanism of electrostatic attraction between metal ions and graphene layer's π electrons. Additionally, oxygen groups present on the carbon surface might behave as ion-exchange sites for the retention of metal ions. Surface functional groups can also act as active chemisorption sites, since oxygen in the surface functional groups possesses a pair of lone electrons (Lewis base) and therefore may coordinate with electron deficient metal ions (Lewis acid) (Vuković et al. 2011).

The surface chemistry of carbon has a great influence on adsorption of organic molecules, both in the case of electrostatic and non-electrostatic interactions. Electrostatic interactions appear when the adsorbate is an electrolyte that is dissociated or protonated in aqueous solution under the experimental conditions used. These interactions, that can be either attractive or repulsive, strongly depend on the charge densities for both the carbon surface and the adsorbate molecule, and on the ionic strength of the solution. The non-electrostatic interactions are always attractive, and can include van der Waals forces, hydrophobic interactions and hydrogen bonding. Aromatic compounds can be physisorbed on carbon materials essentially by dispersion interactions between the π electrons of the aromatic ring and those of the graphene layers. The presence of aromatic rings in an organic compound structure increases the possibility of such interactions due to delocalized π electrons over the ring. Moreover, a branched substituent on the aromatic ring increased the level of organic compound adsorption. On the other hand, electron-acceptor surface groups, present on the carbon sorbent surface, can withdraw π electrons from the graphene layers, decreasing the dispersive interactions and leading to reduced adsorption (Vukcevic et al. 2013). These dispersion interactions, along with the amount and nature of surface functional groups, can be profoundly affected by the functionalization of the carbon sorbent surface. Therefore, adsorption capacity of a carbon sorbent can be improved by changing its surface chemistry by using an appropriate chemical modification method.

The team of researchers from Serbia used carbonized and activated hemp fibers in water purification for the adsorption of heavy metals and pesticides. Adsorption of heavy metal ions was tested through adsorption isotherms and kinetics. The experimental data obtained in this way was used for development of a mathematical model that would describe both the phenomenon of metal ion transport through the porous matrices, and the structure of carbonized hemp fibers. Compared to the adsorption capacities of hemp fibers as biosorbents, carbonized hemp fibers have proved to be much more efficient in heavy metal removal, showing the adsorption capacities for Zn ions of 25.6–27.7 mg/g, Cd ions of 16.7–19.1 mg/g and Pb ions of 28.4–46.6 mg/g.

Table 4.4 Specific surface area, S_{BET}, and amount of pesticides adsorbed on activated hemp fibers surface, q

Sample	S_{BET}, m^2/g	q, μg/g				
		Acetamiprid	Dimethoate	Nicosulfuron	Carbofuran	Atrazine
Ch7	5	7.63	16.41	2.89	47.64	4.94
Ch1	22.64	38.47	50.90	29.01	50.77	30.13
ACh717	272.63	37.72	16.28	9.59	38.30	19.92
ACh727	504.72	49.62	43.70	24.64	44.65	47.64
ACh747	102.2	12.42	39.66	1.50	50.17	4.64
ACh719	432.9	64.45	63.62	63.75	64.13	64.15
ACh729	608.33	66.31	66.31	66.31	66.29	66.24
ACh749	453.76	49.18	49.18	49.18	49.10	49.14
ACh117	158.1	44.88	46.42	26.14	49.66	37.39
ACh127	318.46	40.71	22.55	5.16	48.61	26.40
ACh147	161.18	46.57	28.22	25.70	30.03	33.27
ACh119	261	68.25	68.27	64.42	69.65	66.89
ACh129	456.7	72.25	71.54	71.31	72.19	71.39
ACh149	323.16	60.36	58.13	57.23	58.97	58.74

For the pesticides adsorption study, pesticides with different chemical structure were chosen: acetamiprid, dimethoate, nicosulfuron, carbofuran and atrazine (Table 4.4). It was shown that activated hemp fibers have good adsorption properties toward pesticides. The nature and the amount of surface oxygen groups have a dominant effect on pesticide adsorption, while specific surface area is not the crucial factor. Due to the good adsorption properties toward heavy metals and pesticides, along with the strong antimicrobial effect, the examined materials were successfully used as filter materials in water purification. Activated hemp fibers were successfully applied as a solid-phase sorbent for the preconcentration of pesticides from water, and for same pesticides, recoveries obtained by these cartridges were even higher than recoveries obtained by commercial ones (Vukcevic et al. 2015).

Also, activated hemp fibers were used as thin layer electrodes in order to analyze the representative capacitive performances. It was found that a more pronounced mesoporous character of the surface, along with the higher amount of surface groups, amplified the currents, and consequently increased specific capacitance (Mijailovic et al. 2017).

4.8 Conclusion

Hemp is only a minor crop in Serbia today. However, ambitious plans exist for its revitalization, with special emphasis on classical (threads, twines, ropes, cordage, rough canvas), as well as modern textile products and paper, seed oil, etc.

For the purpose of restoring the production of hemp fibers in Serbia, as well as improving and monitoring their quality, the conducted investigation was focused on the development of adequate methods for testing their chemical composition and mechanical properties. Due to the heterogeneous composition, hemp fiber properties depend both on the chemical composition and the location of the components within the fiber structure. By selecting the appropriate method of chemical modification or/and subsequent processing by plasma and carbonization, fiber characteristics were adjusted to meet the requirements of specific applications. Unmodified and chemically modified short and entangled hemp fibers, obtained as waste from the textile industry, were used as a sustainable and renewable biosorbent for the purification of water contaminated with heavy metals. Furthermore, short and entangled hemp fibers were used as a low-cost precursor for the production of carbon adsorbents for the preconcentration and removal of organic and inorganic contaminants from polluted water. The usage of activated hemp fibers as supercapacitors was also investigated.

The production of innovative hemp-based materials is becoming an important area of research in Serbia, since hemp offers a unique combination of high physical and specific chemical properties, producing a variety of high-value products with low impact on the environment. Disposal of such products does not pose a problem, as is the case with synthetic materials, i.e. they can be composted to improve soil structure, or incinerated with no emission of pollutants, releasing no more carbon than the fiber crops absorbed during their lifetimes.

Based on the rich and long-lasting history of hemp in this region, it is expected that hemp will easily become a major crop in Serbia once again.

References

Aber S, Khataee A, Sheydaei M (2009) Optimization of activated carbon fiber preparation from Kenaf using K_2HPO_4 as chemical activator for adsorption of phenolic compounds. Bioresour Technol 100:6586–6591. https://doi.org/10.1016/j.biortech.2009.07.074

Akin DE (2010) Flax – structure, chemistry, retting and processing. In: Müssig J (ed) Industrial applications of natural fibres: structure, properties and technical applications. Wiley, Chichester, pp 89–108. ISBN 978-0-470-69508-1

Amaducci S, Gusovius H-J (2010) Hemp – cultivation, extraction and processing. In: Müssig J (ed) Industrial applications of natural fibres: Structure, properties and technical applications. Wiley, Chichester, pp 109–134. ISBN 978-0-470-69508-1

Ansell MP, Mwaikambo LY (2009) The structure of cotton and other plant fibers. In: Eichhorn SJ, JWS H, Jaffe M, Kikutani T (eds) Handbook of textile fibre structure, Vol. 2, Natural, regenerated, inorganic and specialist fibres. Woodhead Publishing, Ltd, Cambridge, pp 2–94. ISBN 978-1-84569-730-3 (book)

Baćić-Vukcevic M, Udovičić A, Laušević Z, Perić-Grujić A, Laušević M (2006) Surface characteristics and modification of different carbon materials. Mater Sci Forum 518:217–222. https://doi.org/10.4028/www.scientific.net/MSF.518.217

Bailey SE, Olin TJ, Bricka RM, Adrian DD (1999) A review of potentially low-cost sorbents for heavy metals. Water Res 33:2469–2479. https://doi.org/10.1016/S0043-1354(98)00475-8

Basta AH, Fierro V, El-Saied H, Celzard A (2009) 2-Steps KOH activation of rice straw: an efficient method for preparing high-performance activated carbons. Bioresour Technol 100:3941–3947. https://doi.org/10.1016/j.biortech.2009.02.028

Belgacem MN, Gandini A (2005) The surface modification of cellulose fibres for use as reinforcing elements in composite materials. Compos Interface 12:41–75. https://doi.org/10.1163/1568554053542188

Berenji J, Adamović D, Sikora V, Sabo J (2008) Advances in the production and utilization of alternative crops achieved at Department of hops, sorghum and medicinal plants of Institute of field and vegetable crops. Zbornik radova Instituta za ratarstvo i povrtarstvo 45:145–158. (article language: Serbian). https://scindeks.ceon.rs/article.aspx?artid=0354-76980801145B

Bismarck A, Mishra S, Lampke T (2005) Plant fibers as reinforcement for green composites. In: Mohanty AK, Misra M, Drzal LT (eds) Natural fibers, biopolymers, and biocomposites. CRC Press, Boca Raton, pp 37–108. ISBN 9780849317415

Bledzki AK, Gassan J (1999) Composites reinforced with cellulose based fibres. Prog Polym Sci 24:221–274. https://doi.org/10.1016/S0079-6700(98)00018-5

Bongarde US, Shinde VD (2014) Review on natural fiber reinforcement polymer composites. Int J Eng Sci Innov Technol 3:431–436. https://pdfs.semanticscholar.org/9c49/54b56f7bfd17625207825197e9e0c5f095e1.pdf

Budinova T, Savova D, Tsyntsarski B, Ania CO, Cabal B, Parra JB, Petrov N (2009) Biomass waste-derived activated carbon for the removal of arsenic and manganese ions from aqueous solutions. Appl Surf Sci 255:4650–4657. https://doi.org/10.1016/j.apsusc.2008.12.013

Buschle-Diller G, Fanter C, Loth F (1999) Structural changes in hemp fibers as a result of enzymatic hydrolysis with mixed enzyme systems. Text Res J 69:244–251. https://doi.org/10.1177/004051759906900403

Carter Fox S, Li B, Xu D, Edgar KJ (2011) Regioselective esterification and etherification of cellulose: a review. Biomacromolecules 12:1956–1972. https://doi.org/10.1021/bm200260d

Chowdhury ZZ, Zain SM, Khan RA, Ahmad AA, Islam MS, Arami-Niya A (2011) Application of central composite design for preparation of kenaf fiber based activated carbon for adsorption of manganese (II) ion. Int J Phys Sci 6:7191–7202. https://doi.org/10.5897/IJPS11.1510

Clarke RC (1996) Interview-Dr. Janos Berenji. J Int Hemp Assoc 3:69–70. http://www.hempfood.com/IHA/iha03211.htm

de Lima LS, Araujo MDM, Quináia SP, Migliorine DW, Garcia JR (2011) Adsorption modeling of Cr, Cd and Cu on activated carbon of different origins by using fractional factorial design. Chem Eng J 166:881–889. https://doi.org/10.1016/j.cej.2010.11.062

Demirbas A (2008) Heavy metal adsorption onto agro-based waste materials: a review. J Hazard Mater 157:220–229. https://doi.org/10.1016/j.jhazmat.2008.01.024

Demirbas A, Kucuk MM (1993) Delignification of Ailanthus altissima and spruce orientalis with glycerol or alkaline glycerol at atmospheric pressure. Cellul Chem Technol 27:679–686

Dumanh AG, Windle AH (2012) Carbon fibres from celullosic precursors: a review. J Mater Sci 47:4236–4250. https://doi.org/10.1007/s10853-011-6081-8

El-Hendawy ANA (2009) An insight into the KOH activation mechanism through the production of microporous activated carbon for the removal of Pb^{2+} cations. Appl Surf Sci 255:3723–3730. https://doi.org/10.1016/j.apsusc.2008.10.034

Faruk O, Bledzki A, Fink HP, Sain M (2012) Biocomposites reinforced with natural fibers: 2000–2010. Prog Polym Sci 37:1552–1596. https://doi.org/10.1016/j.progpolymsci.2012.04.003

Faruk O, Bledzki AK, Fink HP, Sain M (2014) Progress report on natural fiber reinforced composites. Macromol Mater Eng 299:9–26. https://doi.org/10.1002/mame.201300008

Febrianto J, Kosasih AN, Sunarso J, Ju YH, Indraswati N, Ismadji S (2009) Equilibrium and kinetic studies in adsorption of heavy metals using biosorbent: a summary of recent studies. J Hazard Mater 162:616–645. https://doi.org/10.1016/j.jhazmat.2008.06.042

Figueiredo JA, Ismael MI, Anjo CMS, Duarte AP (2010) Cellulose and derivatives from wood and fibers as renewable sources of raw-materials. In: Rauter A, Vogel P, Queneau Y (eds) Carbohydrates in sustainable development I. Topics in current chemistry, vol 294. Springer, Berlin/Heidelberg, pp 117–128. https://doi.org/10.1007/128_2010_88

Food and Agriculture Organization of the United Nations (2019) http://faostat3.fao.org/home/E. Accessed on 14 Apr 2019

Garner W (1967) Fibres. In: Textile laboratory manual, vol 5. Heywood Books, London, pp 52–113

Gassan J, Gutowski VS (2000) Effects of corona discharge and UV treatment on the properties of jute-fibre epoxy composites. Compos Sci Technol 60:2857–2863. https://doi.org/10.1016/S0266-3538(00)00168-8

George M, Mussone PG, Bress DC (2014) Surface and thermal characterization of natural fibres treated with enzymes. Ind Crop Prod 53:365–373. https://doi.org/10.1016/j.indcrop.2013.12.037

Giraldo L, Moreno-Pirajan JC (2008) Pb^{2+} adsorption from aqueous solutions on activated carbons obtained from lignocellulosic residues. Braz J Chem Eng 25:143–151. https://doi.org/10.1590/S0104-66322008000100015

Hameed BH, Tan IAW, Ahmad AL (2008) Optimization of basic dye removal by oil palm fibre-based activated carbon using response surface methodology. J Hazard Mater 158:324–332. https://doi.org/10.1016/j.jhazmat.2008.01.088

Hameed BH, Salman JM, Ahmad AL (2009) Adsorption isotherm and kinetic modeling of 2.4-D pesticide on activated carbon derived from date stones. J Hazard Mater 163:121–126. https://doi.org/10.1016/j.jhazmat.2008.06.069

Haque R, Saxena M, Shit SC, Asokan P (2015) Fibre-matrix adhesion and properties evaluation of sisal polymer composite. Fibers Polym 16:146–152. https://doi.org/10.1007/s12221-015-0146-2

Hashem A, Akasha RA, Ghith A, Hussein DA (2007) Adsorbent based on agricultural wastes for heavy metal and dye removal: a review. Energy Educ Sci Technol 19:69–86

Herbert T (2007) Atmospheric-pressure cold plasma processing technology. In: Shishoo R (ed) Plasma technologies for textiles. Woodhead Publishing, Ltd, Cambridge, pp 79–128. https://doi.org/10.1.1.473.3081&rep=rep1&type=pdf . ISBN-13: 978-1-84569-257-5(e-book)

Hernández-Montoya V, Mendoza-Castillo DI, Bonilla-Petriciolet A, Montes-Morán MA, Pérez-Cruz MA (2011) Role of the pericarp of *Carya illinoinensis* as biosorbent and as precursor of activated carbon for the removal of lead and Acid blue 25 in aqueous solutions. J Anal Appl Pyrolsis 92:143–151. https://doi.org/10.1016/j.jaap.2011.05.008

John MJ, Anandjiwala RD (2008) Recent developments in chemical modification and characterization of natural fiber-reinforced composites. Polym Compos 29:187–207. https://doi.org/10.1002/pc.20461

Kafi AA, Magniez K, Cinquemani C, Fox BL (2012) Influence of atmospheric helium plasma on the surface energy of jute fibres and the performance of resulting composites. J Adhes Sci Technol 26:151–162. https://doi.org/10.1163/016942411X569363

Kalia S, Kaith BS, Kaur I (2009) Pretreatments of natural fibers and their application as reinforcing material in polymer composites- a review. Polym Eng Sci 49:1253–1272. https://doi.org/10.1002/pen.21328

Kalijadis AM, Vukcevic MM, Jovanović ZM, Laušević ZV, Laušević MD (2011) Characterization of surface oxygen groups on different carbon materials by the Boehm method and temperature programmed desorption. J Serb Chem Soc 76:757–768. https://doi.org/10.2298/JSC091224056K

Kalijadis A, Đorđević J, Trtić-Petrović T, Vukcevic M, Popović M, Maksimović V, Rakočević Z, Laušević Z (2015) Preparation of boron-doped hydrothermal carbon from glucose for carbon paste electrode. Carbon 95:42–50. https://doi.org/10.1016/j.carbon.2015.08.016

Kennedy LJ, Vijaya JJ, Sekaran G (2004) Effect of two-stage process on the preparation and characterization of porous carbon composite from rice husk by phosphoric acid activation. Ind Eng Chem Res 43:1832–1838. https://doi.org/10.1021/ie034093f

Kessler RW, Becker U, Kohler R, Goth B (1998) Steam explosion of flax – a superior technique for upgrading fiber value. Biomass Bioenergy 14:237–249. https://doi.org/10.1016/S0961-9534(97)10040-X

Khalil HPSA, Ismail H, Rozman HD, Ahmad MN (2001) The effect of acetylation on interfacial shear strength between plant fibres and various matrices. Eur Polym J 37:1037–1045. https://doi.org/10.1016/S0014-3057(00)00199-3

Kostic M, Skundric P, Praskalo J, Pejic B, Medovic A (2007) New functionalities in cellulosic fibers developed by chemical modification. Hemijska Industrija 61:233–237. https://doi.org/10.2298/HEMIND0704233K

Kostic M, Pejic B, Skundric P (2008) Quality of chemically modified hemp fibers. Bioresour Technol 99:94–99. https://doi.org/10.1016/j.biortech.2006.11.050

Kostic MM, Pejic BM, Asanovic KA, Aleksic VM, Skundric PD (2010) Effect of hemicelluloses and lignin on the sorption and electric properties of hemp fibers. Ind Crop Prod 32:169–174. https://doi.org/10.1016/j.indcrop.2010.04.014

Kostic MM, Milanovic JZ, Baljak MV, Mihajlovski K, Kramar AD (2014a) Preparation and characterization of silver-loaded hemp fibers with antimicrobial activity. Fibers Polym 15:57–64. https://doi.org/10.1007/s12221-014-0057-7

Kostic M, Vukcevic M, Pejic B, Kalijadis A (2014b) Hemp fibers: old fibers – new applications. In: Md. Ibrahim M Mondal (ed) Textiles: history, properties and performance and applications. Nova Science Publishers, Inc, New York, pp 399–446. ISBN 163117262X

Kostic M, Pejic B, Vukcevic M (2018) Waste hemp (*Cannabis sativa*) fibers as a biosorbent and a precursor for biocarbon sorbents: influence of their chemical composition on Pb(II) removal. In: Stevanovic T (ed) Chemistry of lignocellulosics: current trends. Taylor & Francis Group/CRC Press, Boca Raton, pp 3–21. ISBN 9781498775694 (hardback) https://www.crcpress.com/Chemistry-of-Lignocellulosics-Current-Trends/Stevanovic/p/book/9781498775694

Kozlowski R, Baraniecki P, Barriga-Bedoya J (2005) Bast fibres (flax, hemp, jute, ramie, kenaf, abaca). In: Blackburn RS (ed) Biodegradable and sustainable fibres. Woodhead Publishing Ltd, Cambridge, pp 36–88. ISBN 9781845690991 (eBook)

Laine JE, Goring DAI (1977) Influence of ultrasonic irradiation on the properties of cellulosic fibers. Cellul Chem Technol 11:561–567

Li Y, Mai Y-W, Ye L (2000) Sisal fibre and its composites: a review of recent developments. Compos Sci Technol 60:2037–2055. https://doi.org/10.1016/S0266-3538(00)00101-9

Lozana-Castelló D, Calo JM, Cazorla-Amoros D, Linares-Solano A (2007) Carbon activation with KOH as explored by temperature programmed techniques, and the effects of hydrogen. Carbon 45:2529–2536. https://doi.org/10.1016/j.carbon.2007.08.021

Lu X, Clarke RC (1995) The cultivation and use of hemp (*Canabis sativa* L.) in ancient China. J Int Hemp Assoc 2:26–32. http://www.druglibrary.org/olsen/hemp/iha/iha02111.html

Mason TJ (1997) Ultrasound in synthetic organic chemistry. Chem Soc Rev 26:443–451. https://doi.org/10.1039/CS9972600443

Mijailovic DM, Vukcevic MM, Stevic ZM, Kalijadis AM, Stojanovic DB, Panic VV, Uskokovic PS (2017) Supercapacitive performances of activated highly microporous natural carbon macrofibers. J Electrochem Soc 164:A1061–A1068. https://doi.org/10.1149/2.0581706jes

Milosavljevic S, Skundric P, Mihajlidi T, Tadic T, Jocic D, Lukic S, Jovancic P, Kostic M, Mihailovic T, Asanovic K, Stankovic S, Radetic M, Topalovic T, Pejic B, Krgovic M, Nikolic S, Borna N, Mijatovic B (eds) (2004) Hemp as raw of the future (Konoplja sirovina buducnosti, in Serbian). Faculty of Technology and Metallurgy, Belgrade. ISBN 86-7401-201-9

Mohamed AR, Mohammadi M, Darzi GN (2010) Preparation of carbon molecular sieve from lignocellulosic biomass: a review. Renew Sustain Energy Rev 14:1591–1599. https://doi.org/10.1016/j.rser.2010.01.024

Mukhopadhay S, Fangueiro R (2009) Physical modification of natural fibers and thermoplastic films for composites—a review. J Thermoplast Compos Mater 22:135–161. https://doi.org/10.1177/0892705708091860

Mwaikambo LY, Ansell MP (2002) Chemical modification of hemp, sisal, jute, and kapok fibers by alkalization. J Appl Polym Sci 84:2222–2234. https://doi.org/10.1002/app.10460
Nor NM, Chung LL, Teong LK, Mohamed AR (2013) Synthesis of activated carbon from lignocellulosic biomass and its applications in air pollution control-a review. J Environ Chem Eng 1:658–666. https://doi.org/10.1016/j.jece.2013.09.017
Oerlikon Textile (2010) The fiber year 2009/10 report, no. 10:88–98. https://www.oerlikon.com/ecomaXL/get_blob.php?name=The_Fibre_Year_2010_en_0607.pdf
Pejic BM, Kostic MM, Skundric PD, Praskalo JY (2008) The effects of hemicelluloses and lignin removal on water uptake behavior of hemp fibers. Bioresour Technol 99:7152–7159. https://doi.org/10.1016/j.biortech.2007.12.073
Pejic B, Vukcevic M, Kostic M, Skundric P (2009) Biosorption of heavy metal ions from aqueous solutions by short hemp fibers: effect of chemical composition. J Hazard Mater 164:146–153. https://doi.org/10.1016/j.jhazmat.2008.07.139
Pejic B, Vukcevic M, Pajic-Lijakovic I, Lausevic M, Kostic M (2011) Mathematical modeling of heavy metal ions (Cd^{2+}, $Zn^{2+}$and Pb^{2+}) biosorption by chemically modified short hemp fibers. Chem Eng J 172:354–360. https://doi.org/10.1016/j.cej.2011.06.016
Pérez S, Rodríguez-Carvajal MA, Doco T (2003) A complex plant cell wall polysaccharide: rhamnogalacturonan II. A structure in quest of a function. Biochimie 85:109–121. https://doi.org/10.1016/S0300-9084(03)00053-1
Pizzi A, Kueny R, Lecoanet F, Massetau B, Carpentier D, Krebs A, Loiseau F, Molina S, Ragoubi M (2009) High resin content natural matrix-natural fibre biocomposites. Ind Crop Prod 30:235–240. https://doi.org/10.1016/j.indcrop.2009.03.013
Potthast A, Kostic M, Schiehser S, Kosma P, Rosenau T (2007) Studies on oxidative modifications of cellulose in the periodate system: molecular weight distribution and carbonyl group profiles. Holzforschung 61:662–667. https://doi.org/10.1515/HF.2007.099
Praskalo J, Kostic M, Potthast A, Popov G, Pejic B, Skundric P (2009) Sorption properties of TEMPO-oxidized natural and man-made cellulose fibers. Carbohydr Polym 77:791–798. https://doi.org/10.1016/j.carbpol.2009.02.028
Princi E, Vicini S, Pedemonte E, Proietti N, Capitani D, Serge AL (2004) Physical and chemical characterisation of cellulose based textiles modified by periodate oxidation. Macromol Symp 218:343–352. https://doi.org/10.1002/masy.200451435
Ragoubi M, Bienaimé D, Molina S, George B, Merlin A (2010) Impact of corona treated hemp fibres onto mechanical properties of polypropylene composites made thereof. Ind Crop Prod 31:344–349. https://doi.org/10.1016/j.indcrop.2009.12.004
Rahman MM, Adil M, Yusof AM, Kamaruzzaman YB, Ansary RH (2014) Removal of heavy metal ions with acid activated carbons derived from oil palm and coconut shells. Materials 7:3634–3650. https://doi.org/10.3390/ma7053634
Ranalli P, Venturi G (2004) Hemp as a raw material for industrial applications. Euphytica 140:1–6. https://doi.org/10.1007/s10681-004-4749-8
Roy SB, Shit SC, Sengupta RA, Shukla PR (2014) A review on bio-composites: fabrication, properties and applications. Int J Innov Res Sci Eng Technol 3:16814–16824. https://doi.org/10.15680/IJIRSET.2014.0310058
Saito T, Isogai A (2004) TEMPO-mediated oxidation of native cellulose: the effect of oxidation conditions on chemical and crystal structures of the water-insoluble fractions. Biomacromolecules 5:1983–1989. https://doi.org/10.1021/bm0497769
Salman JM, Hameed BH (2010) Effect of preparation conditions of oil palm fronds activated carbon on adsorption of bentazon from aqueous solutions. J Hazard Mater 175:133–137. https://doi.org/10.1016/j.jhazmat.2009.09.139
Salman JM, Njoku VO, Hameed BH (2011) Adsorption of pesticides from aqueous solution onto banana stalk activated carbon. Chem Eng J 174:41–48. https://doi.org/10.1016/j.cej.2011.08.026
Sedelnik N (1999) Effect of biological modification of technical flax fibres. Fibers Text East Eur 7:19–21. ISSN 1230-3666

Sharma RK, Wooten JB, Baliga VL, Lin X, Chan WG, Hajaligol MR (2004) Characterization of chars from pyrolysis of lignin. Fuel 83:1469–1482

Shen M, Wang L, Chen F, Long J-J, Rui Y-N (2015) Effect of low-temperature oxygen plasma on the degumming of ramie fabric. J Clean Prod 92:318–326. https://doi.org/10.1016/j.jclepro.2014.12.054

Skundric P, Kostic M, Medovic A, Pejic B, Kuraica M, Vuckovic A, Obradovic B, Mitrakovic D, Puric J (2007) Wetting properties of hemp fibers modified by plasma treatment. J Nat Fibers 4:25–33. https://doi.org/10.1300/J395v04n01_03

Sponner J, Toth L, Franck RR (2005) Hemp. In: Franck RR (ed) Bast and other plant fibres. Woodhead Publishing, Cambridge, pp 176–206. ISBN 9781845690618 (e Book)

Stegmaier T, Dinkelmann A, von Arnim V (2007) Corona and dielectric barrier discharge plasma treatment of textiles for technical applications. In: Shishoo R (ed) Plasma technologies for textiles. Woodhead Publishing, Ltd, Cambridge, pp 129–157. https://doi.org/10.1.1.473.3081&rep=rep1&type=pdf

Stojanović V (2016) Hemp farming development and socioeconomic position of Bačka: example of Odžaci. Geogr Pannonica 20:88–95. https://doi.org/10.18421/GP20.02-05

Sugumaran P, Susan VP, Ravichandran P, Seshadri S (2012) Production and characterization of activated carbon from banana empty fruit bunch and delonix regia fruit pod. J Sustain Energy Environ 3:125–132. https://doi.org/10.1.1.848.1873&rep=rep1&type=pdf

Surina R, Andrassy M (2013) Effect of preswelling and ultrasound treatment on the properties of flax fibers cross-linked with polycarboxylic acids. Text Res J 83:66–75. https://doi.org/10.1177/0040517512452928

Thomas S, Paul SA, Pothan LA, Deepa B (2011) Natural fibres: structure, properties and applications. In: Kalia S, Kaith BS, Kaur I (eds) Cellulose fibers: bio- and nano-polymer composites. Springer, Berlin, pp 3–42. https://doi.org/10.1007/978-3-642-17370-7_1

Tofan L, Paduraru C, Volf I, Balan C (2010) Kinetic and thermodynamic profile of Pb(II) sorption by untreated hemp fibers. Sci Pap Agron Ser 53:146–149. https://www.researchgate.net/publication/268184808_Kinetic_and_thermodynamic_profile_of_PbII_sorption_by_untreated_hemp_fibers

Valente Nabais JM, Gomes JA, Suhas CPJM, Laginhas C, Roman S (2009) Phenol removal onto novel activated carbons made from lignocellulosic precursors: influence of surface properties. J Hazard Mater 167:904–910. https://doi.org/10.1016/j.jhazmat.2009.01.075

Varshney VK, Naithani S (2011) Chemical functionalization of cellulose derived from nonconventional sources. In: Kalia S, Kaith BS, Kaur I (eds) Cellulose fibers: bio- and nano-polymer composites. Springer, Berlin/Heidelberg, pp 43–60. https://doi.org/10.1007/978-3-642-17370-7_2

Vasiljević T, Baćić M, Laušević M, Onjia A (2004) Surface composition and adsorption properties of activated carbon cloth. Mater Sci Forum 453:163–168. https://doi.org/10.4028/www.scientific.net/MSF.453-454.163

Volesky B (2007) Biosorption and me. Water Res 41:4017–4029. https://doi.org/10.1016/j.watres.2007.05.062

Vukcevic M, Kalijadis A, Dimitrijević-Branković S, Laušević Z, Laušević M (2008) Surface characteristics and antibacterial activity of a silver-doped carbon monolith. Sci Technol Adv Mater 9:ID 015006 (7pp). https://doi.org/10.1088/1468-6996/9/1/015006

Vukcevic M, Kalijadis A, Radisic M, Pejic B, Kostic M, Lausevic Z, Lausevic M (2012) Application of carbonized hemp fibers as a new solid-phase extraction sorbent for analysis of pesticides in water samples. Chem Eng J 211-212:224–232. https://doi.org/10.1016/j.cej.2012.09.059

Vukcevic M, Kalijadis A, Babić B, Laušević Z, Laušević M (2013) Influence of different carbon monolith preparation parameters on pesticide adsorption. J Serb Chem Soc 78:1617–1632. https://doi.org/10.2298/JSC131227006V

Vukcevic M, Pejic B, Kalijadis A, Pajić-Lijaković I, Kostic M, Laušević Z, Laušević M (2014a) Carbon materials from waste short hemp fibers as a sorbent for heavy metal ions – Mathematical modeling of sorbent structure and ions transport. Chem Eng J 235:284–292. https://doi.org/10.1016/j.cej.2013.09.047

Vukcevic M, Pejic B, Lausevic M, Pajic-Lijakovic I, Kostic M (2014b) Influence of chemically modified short hemp fiber structure on biosorption process of Zn^{2+} ions from waste water. Fibers Polym 15:687–697. https://doi.org/10.1007/s12221-014-0687-9

Vukcevic MM, Kalijadis AM, Vasiljević TM, Babić BM, Laušević ZV, Laušević MD (2015) Production of activated carbon derived from waste hemp (*Cannabis sativa*) fibers and its performance in pesticide adsorption. Microporous Mesoporous Mater 214:156–165. https://doi.org/10.1016/j.micromeso.2015.05.012

Vukcevic MM, Pejic BM, Pajic-Lijakovic IS, Kalijadis AM, Kostic MM, Lausevic ZV, Lausevic MD (2017) Influence of the precursor chemical composition on heavy metal adsorption properties of hemp (*Cannabis sativa*) fibers based biocarbon. J Serb Chem Soc 82:1417–1431. https://doi.org/10.2298/JSC170310080V

Vuković GD, Marinković AD, Škapin SD, Ristić MĐ, Aleksić R, Perić-Grujić AA, Uskoković PS (2011) Removal of lead from water by amino modified multi-walled carbon nanotubes. Chem Eng J 173:855–865. https://doi.org/10.1016/j.cej.2011.08.036

Wang HM, Postle R, Kessler RW, Kessler W (2003) Removing pectin and lignin during chemical processing of hemp for textile applications. Text Res J 73:664–669. https://doi.org/10.1177/004051750307300802

Wong KK, Tao XM, Yuen CWM, Yueng KW (2001) Wicking properties of linen treated with low temperature plasma. Text Res J 71:49–56. https://doi.org/10.1177/004051750107100108

Yuan X, Jayaraman K, Bhattacharyyaj D (2004) Mechanical properties of plasma-treated sisal fibre-reinforced polypropylene composites. J Adhes Sci Technol 18:1027–1045. https://doi.org/10.1163/1568561041257478

Yun CH, Park YH, Park CR (2001) Effects of pre-carbonization on porosity development of activated carbons from rice straw. Carbon 39:559–567. https://doi.org/10.1016/S0008-6223(00)00163-9

Zhang Q, Jiang Y, Yao L, Jiang Q, Qiu Y (2015) Hydrophobic surface modification of ramie fibers by plasma-induced addition polymerization of propylene. J Adhes Sci Technol 29:691–704. https://doi.org/10.1080/01694243.2014.997380

Zhou JH, Sui ZJ, Zhu J, Li P, Chen D, Dai YC, Yuan WK (2007) Characterization of surface oxygen complexes on carbon nanofibers by TPD, XPS and FT-IR. Carbon 45:785–796. https://doi.org/10.1016/j.carbon.2006.11.019

Chapter 5
Physico-chemical Characterization and Development of Hemp Aggregates for Highly Insulating Construction Building Materials

Yunhong Jiang, Atif Hussain, Davoud M. Heidari, Michael Lawrence, and Martin Ansell

Abstract The natural hemp aggregates and their bio-composite panels which have been developed for sustainable construction with low thermal conductivity and high hygrothermal efficiency. The combination of the hemp aggregates with natural matrix materials results in exceptionally low thermal conductivity and high hygrothermal efficiency compared to conventional materials of construction as a result of their microporosity and breathability. In addition, the developed bio-based composites with nanotechnology improve resistance to liquid water and protect the hemp shiv from biodegradation without impacting the natural ability of the shiv to buffer moisture vapor.

The chapter assesses the physical characteristics of hemp aggregates in terms of their density, microstructure and porosity. Hemp-concrete and novel Hemp-organic composite have been studied and compared. Measurements of the thermal conductivity of hemp-composite panels are described which confirm their highly insulating properties. Hygroscopic testing demonstrates their effectiveness in absorbing and releasing moisture. The thermal and hygroscopic performance of hemp-composite panels in test cells is reported together with their application in construction. The life cycle assessment of hempcrete and hemp organic composite were performed. This chapter is part of the output of the ISOBIO programme supported by the European Union Horizon 2020 program, within the 'Materials for Building Envelopes' call for Energy Efficient Buildings.

Y. Jiang (✉) · A. Hussain · M. Lawrence · M. Ansell
BRE Centre for Innovative Construction Materials, Department of Architecture and Civil Engineering, University of Bath, Bath, UK
e-mail: a.hussain@bath.edu; m.lawrence@bath.ac.uk; m.p.ansell@bath.ac.uk

D. M. Heidari
Interdisciplinary Research Laboratory on Sustainable Engineering and Eco-design (LIRIDE), Faculty of Engineering, Department of Civil and Building Engineering, Université de Sherbrooke, Sherbrooke, QC, Canada

G. Crini, E. Lichtfouse (eds.), *Sustainable Agriculture Reviews 42*,
Sustainable Agriculture Reviews 42, https://doi.org/10.1007/978-3-030-41384-2_5

Keywords Hemp aggregates · Hemp-composite · Hygrothermal properties · Nanotechnology · Life cycle assessment

Abbreviations

AV	Average value
CO_2-eq	Carbon dioxide equivalent
CoV	Coefficient of variation
GHG	Greenhouse gases
GWP	Global warming potential
HDTMS	Hexadecyltrimethoxysilane
IPCC	Intergovernmental panel on climate change
LCA	Life cycle assessment
LCI	Life cycle inventory
LCIA	Life cycle impact assessment
MBV	Moisture buffer value
MIP	Mercury intrusion porosimetry
σ	Standard deviation
RH	Relative humidity
SEM	Scanning electron microscopy
TEOS	Tetraethyl orthosilicate

5.1 Introduction

Using environmental-friendly and sustainable industrial bio-based plant aggregates as raw materials for building insulation and construction is a new approach to address climate change and reduction in carbon dioxide emission. The advantages of bio-based plant materials include a renewable supply chain and significantly reduced carbon footprint through the photosynthetic carbon stored within plant-based materials. The bio-based insulation materials, such as natural fibre batts, offer many benefits in comparison with more established mineral and oil-based alternatives, such as mineral wool and polyurethane rigid form (Binici et al. 2016; Lopez Hurtado et al. 2016). A major advantage of bio-based insulation materials is their ability to form a breathable wall by readily absorbing and desorbing moisture in response to changes in relative humidity (RH) and vapour pressure gradients in the surrounding environment, acting as a hygric buffer, reducing the energy requirements of air conditioning and increasing the comfort of the occupants in the building (Collet et al. 2013; Rahim et al. 2015; Mazhoud et al. 2016). The most commonly available bio-based insulation materials include wood fibre/wool, cellulose, hemp fibre, flax fibre and sheep's wool et al. Their properties have been fully studied by researchers and industrial partners. Table 5.1 shows the typical density and thermal conductivity of the common bio-based

Table 5.1 Properties of the common bio-based insulation materials on the market

Material	Typical thermal conductivity ($W\ m^{-1}\ K^{-1}$)	Typical density ($kg\ m^{-3}$)	Typically embodied energy cradle to the gate ($MJ\ kg^{-1}$)	Manufacturer
Wood fibre	0.038–0.050	160–240	17	Biofib Insulation/FiBRA Natur/Isonat/Steico/Pavatex
Wood wool	0.038–0.040	50	10.8	Biofib Insulation/FiBRA Natur/Isonat/Steico/Pavatex
Paper (cellulose)	0.035–0.040	32	4.9–16.64	Homatherm/Isonat/Fiberlite Technologies/Warmcell
Hemp fibre	0.038–0.040	40	10.5–33	Thermafleece/ Blackmountain/NaturePro/ Pavatex/Hemp Flex/ Lenofon/Nord Tex
Sheep's wool	0.038–0.040	25	12–36.8	Thermafleece/ Blackmountain/NaturePro/ Pavatex/Hemp Flex/ Lenofon/Nord Tex
Flax fibre	0.038–0.040	30–35	11–39.5	Thermafleece/ Blackmountain/NaturePro/ Pavatex/Hemp Flex/ Lenofon/Nord Tex
Cork	0.038–0.070	105–120	26	Amorim/Jelinekcorkgroup/ EnviroNomix/Thermacork/ Agepan
Straw bale	0.052–0.080	100–130	0.24	Stramit/ModCell/Baustron

insulation product on the market. The thermal conductivity of these bio-based products ranges from 0.35 $W\ m^{-1}\ k^{-1}$ to 0.08 $W\ m^{-1}\ k^{-1}$, which are generally more thermally conductive than synthetic materials which tend to range between 0.023 $W\ m^{-1}\ k^{-1}$ (polyurethane) and 0.044 $W\ m^{-1}\ k^{-1}$ (mineral fibre) (Lawrence et al. 2013). Oil derived insulation has the highest embodied the energy of between 95 and 108 $MJ\ kg^{-1}$, and mineral insulation is between 15.7 and 53 $MJ\ kg^{-1}$, while bio-based natural insulation has the lowest embodied energy from cradle to gate between 0.24 and 39.5 $MJ\ kg^{-1}$, as shown in Table 5.1 (Lee and Yeom 2014; Miller and Ip 2013).

However, the bio-based plant aggregates (shiv), which come from the by-product of the stalks of plants, have not to be fully studied as building insulation materials. Bio-based plant aggregates, come from same plant stalk of fibre/seed, are very porous with a low density and have similar hygrothermal properties compared to the product of fibre. Tran et al. (2010) reported the hierarchical cell wall structure of bio-based plant aggregates resulted in good air tightness and minimal thermal bridging of the building. They have huge potential to be used in mainstream building insulation industrial. Recently, several European projects such as ISOBIO, ECO-SEE and HEMPSEC start to create the prefabricate panels using bio-based plant aggregates with an insulation purpose. Among these bio-based plant aggregates, the hemp shiv composite is probably the most widely used and studied in building industry in

Europe. They usually mixed with a lime-based binder to form a hemp concrete, which is known as hemp-lime (Latif et al. 2014; Rahim et al. 2015, 2016).

Collet et al. (2013) compared the hygric behaviour of three hemp concretes, which are a precast compacted hemp concrete, a sprayed hemp concrete and a moulded hemp concrete with fibred hemp shiv. They have a similar hemp/binder mass ratio of 0.5, apparent density between 430 and 460 kg.m^{-3} and a total porosity ranging between 72 and 79%. The results showed the hemp concretes are excellent hygric regulators and the total porosity and manufacturing method have a slight effect on the hygric performance of hemp concrete.

Rahim et al. (2015) reported the excellent hygric properties of hemp-lime concrete and flax-lime concrete according to the classification proposed by the Nordtest project. The density of hemp-lime concrete is around 478 $\pm$ 7 kg.m^{-3}, whereas the density of flax-lime concrete is about 598 $\pm$ 4 kg.m^{-3}. The porosities of hemp-lime concrete and flax-lime concrete are 76.4 $\pm$ 0.1% and 70.6 $\pm$ 0.3%, respectively. Rahim et al. (2016) also reported the hygric properties of rape straw concrete and hemp concrete. They have a similar density between 478 and 487 kg.m^{-3} and total porosity between 75.1 and 76.4%. The moisture buffer value (MBV) for hemp concrete and rape straw concrete are 2.02 and 2.59 g m^{-2}% RH^{-1}, respectively. Mazhoud et al. (2016) studied the hygric and thermal properties of hemp-lime plasters with densities between 723–881 kg.m^{-3} and total porosities ranging from 65.9 to 72%. The MBV of hemp-lime plaster is between 1.23 and 1.64 g/m^{-2} RH %$^{-1}$, which is lower than value above due to the lower hemp to binder ratio in hemp lime plaster than in hemp concrete. The thermal conductivity of hemp-lime plasters is about 0.2 W m^{-1} K^{-1}. They also claimed that the particle size of hemp shiv has a slight impact on the hygric and thermal properties of hemp-lime plaster.

Much of the existing characterization data for natural bio-based aggregate composites relates to their physical (bulk density and complex cell wall structure) and chemical properties (ratio of cellulose, hemicellulose and lignin), pore structure, porosity and pore connectivity (Bismarck et al. 2002; Chundawat et al. 2011; Zhang et al. 2012; Lawrence and Jiang 2017). However, a few studies have explored their scientific characteristics from the benchmark point of bio-based raw aggregates as building insulation materials due to no standardized methods for characterizing such materials. Jiang et al. (2018) studied the cell wall structure, pore size distribution and absolute density of hemp shiv using different approaches. They found the mercury intrusion porosimetry (MIP), which covers a wide range of pore size between 0.003 μm and 100 μm, is a solid method to measure the porosity and absolute density for hemp shiv. Laborel-Preneron et al. (2018) characterised the density, thermal conductivity and sorption-desorption properties of barley straw, hemp shiv and corn cob. They reported that the thermal conductivities are 0.044, 0.051 and 0.096 W m^{-1} K^{-1} respectively for straw (density: 57 kg m^{-3}), hemp shiv (density: 153 kg m^{-3}) and corn cob (density: 497 kg m^{-3}). The RILEM BBM Technical Committee, which set up in 2011, starts to work on the standard protocols for bio-based plant aggregates characterisation (Amziane et al. 2017). The characterisation of the hygrothermal properties of bio-based plant aggregates is still at an early stage.

In this chapter, we fully scientific characterization of hygrothermal and moisture buffer value of the natural hemp aggregates and their bio-composite panels which have been developed for sustainable construction with low thermal conductivity and high hygrothermal efficiency. The bulk density, absolute density, porosity and particle size distribution of these industrial bio-based plant aggregates are analysed using RILEM recommendation methods and MIP method. The microstructure was comparably studied among these aggregates using scanning electronic microscopy (SEM). The thermal and hygroscopic performance of hemp-composite panels in test cells is reported together with their application in construction. The life cycle assessment of hempcrete and hemp organic composite were performed. These data will provide the benchmark data and give a guidance to develop the novel bio-based plant aggregates insulation composites for use in the building insulation industry.

5.2 Characterization of Hemp Aggregates

The SEM microstructures of industrial hemp aggregates are shown in Fig. 5.1. The SEM images showed that the samples have a porous microstructures. It clearly showed the three zones of pith, xylem layer and epidermis from the interior to the exterior of the hemp stem. At higher magnification, the radial arrangement of cells in the xylem layer is visible. A closer view of the pith reveals the foam-like closed cell structure with some voids at the interfaces between cells. The average diameter of parenchyma cells are approximately 5–20 μm. The xylem contains a range of cell diameters the larger cells being termed vessels. The vessels exhibit little variation in size and no clear pore arrangement, which is a diffuse-porous distribution. The vessels are approximately 50 to 100 μm in diameter and are surrounded by relatively thick fibre cells. Thick-walled fibres are located between the vessels with a diameter from 1 μm to 2 μm.

The bulk density and skeleton density of bio-based plant aggregates are given in Table 5.2. The standard deviation between the five measurements of bulk density is very low for all hemp shiv. The accuracy of the measurement was calculated from the characteristics of the balance and from the accuracy of the level, estimated about 0.5 mm. This accuracy is better than 1%. For each material, the measurement is repeated five times. The protocol was based on that developed by the RILEM Technical Committee 236-BBM. Generally, the bulk density of aggregate decreases when the size of aggregate increases. For hemp shiv, the bulk density at dry state ranges from 88 to 104 kg m^{-3}. The skeleton density is similar for all sizes of hemp shiv. The skeleton densities of hemp shiv are close with average values of 1452 kg m^{-3}.

The porosity of the bulk material, including inter-particles and intra-particles porosity, is high. For all the tested hemp shiv, it is higher than 90%. A slight increase of porosity with aggregate size can be noticed, probably as a result of larger voids being created between large particles. The properties of hemp shiv are not only affected by the total pore volume and porosity, but also affected by the pore size

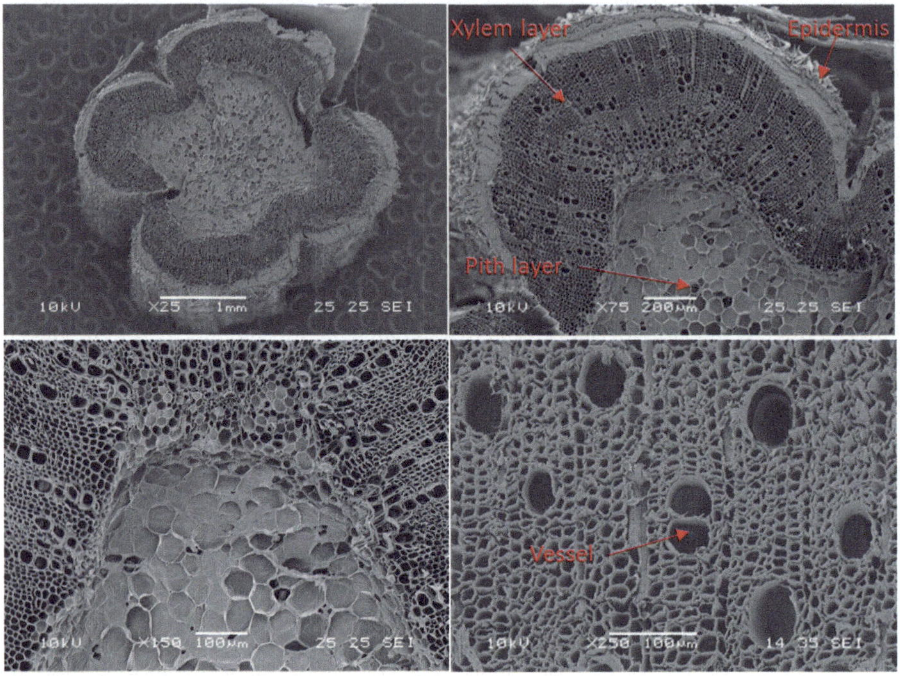

Fig. 5.1 Scanning electron microscopy micrographs of hemp shiv. The industrial bio-based plant aggregates used in this study was sourced and harvested in North West France. They were produced by a mechanical de-fibring processing removing the fibre, chopping, grading and de-dusting and supplied by the CAVAC cooperative (France). The microstructure of the bio-based plant aggregates was characterised using a scanning electron microscope (JEOL SEM-6480LV, Tokyo, Japan). All images were taken at an accelerating voltage of 10 kV. The sample surfaces were coated with a thin layer of gold using an HHV500 sputter coater (Crawley, UK) to provide electrical conductivity sending electrons to earth

Table 5.2 Bulk density and skeletal density of hemp shiv at dry state; The particle size of hemp shiv are calibrated with different sieving grids using a grader such as G7, G8, G12 and G14

Name		Hemp Shiv			
Grade		G7	G8	G12	G14
Bulk density (kg m^{-3})	Av.	104.01	96.65	90.72	87.89
	σ	1.44	0.62	1.48	2.13
	c. Var.	1.36	0.72	1.56	2.35
Skeleton density (kg m^{-3})	Av.	1452			
	σ	83			
	c. Var.	6.11%			

Av.: Average value; σ: Standard deviation; c. Var.: Coefficient of variation

distribution and pore structure. Figure 5.2 display the relationship between the pore diameter and the mercury intrusion volume and the log differential mercury intrusion volume as a function of the pore diameter and pore size distribution curves of hemp shiv determined by mercury intrusion porosimetry. Because of technical restrictions

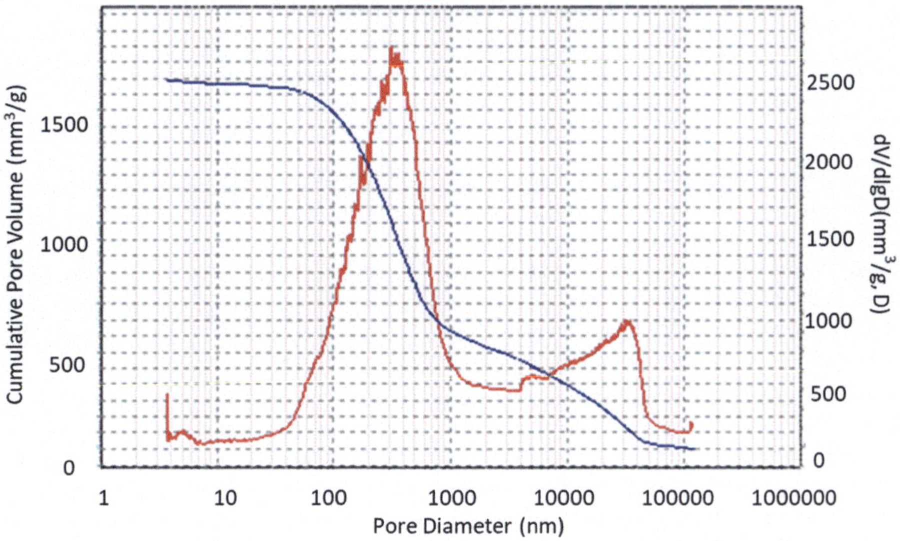

Fig. 5.2 Pore size distribution curves of hemp shiv determined by mercury intrusion porosimetry (PASCAL, Thermo Scientific). Blue line is cumulative pore volume and red line is dV/dlogD. The pore size distribution was presented in the form of cumulative pore volume and logarithmically differential pore volume curves as a function of pore radius. Full details of the methods and measurement process used are as reported by Jiang et al. (2018) (Color figure online)

the measurement of the large tracheids with 100 μm is excluded. Those pores are on the one hand important openings for impregnation but on the other hand easily accessible already without or with low applied pressures. The differential PSD curves clearly show that the aggregates have similar PSDs with representative peaks at pore radii around 0.02–100 μm. However, the magnitude and shape of the peaks differed among different shiv. The vast majority of the pores seen were in the range of 0.1–10 μm. For hemp shiv, they have a similar bi-modal peaks. The results show good agreement with the results of SEM.

Figure 5.3 showed the thermal conductivity of hemp shiv aggregates at dry state using heat flow meter and CT meter methods. The results also compared with the thermal conductivity of their other forms, including fibre and fine powder and plus corn cob. The results measured by the FOX600 are represented by the blue colour and the grey colour represents the results of CT meter. Generally speaking, both methods show relatively similar thermal conductivity on all the bio-aggregates. The thermal conductivity of bio-aggregates ranges from 0.04 W m^{-1} K^{-1} to 0.06 W m^{-1} K^{-1}, except for corn cob, which shows a significantly higher thermal conductivity of 0.08 W m^{-1} K^{-1}. For all the shiv samples, wheat straw G7 and G8 show the lowest thermal conductivity, which is around 0.04 W m^{-1} K^{-1}. Rape shiv shows a slightly lower thermal conductivity than the thermal conductivity of hemp and flax. There are similar results for the thermal conductivity of hemp powder and

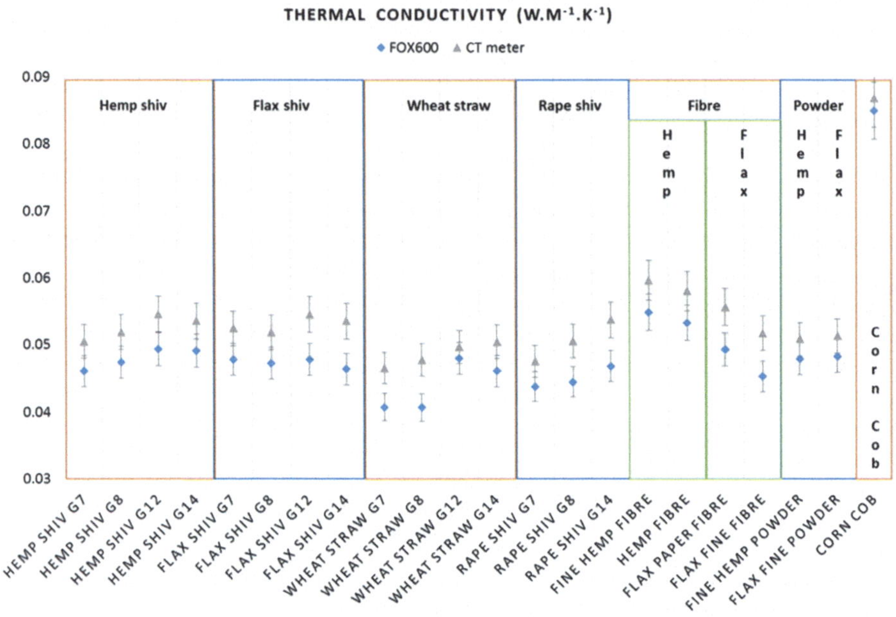

Fig. 5.3 Thermal conductivity of bio-aggregates measured by Fox 600 and CT meter. The industrial bio-based plant aggregates used in this study was sourced and harvested in North West France. They were produced by a mechanical de-fibring processing removing the fibre, chopping, grading and de-dusting and supplied by the CAVAC cooperative (France). The selected bio-based plant aggregates are hemp shiv (grade 7, 8. 12 and 14), flax shiv (grade 7, 8, 12 and 14), rape shiv (grade 7, 8 and 14) and wheat straw (grade 7, 8, 12 and 14)

flax powder. There is no significant difference in thermal conductivity of shiv with different particle sizes, except wheat straw G12 and G14. Wheat straw G12 and G14 show a significantly higher thermal conductivity at lower density due to the additional impact of convection currents. Compared to hemp shiv, both hemp fibre and fine fibre show a slightly higher thermal conductivity. In addition, hemp powder shows a similar thermal conductivity to hemp shiv. It can be concluded that the particle size has little effect on the thermal conductivity of bio-aggregates.

5.3 Binders and Formulation

The hemp organic composites described here was prepared by using a bio-based polysaccharide binder as a matrix binding together the hemp shiv. The bio-based binder was formulated using a starch derivative and a crosslinking agent. Two sets of hemp organic composites were developed for this study consisting of, (i) Untreated hemp composite: raw hemp shiv aggregates and bio-based binder; and (ii) Treated hemp composite: modified hemp shiv aggregates using a hydrophobic pre-treatment and bio-based binder.

The hemp shiv aggregates were pre-treated with a silica-based coating prior to composite preparation for improving the water resistance of the composites. For preparation of the treatment solution, the sol-gel process was followed. 1 M of tetraethyl orthosilicate (TEOS) was added as the silica component to a mixture of 4 M distilled water, 4 M of absolute ethanol and 0.005 M of nitric acid. 0.015 M of hexadecyltrimethoxysilane (HDTMS) was added to the above mixture as the hydrophobic agent. The sol was vigorously stirred for 2 h at 40 °C and atmospheric pressure. The sols were then aged for 96 h in closed container at room temperature before the treatment process. The hemp shiv aggregates were dipped in the sol for 10 min, transferred to an open tray and then dried at room temperature for 1 h. Finally, the hemp shiv aggregates were placed in an oven at 80 °C for 1 h.

Mixing of the constituent materials, hemp shiv aggregates and the bio-based binder, was performed manually to achieve a uniform mixture. The mass of the constituent materials was precalculated to achieve targeted final composite densities (200 kg m^{-3} for the untreated composite and 240 kg m^{-3} for the treated composite). The weight ratio for hemp shiv: binder was 9:1 for both untreated and treated composites. The difference in the final densities of both composites was mainly due to the treatment process depositing silica on the surface of the treated hemp shiv aggregates.

For preparation of the composites, the mixture of the constituent materials was placed into a steel mold of desired dimension and compacted at 0.5 MPa using a hot press (PressMasters 40 T GEM series). The upper and lower plates were then heated to 180 °C and the temperature was maintained for 1 h. The specimens were demolded after cooling down to room temperature and then transferred to a conditioning room at 19 °C and 50% relative humidity.

5.4 Characterisation of Composites

5.4.1 Moisture Buffering

The moisture buffering property of a material corresponds to its ability to absorb and release moisture in response to varying relative humidity levels in internal spaces. This property can have a positive impact on the indoor comfort levels and reduce the energy consumption required for air conditioning (Tran et al. 2010). The moisture buffering performance of construction materials depends on the exposure area, vapour permeability, ventilation rate and surface pre-treatments (Latif et al. 2015).

The Nordtest protocol is the most common method to estimate the moisture buffer value (MBV) of a material (Rode et al. 2005). The kinetics of mass change for the organic hemp composite panels developed under the ISOBIO project is presented in Fig. 5.4. The mass change was calculated per m^2 exposed surface of the samples and plotted against time.

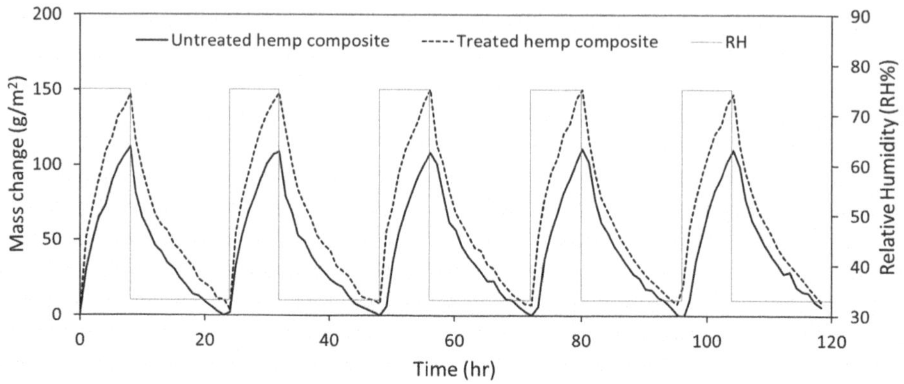

Fig. 5.4 Mass change of ISOBIO hemp shiv composite panels exposed to varying relative humidity at 23 °C

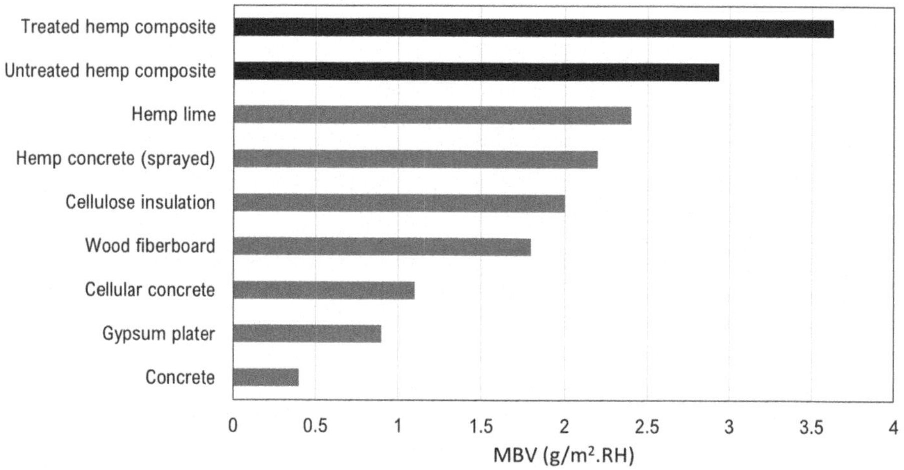

Fig. 5.5 Moisture buffer value of general building materials vs the hemp organic composites (Rode et al. 2005; Collet and Pretot 2012)

The MBV of the usual building materials reported in literature is shown in Fig. 5.5 and compared with the MBV obtained for the organic hemp composites. The materials can be rated according to their MBV as: negligible (MBV ≤ 0.2 g m^{-2} RH), limited (MBV 0.2–0.5 g/m^{2}RH), moderate (MBV 0.5–1.0 g/m^{2}RH), good (MBV 1.0–2.0 g m^{-2} RH) or excellent (MBV ≥ 2.0 g m^{-2} RH). From Fig. 5.5, it can be seen that the prepared hemp organic composites show excellent moisture buffering capacity. The MBV of the both untreated and treated hemp composites is higher than hemp concrete reported in literature (1.75–2.15 g m^{-2} RH) (Tran Le et al. 2010; Collet and Pretot 2012; Dubois et al. 2014). The high hemp shiv binder ratio in these composites result in lower density and thereby enhancing their vapour permeability.

5.4.2 Vapour Permeability

The vapour permeability of a porous material can be expressed by its ability to transfer moisture through a vapour pressure gradient. Moisture transfer takes places in three stages: diffusion such as self-collision of water molecules, effusion including collision of water molecules with the pore walls and liquid transfer associated with capillary condensation (Collet et al. 2013).

Using a vapour permeable wall and maintaining indoor relative humidity levels between 40 and 60% can have a positive impact on wellbeing of residents, controlling respiratory problems and reducing allergies (Maskell et al. 2018). When bio-based materials are used as part of a vapour permeable wall, moisture can penetrate through the fabric of this wall. The risk of moisture build-up is considerably reduced which is expected to reduce bacterial growth, thereby improving the overall indoor air quality (Osanyintola and Simonson 2006; Zhang et al. 2012).

The water vapour permeability and the diffusion resistance factor of thermal insulation materials can be measured using the British Standard BS EN 12086 (BS EN 12086, 2013) under isothermal conditions (23 °C) and at two sets of relative humidity: dry cup and wet cup. The kinetics of mass change of the hemp organic composites presented in Fig. 5.6.

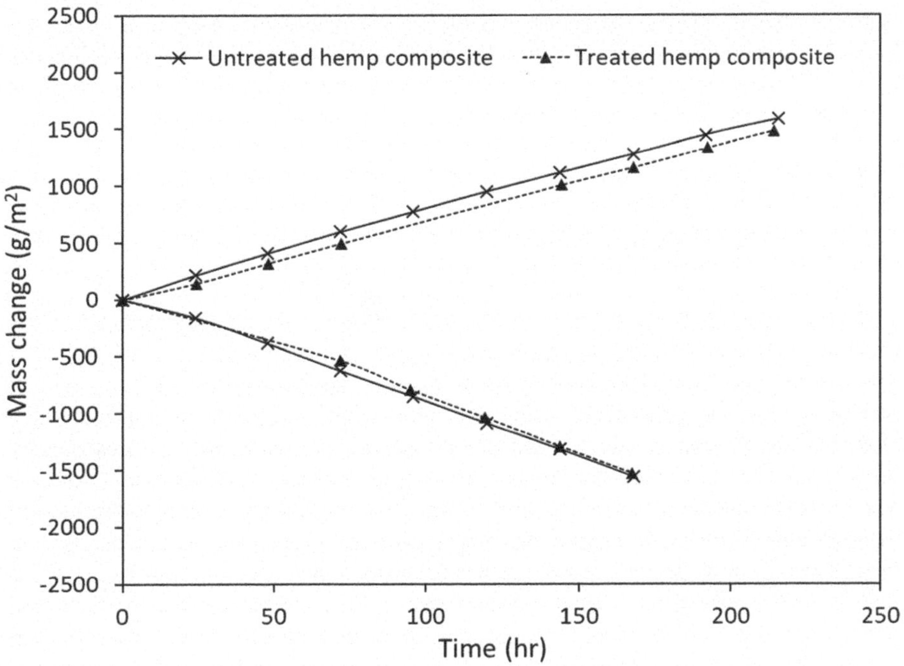

Fig. 5.6 Mass change of the hemp organic composites during the vapour permeability test

Table 5.3 Vapour permeability results of the hemp organic composites

Parameters	Untreated hemp composite		Treated hemp composite	
	Dry cup	Wet cup	Dry cup	Wet cup
Water vapour transmission rate, g (mg h^{-1})	6343.75	9833.33	7224.33	9605.63
Water vapour permeance, W (mg m^{-2} h Pa)	4.72	8.13	5.37	7.94
Water vapour resistance, Z (m^2 h Pa mg^{-1})	0.21	0.12	0.18	0.13
Water vapour permeability δ (mg m^{-1} h Pa)	0.09	0.16	0.11	0.16
Water vapour resistance factor μ	7.54	4.37	6.62	4.48

Table 5.4 Thermal conductivity data of the hemp organic composite panels

Parameter	Untreated hemp composite	Treated hemp composite
Thermal conductivity (W m^{-1} K)	0.053	0.057
Thermal diffusivity (10^{-6}) (m^2 s^{-1})	0.35	0.30
Specific heat capacity (J kg^{-1}·K)	763.20	782.71
Bulk density (kg m^{-3})	200.0	240.0
Volumetric heat capacity (10^6) (J m^{-3} k)	0.15	0.18

The composites show promising results as seen in Table 5.3, behaving as a good vapour permeable material. It can be seen from the results that hydrophobic treatment of hemp shiv did not affect the vapour permeability of the material. Bio-based aggregates such as hemp shiv have very high porosity and as a result they have very low water vapour resistance. On the other hand, solid concrete has a very high water vapour diffusion resistance factor at about 130 while for hemp concrete it ranges between 5 and 12 (Walker and Pavía 2014).

5.4.3 *Thermal Conductivity*

The thermal conductivity of the hemp organic composite panels was measured from transient method using a hand-held measuring instrument ISOMET 2114 at 20 °C. The results are reported in Table 5.4. In general, bio-based materials are light in weight and have low thermal conductivity making them good insulation materials. However, the thermal conductivity of the bio-based composite is also influenced by other factors such as aggregate, binder, aggregate to binder ratio and water content. The density can be affected by the formulation, binder content and production method thereby influencing the thermal conductivity as well (Collet and Pretot 2014; Elfordy et al. 2008). Various studies have showed that as the aggregate content is increased, the thermal conductivity decreases for the composite. This is due to the decrease in density of the composite as usually the aggregate has lower density when compared to the binder. Higher aggregate content increases the overall porosity of the composite (Al Rim et al. 1999).

Table 5.5 Water absorption measurements of the hemp organic composite panels

Sample	Water absorption (kg m^{-2})	Water absorption (%)
Untreated hemp composite	22.11 ± 0.7	221.10 ± 1.3
Treated hemp composite	11.04 ± 0.6	98.02 ± 3.5

5.4.4 Water Absorption

The short term water absorption of the specimens was determined by the British Standard EN 1609 (BS EN 1609, 2013). The water absorption of the composites was calculated in two ways: (i) as change in mass over exposed surface area (WA) using Eq. (5.1) where *WA* is the water absorption (kg/m^2), m_0 is the initial mass of the test specimen (kg), m_{24} is the mass of the test specimen after partial immersion for 24 h (kg), A_p is the bottom surface area of the test specimen (m^2); (ii) as percentage of absorption with respect to initial mass (WA%) using Eq. (5.2) where *WA%* is the water absorption percentage.

$$WA = \frac{m_{24} - m_0}{A_p} \tag{5.1}$$

$$WA\% = \frac{m_{24} - m_0}{m_0} * 100 \tag{5.2}$$

The water absorption results are presented in Table 5.5. The untreated hemp composite shows higher values for WA and WA% due to the absence of hydrophobic silica treatment on hemp shiv in the composites. The hydrophobic treatment significantly reduced the WA by 50% and the WA% reduced by 123%.

Hemp shiv possesses a large water absorption capacity due to its internal pore structure and its chemical composition. The low bulk density of hemp shiv aggregates is due to their high porosity (Jiang et al. 2018) which can trap huge amounts of water. Moreover, the presence of high number of hydroxyl groups (Kidalova et al. 2015) in its structure adds to the retention of water within the bulk of hemp shiv aggregates. High sensitivity to moisture can be responsible for colonial fungal growth leading to cell wall degradation and affecting the durability of the material (Marceau et al. 2017). High water absorption capacity can also affect the manufacturing quality of the final product if it encounters water or is exposed to humid surroundings. The silica treatment used in this research reduced the hydrophilicity of hemp shiv as seen with the water absorption tests making them water resistant and less susceptible to degradation.

5.5 Performance of Panels in the Construction

5.5.1 Hemp Lime Structures

The use of hemp shiv as a bio-aggregate began in France in 1986 when Charles Rasetti, tasked with restoring a medieval timber framed building in Nogent-sur-Seine, made use of chopped hemp stalk – the waste from the manufacture of hemp fibre based paper – mixed with a lime binder as infill in place of wattle and daub. This was found to be highly effective and it use soon spread to the construction of new buildings. Yves Kühn developed the 'Canosmose' system of construction, based on hemp-lime, and by the mid 1990s a hemp specific binder (Tradical®) had been developed by Strasservil to feed the burgeoning demand. In 1998 the professional association 'Construire en Chanvre' was founded, and in 2007 they published 'Règles Professionelles' giving formal guidance on how to build with hemp-lime. By the start of the twenty-first century, hemp-lime construction had been adopted in other European countries as well as in North America, South Africa, Australia and New Zealand. It is now widely used across the world, albeit by enthusiasts rather than main-stream construction companies, but large-scale hemp-lime construction systems are being developed in both France and Australia and expected to make an impact by 2025. In the UK a number of housing developments have been concluded using hemp-lime (Fig. 5.7). Much of the validation for the use of hemp-lime has been based on laboratory based tests rather than measurements taken in actual buildings, from which mainly anecdotal evidence is available. Notable exceptions include an early study done in the UK in 2002 (BRE 2002), which did not identify any particular benefits, and a recent study (Moujalled et al. 2018), which validated a theoretical model over a 4 year period using data from a domestic house. This study demonstrated that hemp-lime is a good hygrothermal regulator, managing internal humidity effectively and showing good thermal inertia.

Housing development in Diss, Norfolk | Housing Development in Swindon, Wiltshire

Fig. 5.7 Examples of housing developments in the United Kingdom

5.5.2 Hemp Organic Composite

The hemp organic composite which is the subject of this chapter was developed as part of the EU funded ISOBIO project and three ISOBIO wall systems and four reference walls were monitored at test facilities in the UK and Spain, for the purpose of measuring thermal transmittance using the Heat Flow Meter method to ISO 9869, and comparing results with calculated U-values to ISO 6946. The results show a measured reduction in thermal transmittance of the ISOBIO New-build panel of 65% in the UK and 54% in Spain, in comparison with their respective reference new-build walls (constructed from traditional materials to current standards). Figure 5.8 below shows a comparison of the measured U-value of the ISOBIO New-build panel, compared with the Reference New-build cavity wall. The results show a 65% reduction in thermal transmittance of the ISOBIO panel compared with the New-build cavity wall.

If we consider the ISOBIO New-build panel's moisture buffering performance and impact on internal relative humidity, we can see that the panel offers superior performance compared to the Reference cavity wall. Figure 5.9 below shows the relative humidity in Cell 4 (ISOBIO New-build panel) compared with Cell 1 (Reference New-build cavity wall), during the same period (24/02–14/03). The ISOBIO panel in Cell 4 helps maintain an average RH of 42% (within the optimum comfort

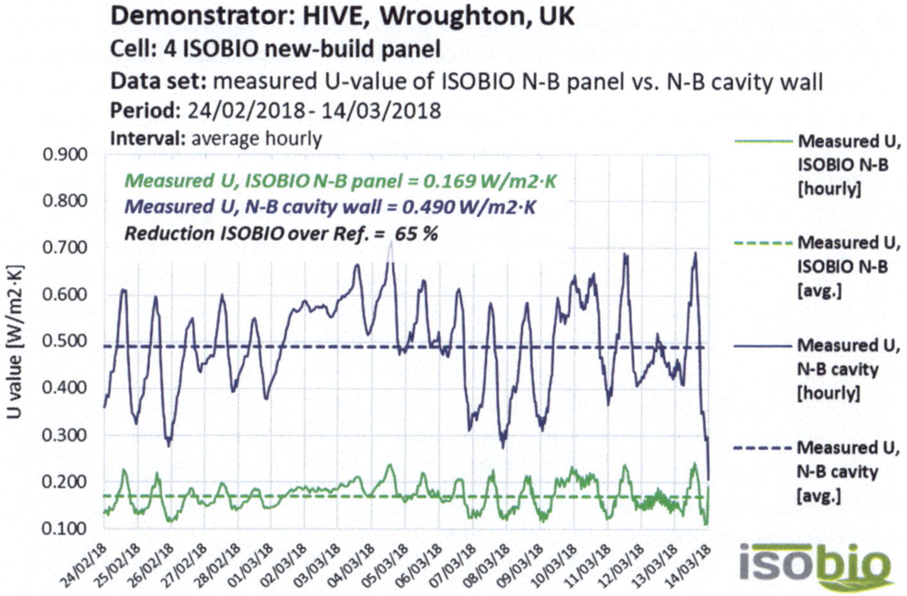

Fig. 5.8 Measured U-value of ISOBIO New-build panel vs. Reference New-build cavity wall, HIVE, United Kingdom

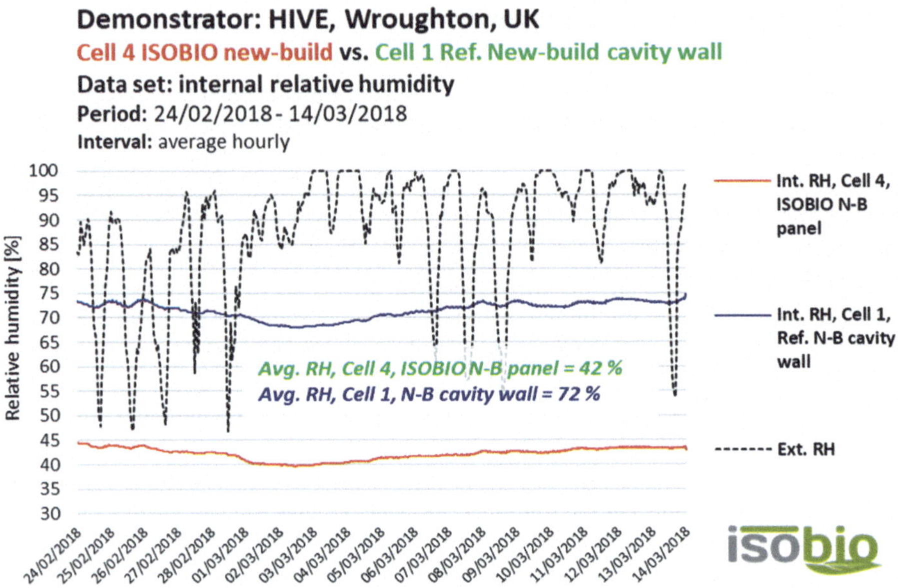

Fig. 5.9 Average internal relative humidity of Cell 1 (ISOBIO New-build panel) vs. internal relative humidity of Cell 1 (Reference New-build cavity wall), HIVE, United Kingdom

range for indoor relative humidity), while in Cell 1, the New-build cavity wall (being vapour impermeable), does not allow water vapour to dissipate, and RH remains at an average of 72%.

The ISOBIO External retrofit system reduced thermal transmittance by 55% (UK) and 71%, (Spain), in comparison with their respective reference walls. The ISOBIO Internal retrofit system reduced thermal transmittance by 77% (UK) and 81% (Spain). The UK monitoring results show that the ISOBIO new-build panel is a high-performance composite wall system, with a very good correlation between calculated and measured values and a near-zero performance gap. The data suggests that bio-based materials, when combined in a structural insulated panel with an air barrier, can provide predictable performance and high thermal resistance, that will reduce heating and cooling energy consumption at building scale.

Monitoring data of the ISOBIO New-build panel during 4 days in August showed an average heat flux of -2.3 W m^{-2}, compared with -7.6 W m^{-2} in the New-build cavity wall, demonstrating that the ISOBIO New-build panel can offer good thermal performance in warm climates and reduced energy consumption from air conditioning. This was confirmed by the calculation of weighted energy savings during the summer period, which showed that the ISOBIO New-build panel generated a 14% saving in cooling energy consumption.

5.6 Life Cycle Assessment of Hemp Shiv Composite

The construction sector is responsible for 40% of materials and energy consumption of the world (Abd Rashid and Yusoff 2015). Concerning the environmental impacts of construction sector such as global warming potential ozone layer depletion, this industry is forced to take in to account new strategies to mitigate its' energy consumption. Whereas in the past the building sector primarily used materials that were local to the construction site with low environmental impact and energy consumption, nowadays the building sector uses global materials with high environmental impact and energy consumption such as concrete, gypsum, steel, aluminum and polyvinyl chloride. Innovative bio-based materials such as hemp shiv are being developed for use in construction, with the goal of being more environmentally friendly during their life cycle. Even though the introduction of innovative bio-based material will reduce the carbon footprint compared with fossil-based materials, it may incur some additional costs such as land use and its related environmental impacts.

Environmental assessment is required to make a strategic decision related to the use of these bio-based materials instead of their fossil-based ones. The growing use of bio-based materials in the construction sector calls for an assessment of their environmental impact. Life Cycle Assessment (LCA) as an environmental assessment technique is used to analyze the environmental impacts of product systems through their life cycle. Some environmental impacts can vary depending on the characteristics of their surroundings, and therefore on the location of the activity. Regionalized LCA is a technique that takes in to account this variability. Previous LCAs has been conducted on a wide range of bio-based materials, including hemp-concrete known as hempcrete (Andrianandraina et al. 2015; Senga Kiesse et al. 2017; Sinka et al. 2018), fiber composites (Zah et al. 2007), starch-based polymers (Patel et al. 2006), Bio-based plastics (Tsiropoulos et al. 2015). A review of the literature shows that climate change is the main indicator that has been considered to analyze the environmental impact of hemp shiv as an aggregate for insulation walls (Miller 2018).

This chapter will assess the damage to terrestrial biodiversity and carbon footprint from the production of hemp shiv treated with sol-gel based materials per functional unit (FU) of 1 kg of hemp shiv produced. Treated hemp shiv as a bio-based composite wall with thermal transmittance U-value of 0.15 W m^{-2} K is compared to two other walls with same U-value, which were a wall made with untreated hemp shiv and a reference cavity wall constructed using traditional masonry techniques. Damage to the terrestrial ecosystem due to land use, acidification, photochemical ozone formation, ecotoxicity, water use and climate change were included. Data on hemp production were obtained from hemp farmers from CAVAC, an agricultural

Table 5.6 Characteristics of walls and materials used

Scenario	Material	Unit	Value	Comment
Reference wall	Solid brick	kg	183.6	Clay brick| market for | cut-off, U
	Air cavity	kg	_	None
	Eco-therm insulation	kg	11.6	Kingspan insulation panel
	Cement block	kg	200.0	Concrete block | market for | cut-off, U
	Gypsum plasterboard	kg	8.7	Gypsum plasterboard | market for | cut-off, U
	Gypsum plaster	kg	2.1	Stucco | market for | cut-off, U
Treated bio-composite wall	Treated hemp shiv	kg	61.8	82% Hemp shiv and 18% Sol-Gel
	Water	kg	101.3	Tap water | market for | cut-off, U
	Bio-based binder	kg	91.2	Tradical ThermO, modelled on SimaPro
	Timber frame	kg	9.3	Plywood, for outdoor use | market for | cut-off, U
	Steel fastening	kg	5.0	Steel, unalloyed | market for | cut-off, U
Untreated bio-composite wall	Untreated hemp shiv	kg	50.7	100% hemp shiv
	Water	kg	152.0	Tap water | market for | cut-off, U
	Bio-based binder	kg	91.2	Tradical ThermO, modelled on SimaPro
	Timber frame	kg	9.3	Plywood, for outdoor use | market for | cut-off, U
	Steel fastening	kg	5.0	Steel, unalloyed | market for | cut-off, U

cooperative based in west France. Potentially disappeared fraction of species (PDF), as a damage to the terrestrial biodiversity, integrated over area and time in m^2 year, was analysed for the functional unit of 1 kg of hemp shiv and sol-gel produced and subsequently per one square meter of insulation wall with a U-value of 0.15 W/ m^{-2} K for 1 year of its service life.

Treated hemp shiv as a bio-composite in a hypothetical building wall was compared to two other hypothetical walls, including a wall with untreated hemp shiv and a cavity wall as a reference wall. The dimensions of the walls were calculated in order to have a U-value of 0.15 W/m^{-2} K. Table 5.6 shows the characteristics of the walls of interest in this study, along with their materials used. The wall with bio-composite has been made up with treated and untreated hemp shiv, bio-based binder and water. A timber frame and steel fixings are used as the load-bearing structural elements. The wall with bio-composite has been made up with treated and untreated hemp shiv, bio-based binder and water. A timber frame and steel fixings are used as the load-bearing structural elements.

Table 5.7 Terrestrial ecosystem damage (Potentially disappeared fraction PDF m^2 year/kg) from the production of 1 kg of hemp shiv, based on hierarchist perspective

	Value	
Impact category (PDF m^2 year/kg)	EA	MA
Land use	0.532	0.863
Terrestrial acidification	0.018	0.027
Ozone formation	0.007	0.011
Terrestrial ecotoxicity	0.001	0.001
Water	0.001	0.001
Global warming	−0.308	−0.292

Economic allocation (EA) and mass allocation (MA) show the results based on economic and mass allocation

Among the environmental impacts, global warming has a negative impact resulting from the sequestered carbon during hemp production (agricultural stage). In spite of the carbon sequestration of hemp production, the agricultural stage is the main contributor to all of the impact categories for hemp shiv production. Table 5.7 shows the total damage to terrestrial biodiversity of 1 kg hemp shiv production. As it can be seen from Table 5.7, depending on allocation methods, the results vary. The environmental damage associated with mass allocation is greater than those analysed based on economic allocation. The carbon footprint of 1 kg hemp shiv was calculated as −1.55 kg CO_2-eq/kg and −1.63 kg CO_2-eq/kg, based on mass and economic allocation respectively. Carbon footprint of common insulation materials for building has been reported as: 7.336 kg CO_2-eq/kg for expanded polystyrene (EPS) foam slab, 1.511 CO_2eq/kg for rock wool, 6.788 CO_2-eq/kg for polyurethane rigid foam, 0.807 kg CO_2-eq/kg for cork slab, 1.831 kg CO_2-eq/kg for cellulose fiber and 0.124 kg CO_2-eq/kg for wood wool (Zabalza Bribian et al. 2011; Lopez Hurtado et al. 2016).

Figure 5.10 shows the total damage to terrestrial biodiversity by using 1 kg of hemp shiv, based on regionalized and generic (world average) characterization factor for hierarchist perspective. Since climate change and terrestrial ecotoxicity results are purely based on generic values. The damage to terrestrial biodiversity based on regionalized characterization factors shows greater values than generic ones. The main differences come from the impact of land use.

Carbon footprint of one square meter of a wall, using treated and untreated hemp shiv was calculated using IPCC 2013 GWP 100a method. Using economic allocation, the total carbon footprint of 1 m^2 of treated hempcrete wall (for its entire service life) is 24.65 kg CO_2-eq and 22.51 kg CO_2-eq based on end of life treatment as composting and landfilling respectively. The results based on mass allocation and untreated hempcrete wall are shown in Fig. 5.11.

Table 5.8 shows the carbon footprint of treated and untreated hempcrete wall of current study compare to similar studies. The current study considered a U-value of 0.15 W/m^2·K for all scenarios which requires a thicker wall compared with similar studies.

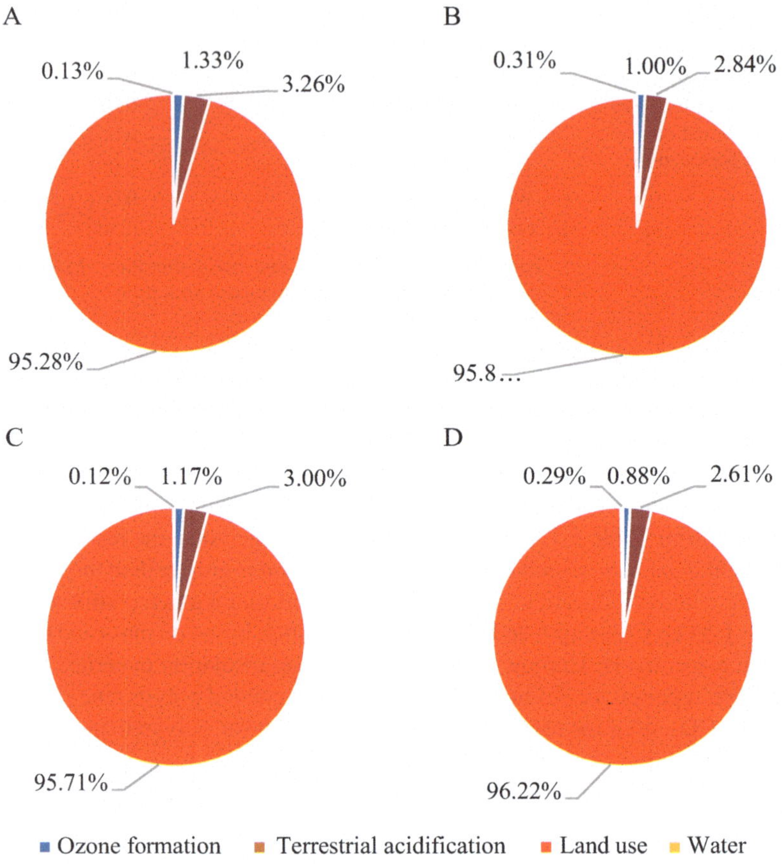

Fig. 5.10 Terrestrial biodiversity of 1 kg of hemp shiv, based on regionalized and generic (world average) characterization factor for hierarchist perspective. (**a**) Regionalized values and economic allocation; (**b**) Generic values and economic allocation; (**c**) Regionalized values and mass allocation; (**d**) Generic values and mass allocation

5.7 Conclusions

A systematic assessment of both commercially available bio-based building products and key raw materials was undertaken as part of the ISOBIO Project. This assessment provides a unique database of biogenic materials, their construction relevant properties and environmental footprints. The assembly of these raw materials into composite core structures is enabling a materials-by-design approach to be developed. Both biogenic and mineral binders are being used to provide a toolkit of composite materials and their key characteristics. Novel chemistries are being developed to provide high levels of water repellence to enable products with durability to external conditions, and to improve the fire resistance of these

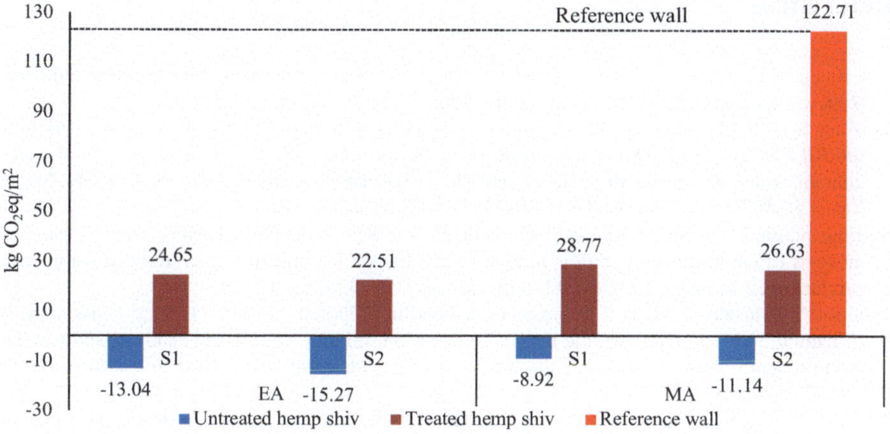

Fig. 5.11 Carbon footprint of 1 m^2 walls for their entire service life. EA and MA show the results based on economic and mass allocation. S1 and S2 are waste scenarios (S1: composting hemp shiv and landfilling the rest, S2: landfilling all the materials)

Table 5.8 Comparison of carbon footprint of similar studies on hempcrete walls with the current study

Type	Wall thickness (mm)	U-value ($W/m^2 \cdot K$)	Carbon footprint (kg CO_2-eq)	References
Hempcrete wall	260	0.42	−35.53	Boutin et al. (2006)
Hemp-lime wall	300	0.19	−36.08	Ip and Miller (2012)
Hempcrete wall	240 + 10[a]	0.36	−1.60	Pretot et al. (2014)
Hempcrete wall	250	0.27	−12.09	Arrigoni et al. (2017)
Untreated hempcrete	507	0.15	−15.27 to −8.92	Current study
Treated hempcrete	507	0.15	22.51–28.77	Current study

[a]This study has 10 mm of coating

composites. Adoption of bio-based materials for the construction sector is dependent on many factors including material performance (real and perceived), availability, cost, regulatory incentive. The ISOBIO project is establishing a database of raw materials and composites, a full environmental assessment of key products and appropriate test methods that enable the advantages of bio-aggregates to be quantitatively measured and compared with conventional materials. These steps are aimed at providing a robust platform to enable the industrial partners to take the emerging products to the market-place by overcoming many of the barriers to adoption that currently exist.

References

Abd Rashid AF, Yusoff S (2015) A review of life cycle assessment method for building industry. Renew Sust Energ Rev 45:244–248. https://doi.org/10.1016/j.rser.2015.01.043

Amziane S, Collet F, Lawrence M, Magniont C, Picandet V, Sonebi M (2017) Recommendation of the RILEM TC 236-BBM: characterisation testing of hemp shiv to determine the initial water content, water absorption, dry density, particle size distribution and thermal conductivity. Mater Struct 50:167. https://doi.org/10.1617/s11527-017-1029-3

Andrianandraina VA, Senga Kiessé T, Cazacliu B, Idir R, van der Werf HMG (2015) Sensitivity analysis of environmental process modeling in a life cycle context: a case study of hemp crop production. J Ind Ecol 19:978–993. https://doi.org/10.1111/jiec.12228

Arrigoni A, Pelosato R, Melià P, Ruggieri G, Sabbadini S, Dotelli G (2017) Life cycle assessment of natural building materials: the role of carbonation, mixture components and transport in the environmental impacts of hempcrete blocks. J Clean Prod 149:1051–1061. https://doi.org/10.1016/j.jclepro.2017.02.161

Binici H, Aksogan O, Demirhan C (2016) Mechanical, thermal and acoustical characterizations of an insulation composite made of bio-based materials. Sustain Cities Soc 20:17–26. https://doi.org/10.1016/j.scs.2015.09.004

Bismarck A, Aranberri-Askargorta I, Springer J, Lampke T, Wielage B, Stamboulis A, Shenderovich I, Limbach H (2002) Surface characterization of flax, hemp and cellulose fibers; surface properties and the water uptake behavior. Polym Compos 23:872–894. https://doi.org/10.1002/pc.10485

Boutin MP, Flamin C, Quinton S, Gosse G (2006) Étude des caractéristiques environnementales du chanvre par l'analyse de son cycle de vie (in French)

BRE (2002) Final report on the construction of the hemp houses at Haverhill, Suffolk. BRE Construction Division

BS EN 12086 (2013) BSI Standards Publication Thermal insulating products for building applications – determination of water vapour transmission properties

BS EN 1609 (2013) BS EN 1609: 2013 BSI Standards Publication Thermal insulating products for building applications – determination of short term water absorption by partial immersion

Chundawat SPS, Donohoe BS, Sousa LC, Elder T, Agarwal UP, Lu F, Ralph J, Himmel ME, Balan V, Dale BE (2011) Multi-scale visualization and characterization of lignocellulosic plant cell wall deconstruction during thermochemical pretreatment. Energy Environ Sci 4:973–984. https://doi.org/10.1039/C0EE00574F

Collet F, Pretot S (2012) Experimental investigation of moisture buffering capacity of sprayed hemp concrete. Constr Build Mater 36:58–65. https://doi.org/10.1016/j.conbuildmat.2012.04.139

Collet F, Pretot S (2014) Thermal conductivity of hemp concretes: variation with formulation, density and water content. Constr Build Mater 65:612–619. https://doi.org/10.1016/j.conbuildmat.2014.05.039

Collet F, Chamoin J, Pretot S, Lanos C (2013) Comparison of the hygric behaviour of three hemp concretes. Energ Buildings 62:294–303. https://doi.org/10.1016/j.enbuild.2013.03.010

Dubois S, Evrard A, Lebeau F (2014) Modeling the hygrothermal behavior of biobased construction materials. J Build Phys 38:191–213. https://doi.org/10.1177/1744259113489810

Elfordy S, Lucas F, Tancret F, Scudeller L, Goudet L (2008) Mechanical and thermal properties of lime and hemp concrete ("hempcrete") manufactured by a projection process. Constr Build Mater 22:2116–2123. https://doi.org/10.1016/j.conbuildmat.2007.07.016

Ip K, Miller A (2012) Life cycle greenhouse gas emissions of hemp–lime wall constructions in the UK. Resour Conserv Recycl 69:1–9. https://doi.org/10.1016/j.resconrec.2012.09.001

Jiang Y, Lawrence M, Ansell MP, Hussain A (2018) Cell wall microstructure, pore size distribution and absolute density of hemp shiv. R Soc Open Sci 5:171945. https://doi.org/10.1098/rsos.171945

Kidalova L, Stevulova N, Terpakova E (2015) Influence of water absorption on the selected properties of hemp hurds composites. Pollack Periodica 10:123–132. https://doi.org/10.1556/Pollack.10.2015.1.12

Laborel-Préneron A, Magniont C, Aubert JE (2018) Characterization of barley straw, hemp shiv and corn cob as resources for bioaggregate based building materials. Waste Biomass Valor 9:1095–1112. https://doi.org/10.1007/s12649-017-9895-z

Latif E, Ciupala MA, Wijeyesekera DC (2014) The comparative in situ hygrothermal performance of Hemp and Stone Wool insulations in vapour open timber frame wall panels. Constr Build Mater 73:205–213. https://doi.org/10.1016/j.conbuildmat.2014.09.060

Latif E, Lawrence M, Shea A, Walker P (2015) Moisture buffer potential of experimental wall assemblies incorporating formulated hemp-lime. Build Environ 93:199–209. https://doi.org/10.1016/j.buildenv.2015.07.011

Lawrence M, Jiang Y (2017) Porosity, pore size distribution, micro-structure. In: Amziane S, Collet F (eds) Bio-aggregates based building materials. RILEM state-of-the-art reports, vol 23. Springer, Dordrecht. https://doi.org/10.1007/978-94-024-1031-0_2

Lawrence M, Shea A, Walker P, De Wilde P (2013) Hygrothermal performance of bio-based insulation materials. Proc Inst Civil Eng Constr Mater 166(4):257–263. https://doi.org/10.1680/coma.12.00031

Lee K, Yeom D (2014) Experimental research on the indoor environment control of natural insulation materials: carbonized rice hull and straw bales. Indoor Built Environ 24:729–739. https://doi.org/10.1177/1420326x14534866

Lopez Hurtado P, Rouilly A, Vandenbossche V, Raynaud C (2016) A review on the properties of cellulose fibre insulation. Build Environ 96:170–177. https://doi.org/10.1016/j.buildenv.2015.09.031

Marceau S, Glé P, Guéguen-Minerbe M, Gourlay E, Moscardelli S, Nour I, Amzian S (2017) Influence of accelerated aging on the properties of hemp concretes. Constr Build Mater 139:524–530. https://doi.org/10.1016/j.conbuildmat.2016.11.129

Maskell D, Thomson A, Walker P, Lemke M (2018) Determination of optimal plaster thickness for moisture buffering of indoor air. Build Environ 130:143–150. https://doi.org/10.1016/j.buildenv.2017.11.045

Mazhoud B, Collet F, Pretot S, Chamoin J (2016) Hygric and thermal properties of hemp lime plasters. Build Environ 96:206–216. https://doi.org/10.1016/j.buildenv.2015.11.013

Miller SA (2018) Natural fiber textile reinforced bio-based composites: mechanical properties, creep, and environmental impacts. J Clean Prod 198:612–623. https://doi.org/10.1016/J.JCLEPRO.2018.07.038

Miller A, Ip K (2013) Sustainable construction materials. In: Yao R (ed) Design and management of sustainable built environment. Springer, London, pp 341–358

Moujalled B, Aït Oumeziane Y, Moissette S, Bart M, Lanos C, Samri D (2018) Experimental and numerical evaluation of the hygrothermal performance of a hemp lime concrete building: a long term case study. Build Environ 136:11–27. https://doi.org/10.1016/j.buildenv.2018.03.025

Osanyintola OF, Simonson CJ (2006) Moisture buffering capacity of hygroscopic building materials: experimental facilities and energy impact. Energ Buildings 38:1270–1282. https://doi.org/10.1016/j.enbuild.2006.03.026

Patel MK, Crank M, Dornburg V, Hermann BG, Roes L, Husing B, Overbeek L, Teraggni F, Recchia E (2006) Mediumand long-term opportunities and risks of the biotechnological production of bulk chemicals from renewable resources – the potential of white biotechnology. The BREW Project – Final Report

Pretot S, Collet F, Garnier C (2014) Life cycle assessment of a hemp concrete wall: impact of thickness and coating. Build Environ 72:223–231. https://doi.org/10.1016/j.buildenv.2013.11.010

Rahim M, Douzane O, Tran Le AD, Promis G, Laidoudi B, Crigny A, Dupre B, Langlet T (2015) Characterization of flax lime and hemp lime concretes: hygric properties and moisture buffer capacity. Energ Buildings 88:91–99. https://doi.org/10.1016/j.enbuild.2014.11.043

Rahim M, Douzane O, Tran Le AD, Promis G, Langlet T (2016) Characterization and comparison of hygric properties of rape straw concrete and hemp concrete. Constr Build Mater 102:679–687. https://doi.org/10.1016/j.conbuildmat.2015.11.021

Rim Al, Ledhem A, Douzane O, Dheilly RM, Queneudec M (1999) Influence of the proportion of wood on the thermal and mechanical performances of clay-cement-wood composites. Constr Build Mater 21:269–276. https://doi.org/10.1016/S0958-9465(99)00008-6

Rode C, Peuhkuri R, Mortensen LH, Hansen KK, Gustavsen A, Ojanen T, Arfvidsson J (2005) Moisture buffering of building materials. Retrieved from http://orbit.dtu.dk/fedora/objects/orbit:75984/datastreams/file_2415500/content

Senga Kiessé T, Ventura A, van der Werf HMG, Cazacliu B, Idir R, Andrianandraina IR (2017) Introducing economic actors and their possibilities for action in LCA using sensitivity analysis: application to hemp-based insulation products for building applications. J Clean Prod 142:3905–3916. https://doi.org/10.1016/J.JCLEPRO.2016.10.069

Sinka M, Van den Heede P, De Belie N, Bajare D, Sahmenko G, Korjakins A (2018) Comparative life cycle assessment of magnesium binders as an alternative for hemp concrete. Resour Conserv Recycl 133:288–299. https://doi.org/10.1016/J.RESCONREC.2018.02.024

Tran Le AD, Maalouf C, Mai TH, Wurtz E, Collet F (2010) Transient hygrothermal behaviour of a hemp concrete building envelope. Energ Buildings 42:1797–1806. https://doi.org/10.1016/j.enbuild.2010.05.016

Tsiropoulos I, Faaij APC, Lundquist L, Schenker U, Briois JF, Patel MK (2015) Life cycle impact assessment of bio-based plastics from sugarcane ethanol. J Clean Prod 90:114–127. https://doi.org/10.1016/J.JCLEPRO.2014.11.071

Walker R, Pavía S (2014) Moisture transfer and thermal properties of hemp-lime concretes. Constr Build Mater 64:270–276. https://doi.org/10.1016/j.conbuildmat.2014.04.081

Zabalza Bribián I, Valero Capilla A, Aranda Usón A (2011) Life cycle assessment of building materials: comparative analysis of energy and environmental impacts and evaluation of the eco-efficiency improvement potential. Build Environ 46:1133–1140. https://doi.org/10.1016/J.BUILDENV.2010.12.002

Zah R, Hischier R, Leão AL, Braun I (2007) Curauá fibers in the automobile industry – a sustainability assessment. J Clean Prod 15:1032–1040. https://doi.org/10.1016/J.JCLEPRO.2006.05.036

Zhang H, Yoshino H, Hasegawa K (2012) Assessing the moisture buffering performance of hygroscopic material by using experimental method. Build Environ 48:27–34. https://doi.org/10.1016/j.buildenv.2011.08.012

Chapter 6
Modelling the Hygrothermal Behaviour of Hemp Concrete: From Material to Building

Yacine Aït Oumeziane, Florence Collet, Christophe Lanos, and Bassam Moujalled

Abstract Global warming, the scarcity of natural resources, the polluting emission is a major concern for the human community. The construction sector especially has a significant impact on the environment and has therefore a role to play in the development of innovative sustainable solutions. Bio-based materials are known to be an interesting solution to address energy and environmental issues. In particular, hemp, fast growing renewable raw vegetal, has the qualities to be a serious alternative to modern insulation solutions. Hemp wool using the fibres and hemp concrete using the shivs of the plant have interesting hygrothermal properties and a good thermal insulation level. Their porous, hygroscopic and permeable structure gives them high moisture transfer and storage capacities, improving the hygrothermal comfort felt by the inhabitants.

In order to answer to the future environmental and energy regulations, expanding the use of hemp in the building sector depends on a better knowledge of its hygrothermal behaviour and its response to climatic variations.

Hemp concrete presents a significant hysteretic behaviour. This complex behaviour influences the evolution of the moisture content inside the material which is a key factor of the evolution of the hygrothermal properties and transfer. This chapter reports that the recent consideration of numerical models suited to the hemp concrete hygric behaviour implemented in a heat, air, and moisture transfer model has improved the knowledge and prediction of the hemp concrete hygrothermal response. Especially, we show that the modelling of the temperature-dependence

---

Y. Aït Oumeziane (✉)
FEMTO-ST Institute, University Bourgogne Franche-Comté, CNRS, Belfort, France
e-mail: yacine.ait_oumeziane@univ-fcomte.fr

F. Collet · C. Lanos
University of Rennes, LGCGM EA 3913, Rennes, France
e-mail: florence.collet@univ-rennes1.fr; christophe.lanos@univ-rennes1.fr

B. Moujalled
Cerema, Equipe-projet BPE, L'Isle d'Abeau, France
e-mail: Bassam.Moujalled@cerema.fr

G. Crini, E. Lichtfouse (eds.), *Sustainable Agriculture Reviews 42*,
Sustainable Agriculture Reviews 42, https://doi.org/10.1007/978-3-030-41384-2_6

of sorption mechanism allows to better represent the effective response of hemp concrete subjected to real weather conditions. The predicted local daily variations of temperature and relative humidity through a wall are found to be consistent with the experimental ones. Moreover, we review experimental campaigns lead *in situ* which show that hemp concrete helps to maintain a good level of hygrothermal comfort.

Keywords Hemp · HAM transfer · Hygrothermal modelling · Building

Abbreviations

BET	Brunauer, Emmet and Teller
DSC	Differential scanning calorimetry
DTA	Differential thermal analysis
DVS	Dynamic vapour sorption
GAB	Guggenheim, De Boer and Anderson (at the origin of GAB formalism)
HAM	Heat, air and moisture
HLC	Hemp-lime concrete
MBV	Moisture buffer value
REV	Representative elementary volume
RH	Relative humidity
SEM	Scanning electron microscopy

Nomenclature

Latin Symbols

c_p	specific heat capacity [J.kg^{-1}.K^{-1}]
C_φ	Milly's coefficient [K^{-1}]
D_l	isothermal liquid diffusion coefficient [m^2.s^{-1}]
D_l^φ	liquid diffusion coefficient under relative humidity gradient [kg.m^{-1}.s^{-1}]
$D_{l,T}$	liquid diffusion coefficient under temperature gradient [kg.K^{-1}.m^{-1}.s^{-1}]
D_l^T	total liquid diffusion coefficient under temperature gradient [kg.K^{-1}.m^{-1}.s^{-1}]
$D_{l,w}^T$	isothermal liquid diffusion coefficient under moisture content gradient [m^2.s^{-1}]
D_v	vapour diffusion coefficient [kg.K^{-1}.m^{-1}.s^{-1}]
D_t	total moisture diffusion coefficient [m^2.s^{-1}]
e	thickness [m]
E	molar bond energy [J.mol^{-1}]
g	mass flow density [kg.m^{-2}.s^{-1}]
h	capillary height [m]
H	enthalpy [J.m^{-3}]
k_g	effective moist air permeability [m^2]

K_l hydraulic conductivity [kg.m^{-1}.s^{-1}.Pa^{-1}]
l_v latent heat of condensation [J.kg^{-1}]
m mass [kg]
M molar mass [g.mol^{-1}]
n_0 open porosity [%]
n total porosity [%]
N_A Avogadro's number [6.02 10^{23} mol^{-1}]
p pressure [Pa]
q_{st} isosteric heat [J.kg^{-1}]
q heat flow density [W.m^{-2}]
r pore radius [m]
r_m mixing rate [–]
R perfect gas constant [8.314 J.mol^{-1}.K^{-1}]
S specific surface [m^2.kg^{-1}]
S_e effective saturation [–]
t time [s]
T temperature [K]
u mass moisture content [kg.kg^{-1}]
v air velocity [m.s^{-1}]
V volume [m^3]
w moisture content [kg.m^{-3}]

Greek Symbols

α thermal diffusivity [m^2.s^{-1}]
γ dynamic viscosity [Pa.s]
δ_a air permeability [kg.m^{-1}.s^{-1}.Pa^{-1}]
δ_p vapour permeability [kg.m^{-1}.s^{-1}.Pa^{-1}]
$\delta_{p,a}$ air vapour permeability [kg.m^{-1}.s^{-1}.Pa^{-1}]
θ contact angle [°]
λ thermal conductivity [W.m^{-1}.K^{-1}]
μ vapour diffusion resistance factor [–]
ρ bulk density [kg.m^{-3}]
σ surface tension [J.m^{-2}]
φ relative humidity [%RH]

Index

$*$ equivalent
$**$ fictitious
0 dry

a	air
o	open
ads	main adsorption
adv	advection
cond	conduction
conv	convection
cr	critical
diff	diffusion
des	main desorption
g	gas phase
l	liquid water
lat	latent
m	molecular
p	pore
sat	saturated
sen	sensitive
suc	succion
t	total
v	vapour water
w	water

6.1 Introduction

Both in France and in Europe, the building sector represents more than 40% of final energy consumption and about 25% of CO_2 emission (Eurostat 2018; CGDD 2018). The actors of sustainable development need to conjointly think about energy consumption and carbon footprint reduction, hygrothermal comfort for the inhabitants and indoor air quality. In this framework, the choice of environmentally friendly and healthy materials like bio-based materials appears is an interesting solution to address these energy and environmental issues (Miller 2018; Spierling et al. 2018). As a construction material, hemp – fast growing renewable raw vegetal – has the potential to radically decrease the carbon footprint of buildings (Pretot et al. 2014; Pittau et al. 2018).

Hemp has been used in the building sector for long time but has fallen into disuse because of the development of modern construction materials. Nevertheless, the research and development on traditional bio-based materials have significantly increased in the last two decades thanks to incentivizing environmental policies. The fibres of hemp are valued in hemp wool used as insulation material. Many studies on hemp wools have shown their interesting hygrothermal properties (Collet et al. 2011; Korjenic et al. 2011; Latif et al. 2014). Associated with a mineral binder, the hemp shivs are used as aggregates in hemp concrete formulations. The most frequently used binder is lime, which has a lower embodied energy than cement

Fig. 6.1 View of the south façade of a 2-floor single-detached house located in south west of France. (From Moujalled et al. 2018). The walls are made of 30-cm thick coated sprayed hemp-lime concrete supported by a timber frame structure

(Pittau et al. 2018). Moreover, in actual hemp concrete formulations, lime is the element which has the strongest environmental impacts. Some recent research have shown the environmental benefit to replace part of the lime by earth fines while maintaining the mechanical and hygrothermal properties of hemp-lime concretes (Mazhoud et al. 2017).

Different compositions and manufacturing methods allow hemp concrete to have a variety of use in walls, floors and roofs (Evrard 2008). In walls, hemp concrete is usually precast, moulded in place or sprayed. It has a compressive strength far below 2 MPa and is consequently not load-bearing (Elfordy et al. 2008; Arnaud and Gourlay 2012; Niyigena et al. 2016). Because of its low mechanical properties, hemp concrete is mainly used as filling material supported on a timber frame. However, its porous, hygroscopic and permeable structure results in high moisture transfer and storage capacities. Its capacity to store moisture by adsorption and remove moisture by desorption gives the ability to hemp concrete walls to moderate the amplitude of indoor relative humidity and temperature. Indeed, relative humidity and temperature are two essential parameters conditioning indoor environmental quality, hygrothermal comfort and by correlation energy consumption. As an illustration, Fig. 6.1 shows the south façade view of a 2-storey single-detached house located in south west of France. Its walls are made of 30-cm thick coated sprayed hemp-lime concrete (HLC) supported by a timber frame structure.

However, hemp concrete presents a significant hysteretic behaviour experimentally highlighted in (Collet et al. 2013; Promis et al. 2018; Seng et al. 2018b). It means that hemp concrete does not release moisture in the same way that it stores it. This complex behaviour influences the evolution of the moisture content within the material. For hygroscopic materials, recent works led on gypsum board have shown that moisture content is the key parameter which governs at the first order the evolution of the hygrothermal properties and influences the hygrothermal transfer (Kwiatkowski et al. 2009; Van Belleghem et al. 2010). Due to the hysteretic effect, the equilibrium moisture content depends not only on relative humidity but also on moisture history. The temperature also influences the equilibrium moisture content. Hysteresis phenomenon and temperature effects on sorption process are most often neglected for modelling the moisture content evolution in heat and moisture transfer models. This can cause significant discrepancies to predict the hygrothermal response of a material subjected to climatic variations as shown by Zhang et al. (2016) in the case of wood materials.

Moreover, the latest research on the transient hygrothermal response of hemp lime concrete have essentially been performed under laboratory-controlled conditions. The experimental feedback on well-monitored wall or buildings in real weather conditions are still scarce in literature. However, this work aims at giving some information about the effective hemp concrete behaviour under real climatic conditions.

6.2 Heat, Air and Moisture Transfer in Porous Media

6.2.1 General Framework

This section aims to briefly mathematically describe the physical phenomena which occur in hemp concrete. As illustrated in Fig. 6.2, hemp concrete is considered as a porous medium composed of a rigid solid phase and pores which can be filled by a liquid and/or gas phase.

Under the assumption of perfect gas for the water vapour phase, it is shown that the diffusion of the vapour phase is initiated by a vapour pressure gradient. The vapour transport is governed by three modes depending on the average free path of a molecule of water through the porous structure:

- effusion determined by the number of collisions of the water molecules with the solid matrix
- free molecular diffusion determined by the number of collisions between the particles themselves
- mixed diffusion determined by both particle/solid collisions and particle collisions

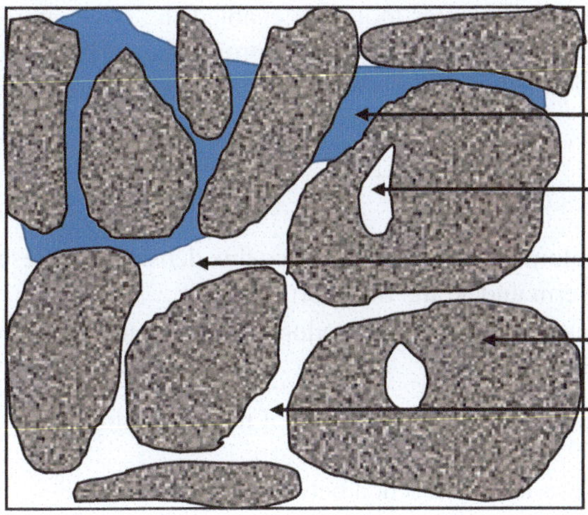

Fig. 6.2 Schematic microscopic representation of a porous medium

The transport of liquid water is governed by the capillary pressure gradient, microscopically described by the Poiseuille's law.

However, the determination of the properties of a material is based on the measurement of macroscopic parameters resulting from the combination of phenomena at the microscopic scale. Whitaker has proposed a volume averaging method to link the microscopic and macroscopic approaches to describe the transfer laws (Whitaker 1986a, b). Some precautions need to be considered to pass from laws defined on the microscopic scale to laws defined on the macroscopic scale:

- as for the experimental determination of the physical characteristics of the material, the medium is considered macroscopically homogeneous which supposes a studied volume higher than the representative elementary volume (REV)
- the Fick's law used to locally describe the diffusive transport must be readapted to the case of a porous medium by considering the global structure of the porous space
- the definition of a macroscopic liquid phase transfer coefficient can be only determined either experimentally, from empirical models, or based on the porous structure because of the difficulty to differentiate vapour and liquid flows

Other authors have built their model from a phenomenological approach based on a macroscopic description of the phenomena (Philip and Vries 1957; Taylor and Cary 1964).

These two approaches, microscopic and macroscopic, lead to a continuous formulation of heat, air and moisture (HAM) transfer. Therefore, they are rigorously equivalent.

In the context of this work, the physical phenomena which occur in hemp concrete are macroscopically described.

The continuing growth of the computer tools has enabled the development of increasingly complex, fast and reliable HAM models in porous media. These models are all based on the conservation principle:

$$\frac{\partial X}{\partial t} = -\nabla \mathrm{j} + \mathrm{S} \tag{6.1}$$

where X is homogenous to a volume density, j to a surface flux density. S represents a source term, $\frac{\partial X}{\partial t}$ the unsteady term and ∇j the transport term.

The modelling assumptions are those usually adopted in literature (Philip and Vries 1957; Luikov 1975; Whitaker 1977):

- the material is isotropic and the considered volume of material is higher than the REV
- the thermodynamic balance is locally reached between the three phases
- the moist air phase is an incompressible perfect fluid
- no chemical reaction occurs in the material
- the radiative transfer is neglected
- the gravity effect on moisture transfer is neglected

Under these conditions, moisture transfer is governed by vapour and liquid diffusion and vapour advection transport through the material. Heat transfer is carried out by conduction, advection and by enthalpy transport of sensible and latent heat due to phase change in pores.

Equations 6.2, 6.3 and 6.4 respectively describe the heat, air and moisture conservation equations.

$$\frac{\partial H}{\partial t} = -\nabla(q_{cond} + q_{adv} + q_{sen} + q_{lat}) + S_h \tag{6.2}$$

H [J.m^{-3}] is the enthalpy, t [s] the time, q_{cond}, q_{adv}, q_{lat} and q_{sen} [W.m^{-2}] are respectively the conduction, advection, latent and sensitive heat flux densities.

$$\frac{\partial w_a}{\partial t} = -\nabla\left(g_{a,diff} + g_{a,adv}\right) + S_a \tag{6.3}$$

w_a [kg.m^{-3}] is the air content inside the porous medium and $g_{a,diff}$ and $g_{a,adv}$ [kg.m^{-2}.s^{-1}] represent respectively the air diffusion and advection flux densities.

$$\frac{\partial w}{\partial t} = -\nabla\left(g_{t,diff} + g_{t,adv}\right) + S_w \tag{6.4}$$

w [kg.m^{-3}] represents the moisture content under vapour and liquid phases inside the porous medium. $g_{t,diff}$ and $g_{t,adv}$ [kg.m^{-2}.s^{-1}] respectively represent the total moisture diffusion and advection flux densities. S_a [kg.m^{-3}.s^{-1}], S_h [W.m^{-3}] and S_w [kg.m^{-3}.s^{-1}] are the source terms.

6.2.2 Air Transfer

The air flow in porous media is generated by total pressure gradients following Poiseuille's law of proportionality (Hens 2007), which relates the total pressure gradient to air flow velocity v [m.s^{-1}]:

$$\mathrm{v} = \frac{k_g}{\gamma_g} \nabla p_t \tag{6.5}$$

k_g [m^2] stands for the effective moist air permeability, μ_g [Pa.s] the air dynamic viscosity and p_t [Pa] the total pressure.

In building physics applications, air is considered incompressible due to very low air velocities. As described in Tariku et al. (2010), the air mass conservation can be written according to Eq. 6.6.

$$0 = -\nabla \left(g_{a,adv+} g_{a,diff} \right) \tag{6.6}$$

In the case of an application to hemp concrete walls, temperature and pressure changes are usually supposed to be too low to create a diffusive or advective air flow through the material. Thus, it is commonly accepted that air advection and diffusion are neglected.

$$g_{a,adv} = \rho_a \mathrm{v} = 0 \tag{6.7}$$

with δ_a [kg.m^{-1}.s^{-1}.Pa^{-1}] the air permeability of the material and:

$$g_{a,diff} = -\delta_a \nabla w_a = 0 \tag{6.8}$$

The advective vapour flow density expressed in Eq. 6.9 is generated by moist air flow through the material:

$$g_{v,adv} = \rho_a r_m \mathrm{v} = 0 \tag{6.9}$$

with r_m [−] the mixing rate:

$$r_m = \frac{M_v}{M_a} \frac{p_v}{(p_{atm} - p_v)} \tag{6.10}$$

M_v [0.018 kg.mol^{-1}] and M_a [0.029 kg.mol^{-1}] are respectively the vapour and dry air molar mass.

Consequently, the heat transfer by air and vapour advection are neglected in the rest of this work.

Nevertheless, recent works have shown that an apparent air permeability of about 10^{-10} m^2 can be considered for hemp concrete with an apparent dry density of about 460 kg.m^{-3} (Seng et al. 2018a). As underlined in Aït Oumeziane (2013), this new experimental data could lead to not neglect vapour advection in heat and moisture transfer numerical simulations even under low total pressure gradients lower than 10 Pa.

6.2.3 Moisture Transfer

Porous hygroscopic materials have the ability to store and release the moisture of the ambient air on the pores' surface. The hygric state of a porous medium depends on its moisture content u [kg.kg^{-1}] or w [kg.m^{-3}] defined in Eq. 6.11.

$$u = \frac{w}{\rho_0} = \frac{m_w}{m_0} \tag{6.11}$$

m_w [kg] stands for the mass of water inside the material and m_O [kg] for the mass of the dry material. ρ_O [kg.m^{-3}] is the dry bulk density of the material. The saturation moisture content u_{sat} [kg.kg^{-1}] is reached at 100%RH. The description of the sorption mechanism is covered in more details in the following section of this chapter.

At low moisture contents, moisture transport in porous media comes from vapour diffusion. The isothermal transport of a fluid, gas or liquid, in a porous medium is described by the general form of the Darcy's law (Hens 2007). The vapour diffusion process is described by the Fick's law in Eq. 6.12, which can be considered as a special application of the Darcy's law. Vapour pressure gradients are the driving forces of vapour phase.

$$g_{v,diff} = -D_v \nabla (p_v) \tag{6.12}$$

$g_{v,diff}$ [kg.m^{-2}.s^{-1}] is the vapour diffusion flow density, D_v [kg.m^{-1}.s^{-1}.Pa^{-1}] the vapour diffusion coefficient and p_v [Pa] the partial vapour pressure. The partial vapour pressure depends on the number of water moles. It is related to the saturation vapour pressure p_{sat} [Pa], which is function of temperature T [K] by the relative humidity φ [%RH].

$$p_v = \varphi p_{sat}(T) \tag{6.13}$$

The vapour diffusion coefficient D_v is called vapour permeability δ_p [kg.m^{-1}.s^{-1}.Pa^{-1}] and is expressed in Eq. 6.14.

$$D_v = \delta_p = \frac{\delta_{p,a}}{\mu} \tag{6.14}$$

μ [−] is the vapour diffusion resistance factor and $\delta_{p,a}$ [kg.m^{-1}.s^{-1}.Pa^{-1}] the air vapour permeability:

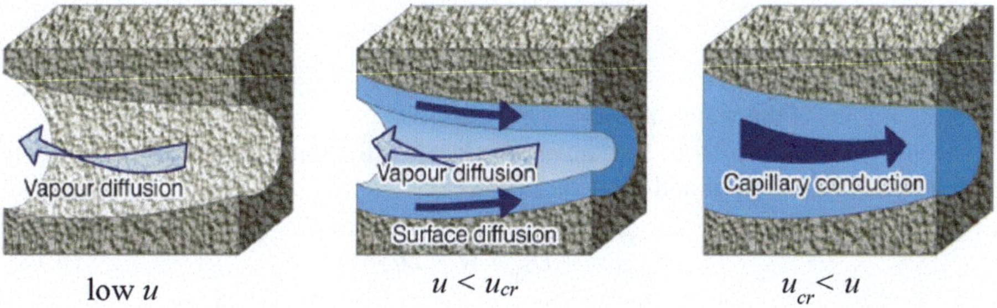

Fig. 6.3 Physical representation of the moisture diffusion evolution. (From Evrard 2008)

$$\delta_{p,a} = 2.31.10^{-5}\frac{M_v}{RT}\left(\frac{T}{273.15}\right)^{1.81} \tag{6.15}$$

with R [8.314 J.mol^{-1}.K^{-1}] the perfect gas constant.

Furthermore, the existence of a temperature gradient through a porous medium creates a migration of the water vapour. This phenomenon is called thermodiffusion or Soret effect. Nevertheless, Krus (1996) showed that thermodiffusion contributes to only around 0.05% of the total vapour diffusion transport. Janssen (2011) has confirmed this result from an experimental campaign on several construction materials. This can be explained by the low temperature gradients which occur in building physics.

As illustrated in Fig. 6.3, liquid transport is considered to be initiated when moisture content exceeds a critical moisture content u_{cr}. Moisture transport by liquid surface diffusion occurs before capillary condensation and is very significant in hygroscopic mineral building materials at relative humidities above 50%RH (Krus 1996). Künzel (1995) reported that liquid surface diffusion in paper products begins to be noticed from 30%RH and in sandstone at about 60%RH. Indeed, when moisture content continues to increase, the thickness of the adsorbed layers becomes higher and thus more mobile. As explained by the BET theory (from the authors Brunauer, Emmet and Teller), the reducing influence of the adsorbate at higher moisture contents, explained by an increasing distance between the water molecules and the adsorbate surface, decreases the bond energy (Brunauer et al. 1938).

The liquid flux density by capillary conduction can be expressed from the Darcy's law in Eq. 6.16. The driving force of the capillary conduction process is the capillary pressure gradient.

$$g_{l,diff,cond} = -K_l \nabla (p_{suc}) \tag{6.16}$$

K_l [kg.m^{-1}.s^{-1}.Pa^{-1}] is the hydraulic conductivity and p_{suc} [Pa] the capillary pressure. p_{suc} can be expressed in function of relative humidity from the Kelvin's law in Eq. 6.17:

$$p_{suc} = -\rho_w \frac{R}{M_l} Tln(\varphi) \tag{6.17}$$

with ρ_w [kg.m^{-3}] the water density.

From Philip and De Vries formalism, the liquid diffusion transport can be written in a general form (Philip and De Vries 1958):

$$g_{l,diff} = -D_{l,w}^T \nabla w - D_{l,T} \nabla T \tag{6.18}$$

$D_{l,w}^T$ [m^2.s^{-1}] is the isothermal liquid diffusion coefficient under moisture content gradient and $D_{l,T}$ [kg.K^{-1}.m^{-1}.s^{-1}] the liquid diffusion coefficient under temperature gradient.

The development of Eq. 6.18 in function of relative humidity and temperature leads to:

$$\begin{aligned} g_{l,diff} &= -D_{l,w}^T \left[\left.\frac{\partial w}{\partial \varphi}\right|_T \nabla \varphi + \left.\frac{\partial w}{\partial T}\right|_\varphi \nabla T \right] - D_{l,T} \nabla T \\ &= -D_l^\varphi \nabla \varphi - D_l^T \nabla T \end{aligned} \tag{6.19}$$

D_l^φ [kg.m^{-1}.s^{-1}] stands for the liquid conduction coefficient under relative humidity gradient and D_l^T [kg.K^{-1}.m^{-1}.s^{-1}] the liquid diffusion coefficient under temperature gradient. They are expressed as:

$$\begin{aligned} D_l^\varphi &= D_{l,w}^T \left.\frac{\partial w}{\partial \varphi}\right|_T \\ D_l^T &= D_{l,w}^T \left.\frac{\partial w}{\partial T}\right|_\varphi + D_{l,T} \nabla T \end{aligned} \tag{6.20}$$

As for vapour transport, the liquid thermodiffusion can be neglected because of low temperature gradients in building physics applications which means that D_l^T is expressed as:

$$D_l^T = D_{l,w}^T \left.\frac{\partial w}{\partial T}\right|_\varphi \tag{6.21}$$

Finally, the moisture transfer conservation equation is given by Eq. 6.22.

$$\frac{\partial w}{\partial t} = -\nabla \left(D_v \nabla p_v + D_l^\varphi \nabla \varphi + D_l^T \nabla T \right) \tag{6.22}$$

6.2.4 Heat Transfer

Heat conduction through a porous media can be expressed using Fourier's law, which relates the heat conduction flow to the temperature gradient:

$$q_{cond} = -\lambda^* \nabla T \tag{6.23}$$

λ^* [W.m^{-1}.K^{-1}] is the equivalent thermal conductivity of the moist material which depends on moisture content.

The enthalpic flow density by sensible heat q_{sen} [W.m^{-2}] is expressed in Eq. 6.24:

$$q_{sen} = c_v\left(T - T_{ref}\right)g_{v,diff} + c_l\left(T - T_{ref}\right)g_{l,diff} \tag{6.24}$$

with T_{ref} [K] the energetic reference temperature. c_v [J.kg^{-1}.K^{-1}] and c_l [J.kg^{-1}.K^{-1}] are respectively the vapour and liquid specific heat capacities at the reference temperature.

The heat flow density by latent heat q_{lat} [W.m^{-2}] is generated by the isothermal phase change by evaporation and condensation on the adsorbate surface and is expressed in Eq. 6.25.

$$q_{lat} = l_v g_v = l_v\left[g_{v,diff} + g_{v,adv}\right] = -l_v[D_v \nabla p_v] \tag{6.25}$$

The conservation equation for heat balance is finally given in Eq. 6.26.

$$= \frac{\frac{\partial\left[\left(\rho_0 c_{p,0} + w c_l\right)\left(T - T_{ref}\right)\right]}{\partial t}}{\nabla[\lambda^* \nabla T] + c_v\left(T - T_{ref}\right)g_{v,diff} + c_l\left(T - T_{ref}\right)g_{l,diff} + l_v[D_v \nabla p_v]} \tag{6.26}$$

6.3 Hygrothermal Properties of Hemp Concrete

6.3.1 Physical Properties

Compared to traditional concretes, hemp concrete is lightweight with a dry bulk density between 300 and 700 kg.m^{-3} depending on its composition and compaction method. This range of values is commonly observed in literature (De Brujin and Johansson 2013; Collet et al. 2013; Chamoin 2013; Amziane and Arnaud 2013; Walker and Pavía 2014).

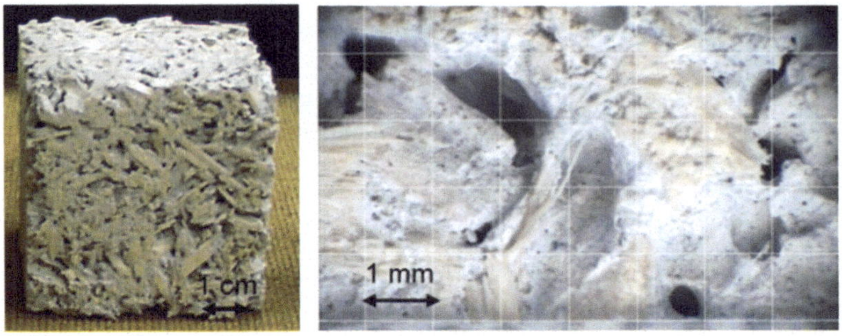

Fig. 6.4 View of the macro porosity of a hemp concrete sample. (From Collet et al. 2013)

Hemp concrete has a highly porous structure, which explains its physical, hygrothermal, mechanical and acoustical characteristics.

The total porosity n [%] is defined as the rate between the pores volume V_p [m^3] ant the total volume V_t [m^3] of the material (Eq. 6.27). This definition includes the consideration of trapped pores (Fig. 6.2). The total porosity is usually experimentally measured with the pycnometer method.

$$n = \frac{V_p}{V_t} \tag{6.27}$$

The open porosity n_O [%] characterizes the interconnection of the pores and is defined as the rate between the open pores volume $V_{p,o}$ [m^3] ant the total volume V_t [m^3] of the material in Eq. 6.28. Experimentally, it can be determined with the hydrostatic weighing method or by mercury porosimetry.

$$n_o = \frac{V_{p,o}}{V_t} \tag{6.28}$$

The works of literature showed that hemp concrete presents a total porosity which ranges from 70% to 85% and an open porosity ranging between 60% and 75% for the lightest compositions. As explained in Collet et al. (2013), the open porosity includes a wide range of interconnected pores. At a macroscopic scale, the porosity due to the arrangement between the hemp shiv and the hemp–binder adhesion can reach a millimetric width (Fig. 6.4). In the lime binder matrix, the porosity due to the trapped air has a width up to 50 μm and the porosity due to hydrates arrangement is lower than 0.01 μm (Fig. 6.5a). In hemp shiv, the width of the pores goes from 5 to 50 μm (Fig. 6.5b, c).

Recently, X-Ray tomography analysis have confirmed these observations (Bennai et al. 2018).

Fig. 6.5 SEM view of a hemp concrete sample (a) and of a hemp shiv (b: length view, c: cross view). (From Collet et al. 2013)

6.3.2 Hygric Properties

6.3.2.1 Sorption Isotherms Modelling

The sorption isotherm relates the amount of equilibrium moisture content to the ambient relative humidity at a given temperature. Sorption isotherms can be measured according to continuous or discontinuous methods (Sing et al. 1986). The continuous method consists of measuring the sorption curve under quasi-equilibrium conditions. The adsorbable fluid is admitted (or removed) at a slow and constant rate. The variation of the adsorbed fluid amount by the specimen with increasing (and then decreasing) pressure is measured by volumetric or gravimetric method. The discontinuous method consists in measuring the amount of fluid adsorbed by the specimen at successive increasing (and then decreasing) pressure stages. It can be performed with desiccators and saturated salt-solutions, with climatic chambers or with a DVS device (Dynamic Vapour Sorption). This latter has been developed to reduce the running time of measurement but can be only used with small specimens that generally do not include REV of the material (Xie et al. 2011). The adsorption and desorption isotherms are determined according to the requirements of standard EN ISO 12571 2000.

At low relative humidities, from the dry state, moisture is gradually adsorbed at the level of the pore surface. The water molecules first settles into one layer (monomolecular adsorption) then, as the moisture increases, over several layers (polymolecular adsorption). This phenomenon is called hygroscopic adsorption. Simultaneously, the phenomenon of liquid surface diffusion occurs at the level of the pore surface as soon as a layer of liquid water molecules is formed. At the highest relative humidities, the phenomenon of capillary conduction appears by the formation of liquid bridges first in the finest pores then in the largest.

At a given temperature, starting from the dry state to the saturated state, the increase of moisture content with relative humidity allows to define a so-called main adsorption curve. Conversely, starting from the saturated state to the dry state, the decrease of moisture content with relative humidity allows to define a so-called main desorption curve. These curves are the physical boundaries of the equilibrium

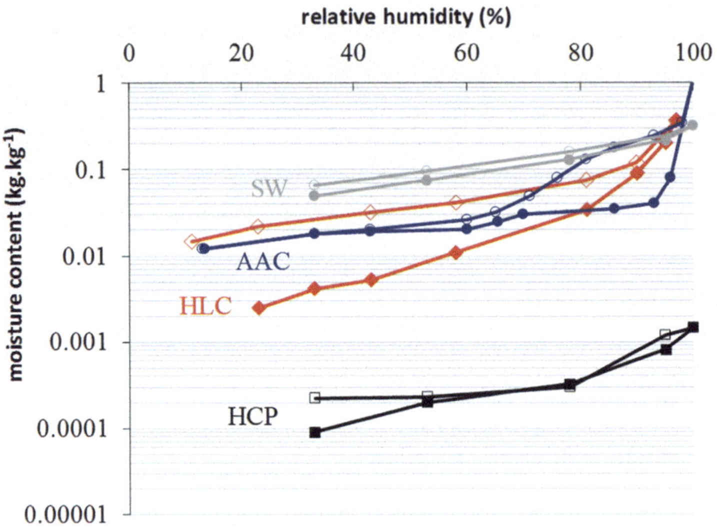

Fig. 6.6 Comparison of the main adsorption and desorption isotherms of different building materials – *HCP* hardened cement paste, *HLC* hemp-lime concrete, *AAC* autoclaved aerated concrete and *SW* spruce wood. Note the strong hysteresis that occurs in HLC and its high hygroscopicity

moisture content evolution in the material. The slope of these curves defines the hygric capacity of the material, i.e. its capacity to store or release moisture.

In order to characterize the hygroscopic behaviour of hemp concrete, the measured sorption isotherms are compared with the results of the literature. As an illustration, in Fig. 6.6, the following curves are represented:

- the adsorption/desorption isotherm of a hemp –lime concrete (HLC) at 23 °C (Collet et al. 2013)
- the adsorption/desorption isotherm of a spruce wood (SW) at 25 °C (Rode and Hansen 2011)
- the adsorption/desorption isotherm of a hardened cement paste with W/C = 0.5 (HCP) at 25 °C (Rode and Hansen 2011)
- the adsorption/desorption isotherm of an autoclaved aerated concrete (AAC) at 25 °C (Trong et al. 2018)

In comparison with the others materials, hemp concrete presents a highly hygroscopic behaviour and a strong hysteresis which occurs between main adsorption and desorption process. The results presented in Fig. 6.6 are very close to those obtained from different characterization works of the literature for different hemp concrete formulations. The results of these different studies are reported in details in (Seng et al. 2018b). Furthermore, Seng et al. (2018b) showed that there are few differences between the results obtained from gradual relative humidity steps and the results obtained with the DVS method.

Mathematically, it exists many models in literature to describe the main sorption isotherms. The historical and most used models in building physics are presented in this section. The first class of models are the phenomenological models. They describe the process of sorption in porous media. In practice, these models are necessarily based on experimental data because the evolution of the moisture content is very sensitive to the nature of the adsorbate. Thus, they tend to lose their predictive character.

The historical Langmuir's theory describes the physical adsorption of the first layer of water molecules (Langmuir 1918). At equilibrium, evaporation and condensation rates on the surface of the solid matrix are assumed to be equal. All the water molecules are bound to the adsorbate by a molar bond energy E_{ads} [J. mol^{-1}]. The Langmuir's model equation is written in Eq. 6.29. This expression is valid only at low relative humidity until 10–20%RH.

$$u(\varphi) = \frac{a_{lan} u_m \varphi}{1 + a\varphi} \tag{6.29}$$

with u_m [kg.kg^{-1}] the molecular moisture content and the parameter a_{lan} [−] is expressed in Eq. 6.30:

$$a_{lan} = e^{\left(\frac{E_{ads}}{RT}\right)} \tag{6.30}$$

Later, from the Langmuir's theory, Brunauer, Emmet and Teller developed a model that describes the polymolecular adsorption (Brunauer et al. 1938, 1940, 1969). This model called BET model in literature considers that molecules of the first layer are bound to the adsorbate by an energy equal to the molar heat of adsorption E_{ads} [J.mol^{-1}] and that the molecules of the other layers are bound between them by an energy equal to the molar latent heat of vaporization E_{lv} [J. mol^{-1}]. The effects of mutual interactions between adsorbed molecules are neglected. In this theory, water is bound to the adsorbate in the form of successive adsorbed layers. The first layer is not necessarily complete before the next one begins to fill. The number of layers that can be adsorbed is not limited by the size of the pores. The moisture content obtained at saturation for 100%RH is infinite. Thus, its application for porous media is considered valid only for relative humidity less than 40%.

$$u(\varphi) = \frac{a_{BET} u_m \varphi}{(1-\varphi)(1-\varphi + a_{BET}\varphi)} \tag{6.31}$$

with u_m [kg.kg^{-1}] the molecular moisture content and the parameter a_{BET} [−] is expressed in Eq. 6.32:

$$a_{BET} = e^{\left(\frac{E_{ads}-E_{lv}}{RT}\right)} \tag{6.32}$$

Guggenheim, Anderson and De Boer independently developed a model based on the physical description of the sorption mechanism (Anderson 1946; Anderson and Hall 1948; De Boer 1953; Guggenheim 1966). The so-called GAB model, like the BET model, is physically valid only in the absence of capillary condensation, i.e. for low and medium relative humidity levels. By extension of the BET theory, this model considers the variation of the molar heat of adsorption for all water molecular layers $E_{ads,m}$ [J.mol^{-1}] in comparison with the molar heat of adsorption of one water molecular layer $E_{ads,1}$ [J.mol^{-1}].

The general form of the GAB equation is described by Eq. 6.33:

$$u(\varphi) = \frac{a_{GAB} b_{GAB} u_m \varphi}{(1 - a_{GAB}\varphi)[1 + (a_{GAB} - 1) b_{GAB}\varphi]} \tag{6.33}$$

with:

$$\begin{aligned} a_{GAB} &= e^{\left(\frac{E_{ads,1} - E_{ads,m}}{RT}\right)} \\ b_{GAB} &= e^{\left(\frac{E_{lv} - E_{ads,m}}{RT}\right)} \end{aligned} \tag{6.34}$$

In practice, the parameters a_{GAB} and b_{GAB} are often derived to fit experimental data from the weighted least square method. Nevertheless, the works of the literature have shown that it is an excellent tool for mathematical smoothing of the sorption isotherms on the whole range of relative humidity. Consequently, its use is widespread in building physics applications.

Both main adsorption and desorption isotherms u_{ads} [kg.kg^{-1}] and u_{des} [kg.kg^{-1}] can be written under a single form in Eq. 6.35.

$$u_i(\varphi) = \frac{a_{GAB,i} b_{GAB,i} u_m \varphi}{(1 - a_{GAB,i}\varphi)[1 + (a_{GAB,i} - 1) b_{GAB,i}\varphi]}, i = ads \text{ or } des \tag{6.35}$$

The molecular moisture content u_m, which corresponds to the moisture content needed to cover the adsorbate surface by one layer of water molecules, allows deducing the specific surface of the material. It corresponds to the total surface of the solid matrix and is expressed in Eq. 6.36.

$$S_{GAb} = S_w N_A \frac{u_m}{M_l} \tag{6.36}$$

S_w [$\approx$ 10 Å^2] represents the surface of one adsorbed water molecule, N_A [6.02 10^{23} mol^{-1}] the Avogadro's number. M_l [0.018 kg.mol^{-1}] is the molar water mass.

The moisture content function obtained from GAB model can also be expressed in the following form (Aït Oumeziane et al. 2016a):

$$u_j(\varphi) = \frac{u_{sat}\varphi\left(1 - b_{GAB,j}\right)\left[1 + \left(a_{GAB,j} - 1\right)b_{GAB,j}\right]}{\left(1 - b_{GAB,j}\varphi\right)\left[1 + \left(a_{GAB,j} - 1\right)b_{GAB,j}\varphi\right]}, j = ads\ or\ des \tag{6.37}$$

This relation has the advantage to be expressed in function of u_{sat}, a parameter which can be more easily experimentally characterised.

Other models describe the evolution of moisture content over the whole range of relative humidity (Brooks and Corey 1964; Van Genuchten 1980). Historically developed to describe the evolution of water content in soils, the Van Genuchten's (VG) model equation is expressed in Eq. 6.38:

$$\begin{cases} S_e = \dfrac{u - u_r}{u_{sat} - u_r} = \left(1 + |\alpha h|^{\eta}\right)^{-\left(1-\frac{1}{\eta}\right)}\ for\ \varphi < 100\%RH \\ S_e = 1\ for\ \varphi = 100\%RH \end{cases} \tag{6.38}$$

S_e [−] stands for the effective saturation of the material, u [kg.kg^{-1}] the mass moisture content, u_{sat} [kg.kg^{-1}] the mass saturated moisture content and u_r [kg. kg^{-1}] the mass residual moisture content. α and η are calibration parameters which depend on the nature of the material.
h [m] stands for the capillary height and is expressed in Eq. 6.39.

$$h = \frac{RT}{M_l g}\ ln\left(\varphi\right) \tag{6.39}$$

g [9.81 m^2.s^{-1}] is the gravity acceleration and T [K] the temperature.

6.3.2.2 Moisture Transport Properties Modelling

The vapour permeability is used to assess the ability of a material to transfer water vapour. The principle of its determination consists in imposing a constant monodimensional vapour gradient to a sample. The method recommended by EN ISO 12572 is the dry and wet cup method (EN ISO 12572 2001). The sample is placed in a cup in which a saline solution fixes a relative humidity φ_i. The sample/cup assembly is then placed in another environment with a relative humidity φ_e different from the relative humidity φ_i (Fig. 6.7). The one-dimensional transfer is ensured by the lateral sealing. The experimental device diagram is presented in Fig. 6.7.

At a reference temperature, the relative humidity difference creates a partial vapour pressure gradient, driving force of the vapour flow through the thickness of the sample. The vapour density g_v [kg.m^{-2}.s^{-1}] is expressed in Eq. 6.40.

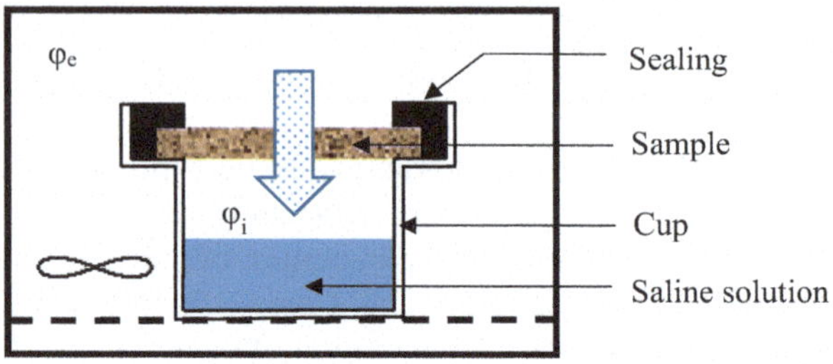

Fig. 6.7 Schematic diagram of the cup test method. The cup is placed in a relative humidity and temperature controlled room

$$g_v = \frac{\delta_p \Delta p_v}{e} \tag{6.40}$$

δ_p [kg.m^{-1}.S^{-1}.Pa^{-1}] is the vapour permeability of the material, p_v [Pa] the vapour pressure and e [m] the thickness of the sample.

The determination of the vapour flow is based on the mass monitoring of the cup/sample system in function of time. The vapour flow density can thus be also expressed by Eq. 6.41.

$$g_v = \frac{\frac{\Delta m}{\Delta t}}{A} \tag{6.41}$$

Δm [kg] is the mass variation of the system during the time period Δt [s] and A [m^2] is the average of the interior and exterior exposed surfaces. Finally, the vapour permeability is deduced in steady state from the relation:

$$\delta_p = \frac{\Delta m}{\Delta t} \frac{e}{A \Delta p_v} \tag{6.42}$$

One can note that the air layers located, both in the test cup between the sample and the saline solution and between the sample and the outside air in the climatic room, create some interface resistances to the water vapour flow. In order to limit this interface resistance, the standard suggests imposing an air flow velocity higher than 2 m.s^{-1}. Generally, this resistance is much lower than that of the sample and is neglected excepted for very permeable materials (EN ISO 12572 2001). For hemp concrete, the inside and outside interface resistances impact the value of the vapour permeability by about 10% (Seng et al. 2018b). Moreover, the vapour permeability is also influenced by the air velocity in the climatic room (Seng et al. 2018b).

Usually, only the relative humidity couple 0/50%RH is tested to measure the vapour permeability. Some values collected in the works of the literature are presented in Table 6.1.

As reported in Seng et al. (2018b), there are some significant discrepancies between the values found in literature. The operating conditions, the method of calculation of the vapour permeability with or without considering the air interface resistances and the shape of the hemp concrete samples can explain this result.

Table 6.1 Values from literature of vapour permeability obtained with the dry cup method for hemp concrete

Vapour permeability [10^{-11} kg.m^{-1}.s^{-1}.Pa^{-1}]	Reference
2.2	Rahim et al. (2016a)
4.0	Evrard (2008)
3.6–3.9	Standberg-De Bruijn and Johansson (2014)
5	Chamoin (2013)
8.6–13.1	Seng et al. (2018a, b)

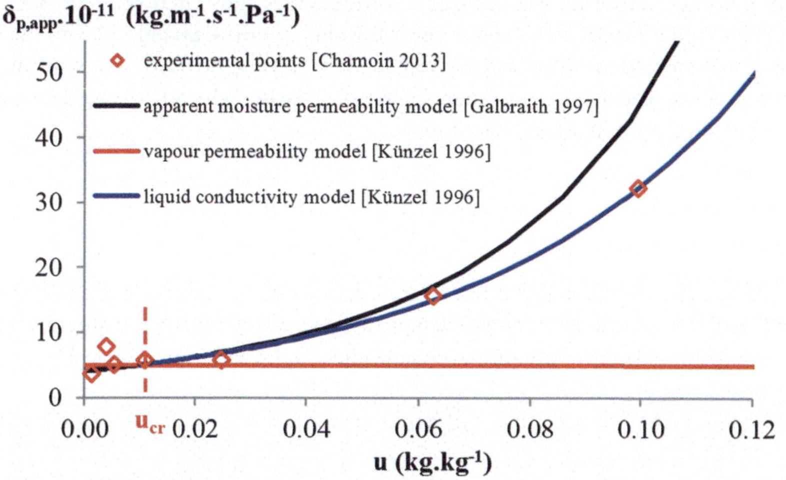

Fig. 6.8 Evolution of the apparent moisture permeability in function of moisture content: experimental results versus models

Furthermore, some experimental works on hemp concrete determined the moisture permeability for many different couples of relative humidities from the cup test method (Collet 2004; Chamoin 2013; Standberg-De Bruijn and Johansson 2014). The results presented in Fig. 6.8 come from the works of Chamoin (2013) and represent the evolution of the apparent moisture permeability in function of the moisture content. The term apparent moisture permeability comes from the fact that at high moisture contents, the moisture transfer is not only due to the vapour phase but also to the liquid phase.

As expected, for low moisture contents, the moisture permeability is found constant for hemp concrete around 5.10^{-11} kg.m^{-1}.s^{-1}.Pa^{-1} for the example presented in Fig. 6.8 (Künzel 1995; Krus 1996; Evrard 2008). Then, from the critical moisture content u_{cr}, the apparent moisture permeability highly increases: for hemp concrete with a dry bulk density $\rho_0 = 320$ kg.m^{-3}, $u_{cr} = u_{ads}$(43%RH); for a hemp concrete with $\rho_0 = 450$ kg.m^{-3}, $u_{cr} = u_{ads}$ (54%RH) (Aït Oumeziane et al. 2016b). In Evrard (2006), the author showed that a hemp concrete made with a low compaction ($\rho_0 = 308$ kg.m^{-3}) has a low vapour diffusion resistance factor of about 3.6. On the contrary, a compacted hemp concrete ($\rho_0 = 466$ kg.m^{-3}) has a high vapour diffusion resistance factor of about 7.7. This means that compaction increases the vapour diffusion resistance factor and induces a reduction of porosity. The influence of compaction on pores connexion is hard to evaluate, but it clearly appears that the open porosity increase favours moisture transport in the material.

The apparent moisture permeability, as a function of the moisture content of the material, can be described from interpolations. Its evolution can thus follow exponential or power laws (Galbraith et al. 1997).

In order to better describe the effective behaviour of the material, a phenomenological approach based on the Künzel's formalism is presented (Künzel 1995). In this approach, the liquid and vapour contributions are dissociated. At low moisture contents, the dry vapour diffusion resistance factor μ_0 can be used as a constant. The fictitious vapour diffusion resistance factor $\mu**$ is introduced to characterize the liquid transport in the hygroscopic region.

$$D_l^{\varphi} = \left(\frac{1}{\mu^{**}(u)} - \frac{1}{\mu_0}\right)\delta_{p,a}p_{sat} \tag{6.43}$$

As previously explained, the liquid transfer is initiated as soon as the moisture content of the material exceeds the value of the critical moisture content u_{cr}. The latter is determined for a moisture content value such that:

$$\mu^{**}(u_{cr}) = \mu_0 \tag{6.44}$$

Figure 6.8 compares the apparent moisture permeability Galbraith's model with the Kunzel's model.

Both models give results close to each other at low moisture contents but the Kunzel's model appears more suited to the experimental data in the hygroscopic region to describe the liquid diffusion.

The liquid transport coefficient D_l [m^2.s^{-1}] can also be determined from the liquid water absorption coefficient A_w [kg.m^{-2}.s$^{-0.5}$] (Künzel 1995). A_w is determined according to EN ISO 15148 (EN ISO 15148 2002). The evolution of D_l in function of moisture content is defined by Eq. 6.45.

$$D_l = 3.8\left(\frac{A_w}{\rho_0 u_{sat}}\right)^2 1000^{\frac{u}{u_{sat}}-1} \tag{6.45}$$

The isothermal total diffusion coefficient D_t [m^2.s^{-1}], sum of the vapour and liquid coefficients, is expressed in Eq. 6.46 and represented in Fig. 6.9:

$$D_t = D_v + D_l == \frac{\delta_p p_{sat}}{\rho_0 \frac{du}{d\varphi}} + \frac{D_{\varphi,l}}{\rho_0 \frac{du}{d\varphi}} \tag{6.46}$$

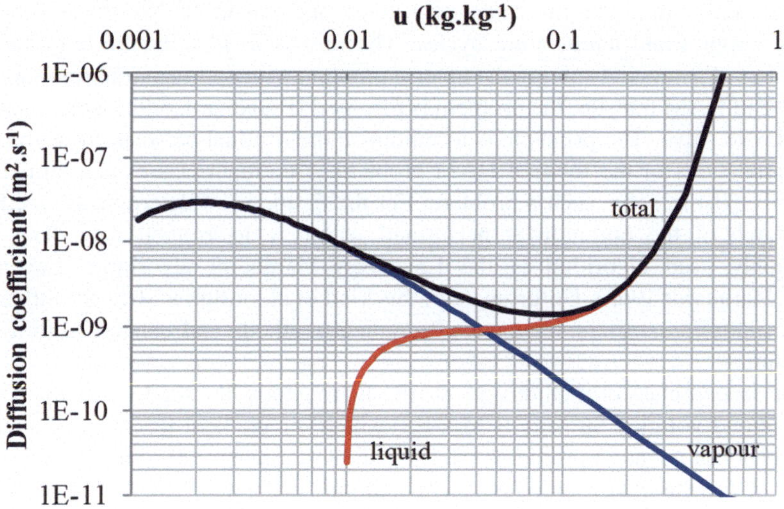

Fig. 6.9 Evolution of the isothermal vapour, liquid and total diffusion coefficient

According to the theory of De Vries (1958), the global evolution of the isothermal diffusion coefficients as a function of the moisture content, presented in Fig. 6.9, can be divided into three main steps. At low moisture contents, below the critical moisture content u_{cr}, moisture is only transferred under vapour phase. The liquid condensed phase is present in adsorbed form or in capillary islands. For higher moisture contents, capillary islands increase in size and in number. The flow section of the vapour decreases and the vapour diffuses through condensation-evaporation mechanisms at the liquid-vapour interfaces. Water in liquid phase also diffuses on the surface of the pores. At high moisture contents, as long as the continuity of the liquid phase is ensured, the transfer into the liquid phase becomes largely predominant and the isothermal liquid diffusion coefficient highly increases.

6.3.2.3 Moisture Buffer Capacity

The Nordtest protocol proposed a methodology to quantify the moisture buffer performance of building materials (Rode 2005). The samples are first sealed on all surfaces except one. After stabilization at 23 °C and 50%RH, the samples are subjected to alternate adsorption/desorption cycles: 8 h at 75%RH followed by 16 h at 33%RH. These conditions are representative of typical daily variations likely to be experienced by a building under isothermal conditions.

Compared to other distributed insulation materials (aerated cellular concrete, earthen bricks), hemp concrete is qualified as an excellent moisture regulator (Collet and Pretot 2012; Collet et al. 2013; Rahim et al. 2015) with a moisture buffer value (MBV) above 2 $g.m^{-2}.\%RH^{-1}$, according to the Nordtest classification (Rode 2005).

In our opinion, it could be interesting to use and extend the Nordtest protocol to others sorption and temperature cycles. The idea is to identify the hygrothermal properties of a material solving the inverse problem with optimization techniques. In the case of HAM transfer, the problem is ill-posed (Hadamard 1902) which makes it complex to solve. The principle is to minimize the residual between measurements and simulations of the forward HAM problem. Two main classes of optimization methods exist to solve such a problem. For the first kind of deterministic methods, the search of the optimum is performed within a determined path. The most commonly used algorithm is the Levenberg-Marquardt algorithm (Levenberg 1944; Marquardt 1963). Unfortunately, this kind of algorithm is strongly influenced by the initial parameters vector as all gradient methods and converges to a local optimum.

The second class of methods are stochastic methods like the Covariance Matrix Adaptation Evolution Strategy (CMA-ES) which is a gradient-free method suited to non-linear optimization problems (Hansen and Ostermeier 2001). It is based on the theory of evolution: new individuals are randomly generated. The evolution strategy consists in randomly generating new individuals from the parents. The recombination and mutation are used to select new individuals with better values to minimize the objective function. The performance of CMA-ES to identify the hygrothermal parameters in the framework of building physics applications has been shown in literature (Rouchier et al. 2015; Rouchier 2018). Even though the use of stochastic methods are longer in term of simulation time than deterministic methods, they are more accurate.

The main advantage of adopting a new protocol could be to faster fully characterize the hygrothermal properties of hemp concrete and building materials than the standard time-consuming experimental methods of characterization. Nevertheless, an accurate description of the hygrothermal transfer and mechanism is needed.

6.3.3 Thermal Properties

6.3.3.1 Thermal Conductivity Modelling

The thermal insulating capacity of a material is characterized by the thermal conductivity λ ($W.m^{-1}.K^{-1}$). There are different experimental methods for determining the thermal conductivity of a material. In general, there is a distinction between measurement methods carried out in steady state and those carried out under transient conditions. In steady state, the measurement consists in imposing constant unidirectional heat flow through a sample and measuring the temperatures in two distinct positions in the flow direction. The application of the Fourier's law thus allows to determine the thermal conductivity. The experimental devices developed in this approach are the well-known guarded or bi-guarded hot plate for example. Nevertheless, these methods do not allow evaluating the evolution of the thermal conductivity under different hygric states as moisture transfer would occur simultaneously to thermal transfer in such conditions because a long period is required to reach steady state.

Under transient conditions, the measurement principle is based on a localized perturbation in time and space of the specimen. The thermal characteristics are then determined from thermograms representing the evolution of temperature as a function of time. The methods used require the simultaneous determination of several parameters. In addition, the transient measurement methods offer greater simplicity of apparatus, a better speed of determination of the parameters and allow measurements in any hygric state of the sample as the solicitation is too short to induce moisture flow during the measurement. For example, the Hot Disk method is a fast and accurate transient plane source widely used to characterize and analyse the evolution of thermal properties of construction materials (Gustafsson 1991). The principle of the method is to deliver a small constant current during a short time in order to increase of a few degrees the temperature of the material. This temperature increase depends on the thermal transport properties of the material surrounding the sensor. By monitoring this temperature increase over a short period after the start of the experiment, it is possible to obtain the equivalent thermal conductivity and the thermal diffusivity α [$m^2.s^{-1}$] expressed in Eq. 6.47:

$$\alpha = \frac{\lambda^*}{\rho c_p^*} \tag{6.47}$$

In this relation, $\lambda*$ [$W.m^{-1}.K^{-1}$] and c_p* [$J.kg^{-1}.K^{-1}$] are respectively the equivalent thermal conductivity and the equivalent specific heat capacity. They depend on the moisture content in the material.

In the case of measurement on hemp concrete, the size of the probe (hot disk or hot wire) must be selected regarding the REV and the shiv size. The representativeness of the measurement must be carefully evaluated.

From the works of literature, Seng et al. (2018a) observed an almost linear trend between the evolution of the thermal conductivity and the dry density. Indeed, the dry thermal conductivity increases from about 0.08 $W.m^{-1}.K^{-1}$ for the lightest hemp concretes under 350 $kg.m^{-3}$ to about an average value of 0.13 $W.m^{-1}.K.^{-1}$ for the heaviest above 650 $kg.m^{-3}$. It is nevertheless good to remain cautious in the interpretation of these results. Indeed, the measured values found in the literature are comparable only if the compositions of the hemp concrete are identical, which is not the case.

It has been shown in literature that the hemp concrete composition has also a significant influence on the thermal conductivity (Collet and Pretot 2014; Somé et al. 2018). The type of binder, the hemp shiv type (fibered or defibered), the hemp shiv/binder ratio and the method of compaction play a role in the measured values of thermal conductivity.

The thermal conductivity is sensitive to moisture: the higher the moisture content in the material, the higher the thermal conductivity, which weakens its insulating power. The equivalent thermal conductivity of the moist material is usually expressed by Eq. 6.49 and shows the linear dependence of the thermal conductivity to the moisture content (Künzel 1995).

$$\lambda^* = \lambda_0(1 + \beta u) \tag{6.48}$$

λ_0 [W.m^{-1}.K^{-1}] is the dry thermal conductivity and β [−] a fitting parameter.

According to the very first works led on the influence of moisture on thermal conductivity for intermediate dry densities around 450 kg.m^{-3}, the conductivity increases by 10%, going from 0.1 to 0.11 W.m^{-1}.K^{-1} between 0 and 50%RH (Cerezo 2005). It reaches 0.13 Wm^{-1}.K^{-1} for a relative humidity of 75%. The last experimental works confirm that the thermal conductivity increase linearly with moisture content (Collet and Pretot 2014; Gourlay et al. 2017; Seng et al. 2018a; Somé et al. 2018).

A linear relation is experimentally observed between thermal conductivity and temperature (Pierre et al. 2014; Rahim et al. 2016b; Seng et al. 2018a). Moreover, the orientation of the hemp fibres in the binder matrix seems to have of an effect on the measured values of thermal conductivity related to the heat flow direction (Pierre et al. 2014). The increase of thermal conductivity with the temperature measured in Seng et al. (2018a) and Pierre et al. (2014) in "perpendicular" configuration is twice as high as those measured by Rahim et al. (2016b) and Pierre et al. (2014) in "parallel" configuration. According to Tran Le et al. (2019), the anisotropic character of hemp concrete has to be considered. The original model of effective thermal conductivity developed in this study from a multi-scale homogenization approach appears promising to better explain the observed experimental differences (Williams et al. 2017).

6.3.3.2 Specific Heat Capacity Modelling

The specific heat capacity quantifies the ability of the material to store heat. As previously presented, the specific heat capacity can be determined from indirect measurements based on thermal conductivity and thermal diffusivity measurement like the Hot Disk method. It can be also measured experimentally using calorimeter, DTA (Differential Thermal Analysis) or DSC (Differential Scanning Calorimetry) methods.

A large dispersion of the measured dry specific heat capacity of hemp concrete is observed in literature. Table 6.2 provides some values found in literature. These differences are mainly explained by the different calculation methods and the composition of the hemp concrete samples (Seng et al. 2018a).

Otherwise, as reported by Aït Oumeziane (2013), Seng et al. (2018a) suggest to express the dry specific heat capacity of hemp concrete in function of its proportion in hemp shiv and binder.

Table 6.2 Values from literature of dry specific heat capacity obtained with different methods

Dry specific heat capacity [J.kg^{-1}.K^{-1}]	Method	Dry density [kg.m^{-3}]	Reference
1000	DSC	450	Lelievre et al. (2014)
905	DSC	466	Seng et al. (2018b)
911	Calorimeter	466	Seng et al. (2018b)
1240–1350	Calorimeter	508–627	Walker and Pavía (2014)
650–900	Hot disk	340–415	Gourlay et al. (2017)
1560	Calorimeter	440	Evrard (2008)

$$c_{p,HLC,0} = X_{HS}c_{p,HS} + X_B c_{p,B} \tag{6.49}$$

In Eq. 6.49, X_{HS} [−] stands for the mass proportion of hemp shiv and X_B [−] the mass proportion of binder. $c_{p,HS}$ [J.kg^{-1}.K^{-1}] the specific heat capacity of hemp shiv is about 1250 J.kg^{-1}.K^{-1} and $c_{p,B}$ [J.kg^{-1}.K^{-1}] the specific heat capacity of the lime is about 900 J.kg^{-1}.K^{-1}.

As highlighted in the last experimental works, the specific heat capacity of hemp concrete increases with moisture content (Pierre et al. 2014; Gourlay et al. 2017).

The equivalent specific heat capacity is usually expressed by Eq. 6.50 (Evrard 2008).

$$c_p^* = c_{p,0} + c_l u \tag{6.50}$$

with $c_{p,0}$ [J.kg^{-1}.K^{-1}] and c_l [4180 J.kg^{-1}.K^{-1}] respectively the dry specific heat capacity of the material and the liquid water specific heat capacity.

A close expression is proposed in (Seng et al. 2018a).

$$c_p^* = \frac{1}{1+u}c_{p,0} + \frac{u}{1+u}c_l \tag{6.51}$$

Finally, from experimental data analysis, the influence of moisture content on specific heat capacity is demonstrated to be predominant compared to that of temperature (Seng et al. 2018a).

6.4 The Hysteresis Phenomenon

6.4.1 Phenomenology and Definitions

As observed for many hygroscopic materials, a significant hysteresis occurs in the sorption process of hemp concrete (Fig. 6.6). At a given relative humidity, various levels of differences are observed between the equilibrium moisture content obtained

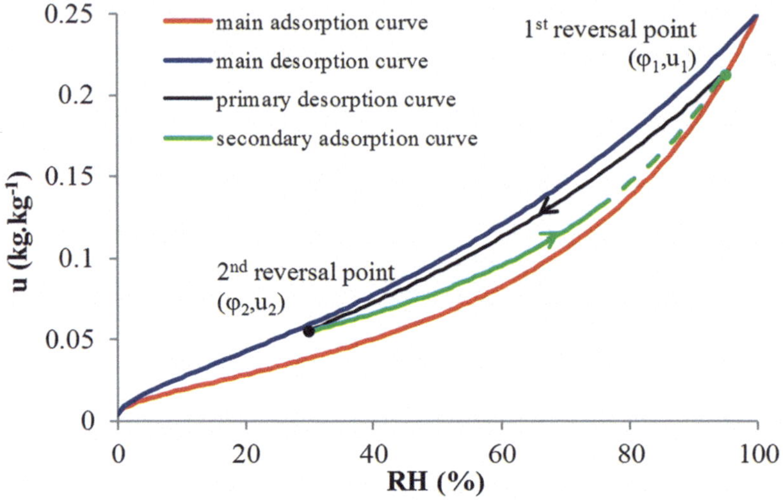

Fig. 6.10 Schematic representation of an intermediate adsorption/desorption cycle. Note that the slope, which characterizes the hygric capacity, of the intermediate scanning curves is reduced in comparison with the slope of the main curves

during adsorption or desorption. This is known as the hysteresis effect. The hysteresis phenomenon is mainly attributed to the pore space morphology ("ink bottle effect"), the effects of spatial arrangement of pores and variation in liquid–solid contact angle.

As an illustration, Fig. 6.10 provides a schematic representation of the dependence of the moisture content using a sequence of relative humidity variations. Starting from any equilibrium state on the main adsorption isotherm, a decrease of relative humidity before reaching the saturated state results in an evolution along a so-called primary scanning curve. Starting from any equilibrium state on this primary scanning curve, an increase of relative humidity before reaching the dry state results in an evolution along a so-called secondary scanning curve. The shift between adsorption and desorption curves is called reversal point.

6.4.2 Modelling of the Hysteresis Phenomenon

Historically, hysteresis models have been first developed in literature for ferromagnetic applications (Preisach 1935; Néel 1942, 1943; Everett 1955). In these works, the medium is considered as an infinite number of independent domains connected in parallel. These models were then adapted for soil physics applications (Poulovassilis 1962). The soils are described as an assemblage of pore space elements which are assumed to be filled or emptied independently of each other.

Topp and Miller (1966) and Everett (1967) have highlighted the limits of the previous models and proposed the dependent domain theory.

In this framework, Mualem (1974) has developed a phenomenological hysteresis model using the ink bottle concept. A special space is introduced and defined by two functions f and a. The function f describes the distribution of the possible sorption sites, and the normalised function a describes the accessibility to the sorption sites. The moisture content function is defined by the integration of the functions f (filled sorption sites) and a in the $f - a$ space also called Mualem's space. This model explains the shape of the scanning curves by means of physical properties such as the distribution of pore radii. Later studies give more accurate results by considering the phenomenon of blockage against air (and water) entry (Mualem 1977; Mualem 1984; Mualem 2009).

Significant progress in modelling the hysteresis of building materials has been made during the last few years. Empirical models like the Pedersen's model use the weighted values of the hydric capacity to model the scanning curves (Pedersen 1990). The implementation of the Pedersen's model requires the determination of the experimental main adsorption and desorption curves as well as primary adsorption and desorption scanning curves. Unlike other mathematical and phenomenological models, the closure condition between desorption and adsorption curves is not respected, whatever the values of the chosen coefficients. The mathematical formalism of the Pedersen's model explains this particularity. For cyclic relative humidity variations, the Pedersen's model leads to an overestimation of the moisture content. Thus, from a given relative humidity, the adsorption scanning curve is found above the complementary desorption scanning curve. It is a mathematical artefact commonly known as "pump effect" which is not physically admissible (Huang et al. 2005; Mualem 2009). The disrespect of this closure condition raises a significant problem in justifying the use of the Pedersen's model. Carmeliet et al. (2005) pointed out that like any empirical approach, this model does not have physical background and shows a lack of flexibility.

In their approach, Carmeliet et al. (2005) have adapted the Mualem's model to the case of building materials. The moisture content depends on relative humidity instead of capillary suction height usually chosen in soil physics. This model requires only the knowledge of the main adsorption and desorption curves.

An adsorption scanning curve (indexed i) after a series of alternating processes of desorption and adsorption is described by:

$$u(\varphi, i) = u_{i-1} + (1 - A(\varphi_{i-1}))(u_{ads}(\varphi) - u_{ads}(\varphi_{i-1})) \tag{6.52}$$

where (φ_{i-1}, u_{i-1}) refers to the relative humidity and moisture content of the reversal point i.

A desorption scanning curve (indexed i) after a series of alternating processes of desorption and adsorption is defined by:

$$u(\varphi, i) = u_{i-1} - (1 - A(\varphi))(u_{ads}(\varphi_{i-1}) - u_{ads}(\varphi)) \tag{6.53}$$

The function *A*, primitive of the previously introduced function *a*, is given by:

$$A(\varphi) = \frac{u_{des}(\varphi) - u_{ads}(\varphi)}{u_{sat} - u_{ads}(\varphi)} \tag{6.54}$$

The subscripts *ads*, *des* and *sat* refer respectively to the main adsorption curve, the main desorption curve and the saturated moisture content.

A second approach proposed by Scott et al. (1983) is based on a mathematical description of the sorption process. It describes the adsorption scanning curves with the same shape parameters as those for the main adsorption curve. The same assumption is done for the desorption process. Kool and Parker (1987) assume that one of the two shape parameters is the same for both adsorption and desorption processes. Parker and Lenhard (1987) and Huang et al. (2005) have developed their models by using this assumption. These models achieve perfect closure at scanning curves reversal points.

Like the Carmeliet's model, the Huang's model has been adapted for building materials by using relative humidity instead of capillary suction. Equations 6.55 and 6.56 describe respectively the adsorption and desorption scanning curves after a series of alternating processes of desorption and adsorption:

$$u(\varphi, i) = u_r(i) + (u_s(i) - u_r(i)) \frac{u_{ads}(\varphi)}{u_{sat}} \tag{6.55}$$

$$u(\varphi, i) = u_r(i) + (u_s(i) - u_r(i)) \frac{u_{des}(\varphi)}{u_{sat}} \tag{6.56}$$

$u_r(i)$ and $u_s(i)$ are respectively residual and saturated moisture contents of the scanning curve indexed *i*. The main adsorption and desorption curves give residual and saturated moisture content values at order *0*: $u_r(0) = 0$ and $u_s(0) = u_{sat}$. The calculation of these parameters is based on the perfect closure of the scanning curve at reversal points. Scanning curve indexed *i* includes the last reversal point (φ_i, u_i) and the penultimate reversal point (φ_{i-1}, u_{i-1}). Calculating $u_r(i)$ and $u_s(i)$ finally leads to solve a linear system of two equations.

Mathematical and phenomenological hysteresis models present common features. Both only require the knowledge of the main adsorption and desorption curves and achieve perfect closure at reversal points during a sequence of relative humidity variations. As shown in Aït Oumeziane (2013), the mathematical expressions of adsorption scanning curves are the same for both models. Two differences are noted. First, Carmeliet's model needs only one reversal point whereas Huang's requires two. Then, some differences are found in the mathematical expressions of desorption scanning curves between both hysteresis models.

6.4.3 Application to Hemp Concrete

The Huang's model and Carmeliet's model associated with the Van Genuchten's model of sorption were compared in (Aït Oumeziane et al. 2014). The original method developed to fit the models to experimental data leads to the identification of the main desorption curve (0–100%RH), considering the experimental dataset (0–97% RH) as a primary scanning curve. This approach is helpful because it makes the experimental determination of the main desorption curve unnecessary. For the studied hemp concrete, the Carmeliet's model fails to reproduce realistic moisture content for the main desorption curve especially at low relative humidities, whereas Huang's model leads to decrease more significantly the hygric capacity. This latter model seems thus to be the most relevant to predict the hysteretic behaviour of hemp concrete in sorption process.

Like Carmeliet et al. (2005), some authors have chosen to use the Mualem's model adapted to building materials to describe the hysteresis phenomenon which occurs in hemp concrete. Associated with the GAB sorption model, they have evaluated the performance of the model under transient conditions (Lelievre et al. 2014; Promis et al. 2018). Contrary to Aït Oumeziane et al. (2014), they did not test its performance to reproduce primary or secondary scanning curves. Nevertheless, their results are discussed in Sect. 6.6.

6.5 Effect of Temperature on Sorption Mechanism

6.5.1 Modelling Approach

The effect of temperature variations on the relative humidity evolution inside hemp concrete is not intuitive. Indeed, while the relative humidity tends to decrease with increasing temperature in air, it tends to increase in hemp concrete (Colinart et al. 2017; Moujalled et al. 2018). This effect is explained by the temperature dependence of the sorption mechanism.

Three main theories try to explain the influence of temperature on sorption isotherms. The modification of the pore structure due to temperature is one explication. Thus, the determination of specific surface areas and porosimetry analysis on cement-based materials were performed at different temperatures (Radjy and Richards 1973; Bray and Sellevold 1973). These studies have shown that a temperature increase between ambient temperature and 90 °C or 100 °C results in a reduction of the specific surface area. The comparison of the pore size distributions showed the presence of pores with larger diameters at 80 °C than at 30 °C. A modification of the microstructure by an enlargement of pores can thus be attributed to a rise of temperature. This can explain the modification of the sorption isotherms shapes until 60 °C (Radjy et al. 2003). Nevertheless, this approach turns out to be insufficient to explain the decrease of moisture content at saturation observed in a large range of temperature (between 20 °C and 80 °C) for cement based materials (Poyet 2009). To overcome the limitations of the microstructure alteration with the increase of temperature, the modification of water thermophysical properties with temperature was also investigated together with coarsening of the pore structure (Bazant and Thonguthai 1978; Bazant and Thonguthai 1979) or without (Milly 1984).

Milly (1984) has developed its model from the differential equation relying on moisture content and temperature:

$$\left.\frac{\partial u}{\partial T}\right|_{\varphi} = \left.\frac{\partial u}{\partial \varphi}\right|_{T} \left.\frac{\partial \varphi}{\partial T}\right|_{u} \tag{6.57}$$

The first term of the second member of Eq. 6.57 stands for the hygric capacity ie. the slope of the sorption isotherm at a given temperature. The term $\left.\frac{\partial \varphi}{\partial T}\right|_{u}$ is calculated for a constant moisture content. Milly (1984) suggests that this term follows an exponential evolution.

$$\varphi_2(u, T_2) = \varphi_1(u, T_1) e^{C_\varphi (T_2 - T_1)} \tag{6.58}$$

with C_φ [K^{-1}]:

$$C_\varphi = \frac{1}{\varphi} \left.\frac{\partial \varphi}{\partial T}\right|_{u} \tag{6.59}$$

The combination of the Young-Laplace and Kelvin laws allows expressing the relative humidity φ as a function of the water thermophysical characteristics (surface tension σ [N.m^{-1}] and density ρ_w [kg.m^{-3}]) and of the structural characteristics of the material (pore radius r [m] and contact angle θ [°]):

$$= e^{-\frac{2\sigma \cos\theta}{\rho_l \frac{RT_{ref}}{M_l} r}} \tag{6.60}$$

According to Milly (1984), the effect of the temperature on the pore structure is assumed to be negligible. The relative humidity/temperature relationship at a given moisture content is only related to the dependence of the water physical properties on temperature:

$$C_\varphi = -\left(\frac{1}{\sigma}\frac{\partial \sigma}{\partial T} - \frac{1}{\rho_l}\frac{\partial \rho_l}{\partial T}\right) \tag{6.61}$$

Poyet (2009) suggested that the microstructure alteration and the water properties evolution had negligible effects and did not fully explain the results obtained for concrete. He proposed a new approach based on the thermodynamic evolution of sorption mechanism. He based his approach on the exothermic process of adsorption (Brunauer 1945) and on the principle of Le Chatelier and the Van't Hoff law. According to these principles, increasing temperature promotes the reverse process of adsorption, namely desorption.

The mathematical formalism is based on the description of the evolution of the sorption heat as a function of the amount of the adsorbed moisture. This method allows, on the basis of the results obtained at only two temperatures, a reliable estimation of sorption isotherms at any other temperature.

The heat involved in a phase change for a system in equilibrium can be described by the Clausius-Clapeyron relation (Brunauer 1945):

$$q_{st} = \frac{Q_{st}}{M_l} = -\frac{R}{M_l}\left[\frac{\partial\, ln\,(p_v)}{\partial\left(\frac{1}{T}\right)}\right]_u \tag{6.62}$$

Q_{st} [J.mol^{-1}] is the molar isosteric heat of sorption, q_{st} [J.kg^{-1}] the mass isosteric heat of sorption. p_v [Pa] the partial vapour pressure. The index u means that this relation is available at constant equilibrium moisture content. The isosteric heat of adsorption Q_{st} stands for the amount of energy involved in the sorption process. The validity of this relation implies that microstructure alteration and the water properties evolution with temperature are negligible.

The integration of Eq. 6.62 between two equilibrium states (p_{v1}, T_1) and (p_{v2}, T_2) gives the expression:

$$q_{st} = \frac{R}{M_l}\frac{T_1 T_2}{T_2 - T_1}\, ln\left[\frac{p_v(T_2, u)}{p_v(T_1, u)}\right] \tag{6.63}$$

The isosteric heat of sorption describes a decreasing curve versus moisture content which tends towards the latent heat of condensation at high moisture contents. A schematic evolution of the isosteric heat is provided in Fig. 6.11. It reaches a maximum value at low moisture content. This maximal isoteric heat gives the "physisorption order of magnitude" of a material which means the strongest interactions between the water molecules and the adsorbate surface (Drouet 2010). According to Poyet and Charles (2009), the isosteric heat decrease corresponds to the decreasing interactions between water molecules and the adsorbate. At high moisture contents, it can be assumed that the adsorbate does not influence the sorption process any more. This explains the value of isosteric heat at high moisture content which tends towards latent heat of condensation. According to Poyet and Charles (2009), the isosteric heat evolution and the maximal isosteric heat are specific characteristics of a material.

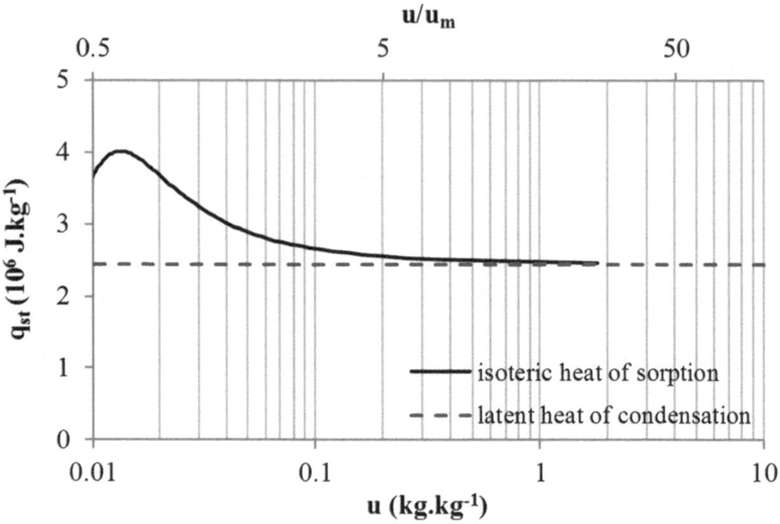

Fig. 6.11 Evolution of the isosteric heat of sorption. The isosteric heat increases at low moisture contents to a maximum value then decreases and tends to the latent heat of condensation at high moisture contents

In practice, the isosteric heat can be experimentally determined by calorimetry (Powers and Brownyard 1948) or sorption isotherms derivation (Poyet and Charles 2009).

Many different approaches can be used to model the isosteric heat of sorption. First approaches suggested to fit an exponential or power law to the experimental data. A full description of the models found in literature is given in (Chen 2006).

The isosteric heat of sorption can be also obtained using the fitted sorption isotherms functions. If a function f describes the sorption isotherm function by the relation $u = f(\varphi)$ at a given temperature T, at two different temperatures T_1 and T_2, corresponding relative humidities can be reciprocally determined as follows:

$$\begin{aligned} at\ T_1, \varphi_1 &= f_1^{-1}(u) \\ at\ T_2, \varphi_2 &= f_2^{-1}(u) \end{aligned} \tag{6.64}$$

The substitution of φ_1 and φ_2 in Eq. 6.63 gives the expression of isosteric heat.

Knowing the evolution of the isosteric heat of sorption, it is then possible to predict the evolution of sorption isotherms for any temperature. Integrating Eq. 6.62 between (T_1, p_{v1}) and (T_2, p_{v2}) at constant equilibrium moisture content gives the relation:

$$\varphi(T_2, u) = \varphi(T_1, u) \frac{p_{sat}(T_1)}{p_{sat}(T_2)} e^{q_{st}(u) \frac{M_l}{R} \frac{T_2 - T_1}{T_1 T_2}} \tag{6.65}$$

In their approach, Staudt et al. (2013a) expressed the parameter a_{BET} of the BET model by an Arrhenius-like temperature function. They adopted a similar approach concerning the GAB model and assumed that only the parameter b_{GAB} depends on temperature and thus can be expressed through an Arrhenius-like expression (Staudt et al. 2013b). The expression of the parameters b_{GAB} is given in Eq. 6.66.

$$a_{GAB} = a_{GAB,0} e^{\frac{Q}{R}\left(\frac{1}{T_1}-\frac{1}{T_2}\right)} \tag{6.66}$$

From the determination of two main sorption isotherms at two different temperatures, the sorption energy Q is calculated by inversing Eq. 6.66.

Finally, Promis et al. (2019) proposed a physically modified version of the GAB model. As introduced in Sect. 6.3, they expressed the temperature-dependency of the GAB parameters in Eq. 6.67.

$$\begin{cases} u_m = u_{m,0} e^{\frac{q_m}{RT}} \\ a_{GAB} = a_{GAB,0} e^{\left(\frac{E_{ads,1}-E_{ads,m}}{RT}\right)} \\ b_{GAB} = b_{GAB,0} e^{\left(\frac{E_{lv}-E_{ads,m}}{RT}\right)} \end{cases} \tag{6.67}$$

The parameters $u_{m,0}$, [kg.kg^{-1}] $a_{GAB,0}$ [−] and $b_{GAB,0}$ [−] are constant factors. q_m is a constant fitted parameter and E_1, E_m and E_{lv} [J.mol^{-1}] are respectively the molar heat of sorption of one layer of water molecules, the multilayers molar heat of sorption and the molar latent heat of condensation. These six unknown parameters are estimated from the knowledge of two main sorption isotherms determined at two different temperatures.

6.5.2 Application to Hemp Concrete

The latest studies performed for cement-based materials and ordinary concrete (Hundt and Kantelberg 1978; Daïan 1988; Poyet 2009; Brue 2009; Drouet 2010) or organic-based products (McLaughlin and Magee 1998; McMinn and Magee 2003; Jamali et al. 2006; Yan et al. 2008) have highlighted that the higher the temperature, the lower the equilibrium moisture content at the same relative humidity.

Otherwise, research on the coupling between temperature-dependence and hysteresis are rather scarce in literature. Rode and Hansen (2011) provided some measurements for several building materials (cement paste, spruce, aerated concrete). For the three temperatures investigated (10 °C, 25 °C and 40 °C), no significant differences were observed for aerated concrete in both adsorption and desorption for a range of relative humidity between 10 and 80%RH. For very high relative humidities over 80%RH and very low below 10%RH, more differences are observed between each sorption isotherms. Furthermore, the results obtained by Rode and Hansen (2011) are in accordance with the hypothesis of Ishida, which assumes that an increase of temperature results in a reduction of the hysteresis loop (Ishida et al. 2007).

As expected, experimental moisture content measurements performed respectively at 10 °C and 23 °C in (Aït Oumeziane et al. 2016a; Promis et al. 2019) and at 23 °C and 40 °C in (Fabbri and McGregor 2017) have shown that a decrease of temperature results in an increase of moisture content. The Ishida's hypothesis is also verified. Moreover, at 10 °C as well as at 23 °C, experimental scanning curves achieved on hemp concrete show that the hygric capacity is significantly reduced along a scanning curve and pointed out that hysteresis and temperature have significant influence on the storage or release of moisture in hemp concrete (Aït Oumeziane et al. 2016a).

Associated with the GAB model, the Poyet's model showed its performance to describe the evolution of the isosteric heat (Aït Oumeziane et al. 2016a; Promis et al. 2019). Promis et al. (2019) have also shown the consistency of the the Staudt's model.

As illustration, the evolution of the isosteric heat is plotted on a dual scale in function of u and of the ratio u/u_m in Fig. 6.11.

The isosteric heat increases from the dry state to a maximal value then decreases and tends to the latent heat of condensation at a rate u/u_m about 5 (Aït Oumeziane et al. 2016a). According to Promis et al. (2019) for very low moisture contents, a first phase of sorption called chemical adsorption or chemisorption is observed. During this phase where water is chemically bound to the adsorbate through strong interactions, the isosteric heat increases until a maximum value observed at low moisture content between about 4 MJ.kg^{-1} (Aït Oumeziane et al. 2016a) and 6.5 MJ.kg^{-1} (Promis et al. 2019). This maximum corresponds to the bond energy between water molecules and the adsorbate surface created by the Van der Walls interactions. This value depends on the nature of the material. The decrease of the isosteric heat highlights the reducing influence of the adsorbate at higher moisture contents explained by an increasing distance between the water molecules and the adsorbate surface. This phase is called physical adsorption or physisorption. Otherwise, a significant difference is found for the isosteric heat evolution between adsorption and desorption phases (Promis et al. 2019). The authors have concluded that the Poyet's and modified GAB models are the most suited to predict the sorption isotherms.

6.6 Applications

6.6.1 Hygrothermal Response Under Laboratory-Controlled Climatic Conditions

Two main kinds of studies are led to evaluate the performance of hemp concrete subjected to transient climatic conditions. First, as for MBV evaluation, the experiments are launched at the scale of the specimen in a climatic room. The main purpose of this test is to follow the variations of the moist mass. Others studies are performed at the scale of a wall placed in a biclimatic room able to impose different climatic conditions on both sides of the wall. This test configuration is usually favoured to analyse the hygrothermal response of hemp concrete in conditions close to reality.

The very first hygrothermal response to climatic variations studies were led at wall scale (Samri 2008; Evrard 2008; Tran Le 2011). Based on the results of experimental characterization campaigns, these authors have first modelled the hemp concrete hygrothermal behaviour. This data allowed them to implement a coupled heat and moisture transfer model based on the conservation principle described in Sect. 6.2. Samri (2008) has used the software Comsol Multiphysics 3.2., Evrard (2008) Wufi 4.1 Pro and Tran Le (2011) the Spark environment.

In parallel to his numerical approach, Samri (2008) also carried out an experimental study at the scale of a wall in order to compare the numerical results with the collected experimental results. Nevertheless, he underlined the insufficiency of his approach especially under transient conditions by the use of constant fitted hygrothermal properties independent of the moisture content evolution and by not considering the hysteresis phenomenon.

Evrard (2008) exclusively focused on a numerical approach in the study of the hygrothermal response of a hemp concrete wall under climatic conditions. He was interested in the temperature and relative humidity distributions inside a hemp concrete wall and compared them with the response of other materials. These walls were studied in a configuration of use, which means that the walls were coated. Evrard (2008) analysed separately the effect of gradients of temperature and relative humidity on the behaviour of the studied materials first daily then yearly. Evrard (2008) favoured a comparative approach to the detriment of an analysis of the effective hygrothermal behaviour of hemp concrete. Although hemp concrete was not distinguished from other materials by its insulating capacity, Evrard (2008) showed that hemp concrete thanks to its hygrothermal regulation performance stands out by the quality of comfort that it allows. Moreover, the simulated moisture buffer capacity presented by Evrard (2008) was obtained from numerical results that did not exactly match the experimental results in terms of moisture mass evolution. The discrepancies observed did not invalidate Evrard's conclusions but demonstrated a poor modelling of the evolution of moisture content of hemp concrete and therefore its hygrothermal behaviour and its response to climatic variations at wall scale.

In order to overpass the limitations of these first works, the later studies focused on the consideration of hysteresis and temperature effect on sorption process in HAM transfer model. Indeed, HAM transfer simulations through building walls often use simplified modelling of sorption process. The relation between moisture content and relative humidity is thus usually modelled by a single curve either the adsorption isotherm or an average of the adsorption and desorption isotherms. Consequently, the hygric capacity of the material is overestimated. This results in inaccurate predictions of the moisture transfer. Steeman et al. (2009) and Van Belleghem et al. (2010) showed that the use of a hysteresis model improves the numerical simulation results providing a more accurate relative humidity evolution in gypsum board.

The influence of hysteresis in hemp concrete was the first analysed. Lelievre et al. (2014) worked on a specimen located in a climatic room. The simulations performed with Mualem's hysteresis model showed a good agreement with the relative humidity and temperature measured within the specimen. According to the authors, some differences still remained between the simulated and measured relative humidity deeper within the material. One reason was that hemp concrete was not homogeneous throughout its thickness due to the manufacturing process. Moreover, the studied hemp concrete specimen was coated with permeable and hygroscopic plaster inside and impermeable plaster outside, creating a multi-layered wall. Nevertheless, these coated layers were not as accurately modelled as hemp concrete and could also explain the differences between experimental ad numerical results.

Colinart et al. (2016) investigated the hygrothermal behaviour of coated hemp concrete at the scale of the specimen and at the scale of the wall. The investigation performed at the scale of the wall revealed that the consideration of hysteresis improves the prediction of the relative humidity distribution within a wall subjected to isothermal and non-isothermal conditions. They have experimentally shown that coating hemp concrete with a less hygroscopic plaster than hemp concrete reduced its moisture buffer capacity. However, they concluded that it could be interesting to consider the hygrothermal history of the wall in order to set initial moisture content.

By comparing experimental and numerical hygrothermal responses with and without hysteresis modelling of a hemp concrete wall located in a biclimatic room, Aït Oumeziane et al. (2016b) confirmed that the simple use of the main adsorption curve is insufficient to describe the effective evolution of the moisture content. Considering the hysteresis phenomenon with a model suited to hemp concrete has significantly improved the numerical results. A more accurate estimation of moisture storage improved both temperature and relative humidity predictions within the wall. The improved evaluation of the relative humidity distribution allowed a better representation of vapour pressure gradients which are driving force of heat transport by evaporation and condensation. Simulated temperature distributions were also in better agreement with experimental ones. Moreover, as underlined by Colinart et al. (2016), the moisture content through the wall is also affected by the moisture content history of the material. In Aït Oumeziane et al. (2016b), the simulations of different hygric scenarii providing initial moisture content gradients gave relevant results throughout the full depth of the wall. Identifying the moisture content distribution through the wall thus seems necessary to predict the hygrothermal response of a wall subjected to climatic variations. Within this framework, numerical simulations are valuable tools. Promis et al. (2018) have also shown that not considering hysteresis leads to significant differences with experimental data, especially in terms of initial moisture content and in terms of variation of moisture content to reach the steady state. Furthermore, they justified the use of a hysteresis model to obtain an accurate estimation of the moisture content especially if high level of relative humidities are reached and an accurate estimation of the moisture transport coefficients is needed.

Moreover, as previously highlighted in Sect. 6.5, on the contrary to the atmosphere, relative humidity in hemp concrete is experimentally observed to increase with temperature (Colinart et al. 2017; Moujalled et al. 2018). This effect is explained by the temperature dependence of the sorption mechanism. Thus, Colinart et al. (2017) have shown that standard modelling approaches failed to reproduce the temperature effect on the relative humidity variations and underlined the necessity of accounting for the temperature dependency of the sorption isotherms in the simulations.

Promis et al. (2019) have worked on a hemp concrete specimen subjected to hygrothermal variations in a climatic chamber. In their study, the Staudt's model appeared to be in poor agreement with the measured experimental data within hemp concrete samples. The Poyet's and Milly's models presented results consistent with experimental data but seemed to overestimate (or underestimate) variations of the moist mass. They identified the modified GAB model as the most relevant to reproduce the experimental results. Nevertheless, they have shown that all these models remain extremely sensitive to sorption isotherms parameters determination.

6.6.2 Hygrothermal Response Under Real Climatic Conditions

A first experimental campaign on a hemp concrete room has been set up on two PASSYS test cells located in Le Bourget du Lac in south east of France (Bejat et al. 2015). The facilities allow to control indoor conditions and tested walls are exposed to real weather on the outside (Fig. 6.12). The authors have used the commercial software Wufi to compare numerical and experimental results within the wall. Even though the software is able to reproduce the global trends of temperature and relative humidity evolutions, they concluded that the software fails to reproduce daily variations. According to the authors, the improvement of the numerical modelling and the characterization of the material should improve the numerical results.

Otherwise, Shea et al. (2012) have instrumented a little building in Bath, England (Fig. 6.13). First, like Bejat et al. (2015), they highlighted the inability of the commercial software Wufi to reproduce the daily hygrothermal variations within a hemp concrete wall.

Moujalled et al. (2018) also studied the hygrothermal performance of a hemp-lime building located in south west of France during 4 years (Fig. 6.1). The specificity of the case studied by Moujalled et al. is that the building is inhabited. The numerical evolutions of relative humidity and temperature inside the wall have given promising results in comparison with the experimental measurements. Especially, the experimental daily variations of relative humidity increasing with temperature inside the wall were numerically reproduced. Despite a more physical

Fig. 6.12 Experimental setup of the two PASSYS test cells at Le Bourget du Lac, France (Bejat et al. 2015) © CEA

Fig. 6.13 View of the instrumented hemp concrete building 'Hempod' in Bath, England (Shea et al. 2012) © Mike Lawrence, University of Bath

representation of the effective hygric behaviour of hemp concrete, the authors observed that the numerical simulations with hysteresis did not significantly improve the results under the studied hygrothermal conditions. However, the simulations performed showed the relevance to consider the temperature-dependence of the sorption mechanism. As highlighted by Promis et al. (2019), they showed that complementary experimental investigations have to be performed to better identify the main sorption isotherms parameters. Moreover, as underlined by Lelievre et al. (2014) and Colinart et al. (2017), if the effective sorption mechanism are better estimated for hemp concrete, some uncertainties remained also concerning the effective hygrothermal properties of the coating materials.

6.6.3 Hygrothermal Comfort for the Inhabitants and Experience Feedback at Building Scale

Recently, Costantine et al. (2018) have also led an *in situ* experimental campaign on a hemp concrete based building for a 10-month period under real weather conditions (Fig. 6.14). The walls are composed of 20 cm of clay brick and 13 cm of hemp concrete placed on the exterior side. They are coated with a lime-sand plaster on the exterior surface and with a gypsum board on the interior surface.

The temperature and relative humidity measurements collected *in situ* for the different presented cases showed that hemp concrete envelope provides a significant

Fig. 6.14 View of the southern façade of the building located at Fleury la Riviere, France; (Costantine et al. 2018) © Foyer Rémois

amount of attenuation of the exterior climatic variations (Shea et al. 2012; Bejat et al. 2015; Moujalled et al. 2018; Costantine et al. 2018). Indeed, Moujalled et al. (2018) showed the good thermal inertia of 30 cm thick HLC wall, which allowed to dampen the daily temperature and relative humidity variations by 90% and to delay the effects of peak values up to about 12 h. It confirmed the general idea that bio-based materials are good hygrothermal regulators. However, relative humidity values within the wall were observed slightly high and need to be further examined in order to evaluate the risk of mould development.

Building inspection performed with a thermographic camera showed that exterior surface temperature of the walls is homogeneous and no structural thermal bridge due to the timber framework is observed (Fig. 6.15).

Furthermore, the blower door test performed by Moujalled et al. (2018) indicated an air permeability of 1.32 $m^3.h^{-1}.m^{-2}$ under 4 Pa, twice as high as the French regulation maximal value (0.6 $m^3.h^{-1}.m^{-2}$ for a single-family house). According to the authors, this can be explained by the shrinkage of the timber frame structure which creates gaps at the joints between the frame elements and interior finishing of the walls. This problem can be avoided by applying a continuous render over the internal surface of the walls. Indeed, the blower door test performed on the "Hempod" building showed a measured air permeability about 0.55 $vol.h^{-1}$ below the PassivHaus limiting value of 0.6 air changes per hour under 50 Pa (Shea et al. 2012).

The hygrothermal comfort is difficult to analyse. It mainly depends on the relative humidity, operative temperature and air speed. In Moujalled et al. (2018), the hygrothermal comfort during winter and summer period is analysed according to EN 15251 standard (EN 15251 2007). The standard defines four categories of comfort depending on the level of expectation: category I for spaces occupied by very sensitive inhabitants, category II for new buildings, category III for existing buildings, and category IV for values outside the previous categories. The values of category IV correspond to discomfort and are only acceptable for a limited part of the

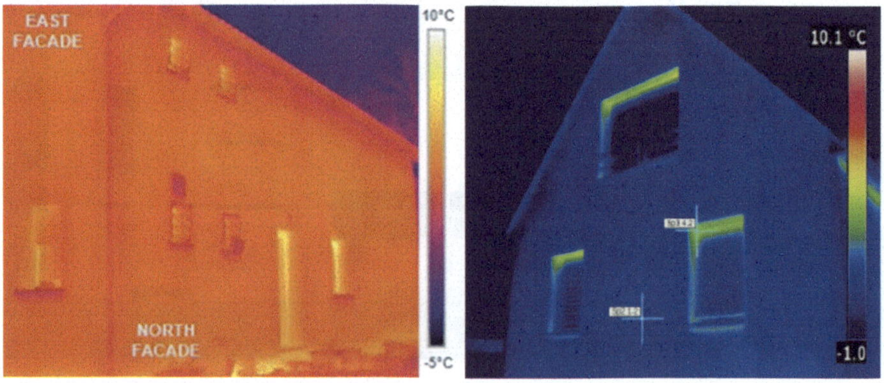

Fig. 6.15 Thermograms of the North and East facades of the building (left: Moujalled et al. (2018), and West façade (right: Costantine et al. (2018)). Note that no structural thermal bridge is observed

measurement period (5% of occupied hours). The *in situ* monitoring of the building showed that hemp concrete helps to maintain good hygrothermal comfort level in winter and in summer, while outside temperature and relative humidity daily variations were up to 15 °C and 50%RH respectively. Relative humidity of indoor air was never observed below 30%RH during the heating period, and indoor air temperature never above 27 °C during summer, even with outdoor air temperature higher than 35 °C. The measurements in different rooms showed a good level of hygrothermal comfort, except in a bedroom during winter where temperatures were observed lower during night. Similar trends are observed in Costantine et al. (2018). They also showed that the inhabitant's behaviour has a significant influence on the indoor relative humidity and temperature (ventilation, cooking activities).

Numerical studies on hemp concrete were also conducted at the room scale in order to analyse the indoor hygrothermal comfort (Tran Le et al. 2010; Tran Le 2011; Maalouf et al. 2014; Costantine et al. 2018).

In their studies of a room made of hemp concrete subjected to climatic variations, Tran Le et al. (2010) and Tran Le (2011) have also considered the impact of complementary systems (heating, ventilation). Tran Le, before presenting his study at the room scale, has validated his model at the scale of the wall from the results of Samri (2008) and thus presented his limits (cf Sect. 6.6.1). However, their conclusions remain very interesting to analyse the hygrothermal comfort and are here discussed. They showed that considering moisture transfer and indoor heat and moisture production sources has a significant effect on indoor relative humidity and energy consumption predictions. Differences of about 65% in indoor relative humidity and 19.5% in energy consumption were found. A sensitivity analysis was also carried out and showed that envelope hygrothermal performance is very sensitive to ventilation rate and to the physical properties of wall coating or external layer. Besides, hemp concrete behaviour was compared to cellular concrete's. It was found that reduction of 45% in energy consumption can be reached with hemp concrete. Finally, different ventilation strategies were compared and it was found that the use of a sensitive relative humidity ventilation strategy with hemp concrete can reduce energy consumption about 15% in comparison with constant flow ventilation system.

Based on the works of Tran Le (2011), simulations were run under summer conditions for three French cities with moderate to hot summers (Maalouf et al. 2014). The results suggested that there is a risk of indoor overheating due to the hemp concrete low effusivity. According to the authors, external solar shadings and night ventilation technique should be used in order to improve indoor thermal comfort conditions. In southern France, these techniques can be insufficient and coupling hemp concrete with a higher thermal inertia component should be used.

Finally, if Costantine et al. (2018) are not interested in the effective hygrothermal response of hemp concrete inside the wall, their simulations showed a quite good agreement of the indoor temperature with the experimental data. However, they highlighted the necessity to have fine knowledge of the hygrothermal properties of the materials. At the scale of the room, they showed that the implemented inhabitants' occupancy scenario have a great influence. As highlighted by Moujalled et al. (2015), Costantine et al. (2018) showed that the indoor and outdoor surface exchange coefficients have a significant influence on temperature and relative humidity evolutions. Indeed, under Neumann boundary conditions, the outdoor and indoor heat surface exchange coefficients are, in a great majority of the simulations, respectively set to 25 $W.m^{-2}.K^{-1}$ and 8 $W.m^{-2}.K^{-1}$, which correspond to the standard values. Moisture transfer coefficients are deduced from the heat exchange coefficients with the Lewis relation (Lewis 1922). Nevertheless, these coefficients do not take into account the variations of the climatic conditions caused by wind or solar radiation. Wind-driven rain can also participate in surface exchange by increasing the rate of moisture entering in the wall. These phenomena have to be considered in more details in future works.

6.7 Conclusion

The knowledge of the temperature, relative humidity and moisture content evolutions within hemp concrete walls is fundamental to address many issues. Indeed, these parameters define and influence the hygrothermal comfort and indoor air quality. They allow to evaluate the hygrothermal performance of a building and demonstrate the energy performance and environmental benefit of a chosen constructive solution. More broadly, in building physics, an accurate estimation of the moisture content through the material is fundamental to predict and avoid the different damages linked to moisture (moulds, mechanical resistance decrease, shelf life, etc.). The strong coupling between moisture and heat requires an efficient modelling of the hygrothermal behaviour of hemp concrete to assess the energy performance of a building.

Even though the works are rather scarce in literature compared to other building materials, recent research have improved the knowledge about the characterization and the modelling of the hygrothermal behaviour of hemp concrete. Especially, the consideration of the hysteresis phenomenon and of the temperature-dependence of the sorption process in HAM model has given promising results in transient climatic conditions (Aït Oumeziane et al. 2014, 2016a; Colinart and Glouannec 2017; Colinart et al. 2017; Moujalled et al. 2018; Promis et al. 2018, 2019).

The standard experimental methods of characterization are based on a static approach, which assume that the evolution of moisture content is applicable under transient conditions. Dynamic processes are thus assumed not to affect the moisture content evolution. Nevertheless, this simplification is demonstrated to not be valid (Janssen et al. 2016). Introduce a sorption kinetic appears full of interest on the case of hemp concrete as discussed by Reuge et al. 2019. The definition of a new protocol to fully characterize the hygrothermal properties of hemp concrete under various temperature and relative humidity transient cycles constitutes a major scientific objective.

The recent experimental campaigns led *in situ* have shown the performance of hemp concrete to ensure a good level of hygrothermal comfort for the inhabitants. This is especially explained by the moisture buffer ability of hemp concrete.

At building scale under real weather conditions, the modelling of the surface exchange and inhabitants influence need to be improved (Costantine et al. 2018). In addition, in real configurations, the influence of the coating materials has been demonstrated (Colinart et al. 2016; Moujalled et al. 2018). In this way, further investigations should be done in the future to better estimate their effective hygrothermal behaviour.

References

Aït Oumeziane Y (2013) Evaluation des performances hygrothermiques d'une paroi par simulation numérique: application aux parois en béton de chanvre. PhD thesis [In French]. Université Européenne de Bretagne – Institut National des Sciences Appliquées de Rennes, France

Aït Oumeziane Y, Bart M, Moissette S, Lanos C (2014) Hysteretic behaviour and moisture buffering of hemp concrete. Transp Porous Media 103:515–533. https://doi.org/10.1007/s11242-014-0314-7

Aït Oumeziane Y, Moissette S, Bart M, Lanos C (2016a) Influence of temperature on sorption process in hemp concrete. Constr Build Mater 106:600–607. https://doi.org/10.1016/j.conbuildmat.2015.12.117

Aït Oumeziane Y, Moissette S, Bart M, Collet F, Pretot S, Lanos C (2016b) Influence of hysteresis on the transient hygrothermal response of a hemp concrete wall. J Build Perform Simul 10:1–16. https://doi.org/10.1080/19401493.2016.1216166

Amziane S, Arnaud L (2013) Bio-aggregate-based building materials: applications to hemp concretes. ISTE Ltd./Wiley, London/Hoboken

Anderson B, Hall WK (1948) Modifications of the Brunauer, Emmett and Teller, equation II. J Am Chem Soc 70:1727–1734. https://doi.org/10.1021/ja01185a017

Anderson RB (1946) Modifications of the Brunauer, Emmett and Teller. J Am Chem Soc 68:686–691. https://doi.org/10.1021/ja01208a049

Arnaud L, Gourlay E (2012) Experimental study of parameters influencing mechanical properties of hemp concretes. Constr Build Mater 28:50–56. https://doi.org/10.1016/j.conbuildmat.2011.07.052

Bazant ZP, Thonguthai W (1978) Pore pressure and drying of concrete at high temperature. J Eng Mech Div ASCE 104:1059–1079

Bazant ZP, Thonguthai W (1979) Pore pressure in heated concrete walls – theoretical prediction. Mag Concr Res 31:67–76. https://doi.org/10.1680/macr.1979.31.107.67

Bennai F, El Hachem C, Abahri K, Belarbi R (2018) Microscopic hydric characterization of hemp concrete by X-ray microtomography and digital volume correlation. Constr Build Mater 188:983–994. https://doi.org/10.1016/j.conbuildmat.2018.08.198

Bejat T, Piot A, Jay A, Bessette L (2015) Study of two hemp concrete walls in real weather conditions. Energy Procedia 78:1605–1610. https://doi.org/10.1016/j.egypro.2015.11.221

Bray WH, Sellevold EJ (1973) Water sorption properties of hardened cement paste cured or stored at elevated temperatures. Cem Concr Res 3(6):723–728. https://doi.org/10.1016/0008-8846(73)90007-0

Brooks RH, Corey AT (1964) Hydraulic properties of porous media, Hydrology papers 3. Colorado State University, Fort Collins

Brue F (2009) Rôles de la température et de la composition sur le couplage thermohydromécanique des bétons. PhD thesis [In French]. Ecole Centrale de Lille, France

Brunauer S, Emmet PH, Teller E (1938) Adsorption of gases in multimolecular layers. J Am Chem Soc 60:309–319

Brunauer S, Deming LS, Deming WS, Teller E (1940) On a theory of Van der Walls adsorption of gases. J Am Chem Soc 62:1723–1732. https://doi.org/10.1021/ja01864a025

Brunauer S (1945) The adsorption of gases and vapors, Physical adsorption, vol I. Oxford University Press, London

Brunauer S, Skalny, Bodor EE (1969) Adsorption on nonporous solids. J Colloid Interface Sci 30:546–552. https://doi.org/10.1016/0021-9797(69)90423-8

Carmeliet J, De Wit M, Janssen H (2005) Hysteresis and moisture buffering of wood. In: Proceedings of 7th Nordic symposium on building physics, Reykjavik

Cerezo V (2005) Propriétés mécanique, thermiques et acoustiques d'un matériau à base de particules végétales: approche expérimentale et modélisation théorique. PhD thesis [In French]. ENTPE Lyon, France

CGDD (Commissariat Général au Développement Durable) (2018) Bilan énergétique de la France pour 2017 – Données définitives. Datalab essentiel 162. ISSN: 2557-8510

Chamoin J (2013) Optimisation des propriétés (physiques, hydriques et mécaniques) de bétons de chanvre par la maîtrise de la formulation. PhD thesis [In French]. Université Européenne de Bretagne – Institut National des Sciences Appliquées de Rennes, France

Chen CC (2006) Obtaining the isosteric sorption heat directly by sorption isotherm equations. J Food Eng 74:178–185. https://doi.org/10.1016/j.jfoodeng.2005.01.041

Colinart T, Glouannec P (2017) Temperature dependence of sorption isotherm of hygroscopic building materials. Part 1: xxperimental evidence and modeling. Energ Buildings 139:360–370. https://doi.org/10.1016/j.enbuild.2016.12.082

Colinart T, Glouannec P, Bendouma M, Chauvelon P (2017) Temperature dependence of sorption isotherm of hygroscopic building materials. Part 2: influence on hygrothermal behavior of hemp concrete. Energ Buildings 152:42–51. https://doi.org/10.1016/j.enbuild.2017.07.016

Colinart T, Lelievre D, Glouannec P (2016) Experimental and numerical analysis of the transient hygrothermal behavior of multilayered hemp concrete wall. Energ Buildings 112:1–11. https://doi.org/10.1016/j.enbuild.2015.11.027

Collet F (2004) Caractérisation hydrique et thermique de matériaux de Génie Civil à faibles impacts environnementaux. PhD thesis [In French]. Institut National des Sciences Appliquées de Rennes, France

Collet F, Achchaq F, Djellab K, Marmoret BH, Beji H (2011) Water vapor properties of two hemp wools manufactured with different treatments. Constr Build Mater 25:1079–1085. https://doi.org/10.1016/j.conbuildmat.2010.06.069

Collet F, Chamoin J, Pretot S, Lanos C (2013) Comparison of the hygric behaviour of three hemp concretes. Energ Buildings 62:294–303

Collet F, Pretot S (2012) Experimental investigation of moisture buffering capacity of sprayed hemp concrete. Constr Build Mater 36:58–65. https://doi.org/10.1016/j.conbuildmat.2012.04.139

Collet F, Pretot S (2014) Thermal conductivity of hemp concretes: variation with formulation, density and water content. Constr Build Mater 65:612–619. https://doi.org/10.1016/j.conbuildmat.2014.05.039

Costantine G, Maalouf C, Moussa T, Polidori G (2018) Experimental and numerical investigations of thermal performance of a Hemp Lime external building insulation. Build Environ 131:140–153. https://doi.org/10.1016/j.buildenv.2017.12.037

Daïan JF (1988) Condensation and isothermal water transfer in cement mortar Part I – pore size distribution, equilibrium water condensation and imbibition. Transp Porous Media 3:563–589. https://doi.org/10.1007/BF00959103

De Boer JH (1953) The dynamical character of adsorption. Oxford University Press, Oxford
De Brujin P, Johansson P (2013) Moisture fixation and thermal properties of lime-hemp concrete. Constr Build Mater 47:1235–1242. https://doi.org/10.1016/j.conbuildmat.2013.06.006
De Vries DA (1958) Simultaneous transfer of heat and moisture in porous media. Eos Trans AGU 39:909–916. https://doi.org/10.1029/TR039i005p00909
Drouet E (2010) Impact de la température sur la carbonatation des matériaux cimentaires – prise en compte des transferts hydriques. PhD thesis [In French]. Ecole Normale Supérieure de Cachan, France
Elfordy S, Lucas F, Tancret F, Scudeller Y, Goudet L (2008) Mechanical and thermal properties of lime and hemp concrete ("hempcrete") manufactured by a projection process. Constr Build Mater 22:2116–2123. https://doi.org/10.1016/j.conbuildmat.2007.07.016
EN 15251 (2007) Indoor environmental input parameters for design and assessment of energy performance of buildings addressing indoor air quality, thermal environment, lighting and acoustics
EN ISO 12571 (2000) Hygrothermal performance of building materials and products – determination of hygroscopic sorption properties
EN ISO 12572 (2001) Hygrothermal performance of building materials and products – determination of water vapour transmission properties
EN ISO 15148 (2002) Hygrothermal performance of building materials and products – determination of water absorption coefficient by partial immersion
Eurostat (2018) Energy balance Sheets Data 2016. Statistical Books. https://doi.org/10.2785/02631
Everett DH (1955) A general approach to hysteresis, an alternative formulation of the domain model. Trans Faraday Soc 51:1551–1557. https://doi.org/10.1039/TF9555101551
Everett DH (1967) Adsorption hysteresis. The solid-gas interface, vol 2. Marcel Dekker, New York, pp 1055–1113
Evrard A (2006) Sorption behaviour of Lime-Hemp Concrete and its relation to indoor comfort and energy demand. In: 23rd conference on Passive and Low Energy Architecture, Geneva, Switzerland
Evrard A (2008) Transient hygrothermal behaviour of Lime-Hemp materials. PhD thesis. Université Catholique de Louvain, Belgium
Fabbri A, McGregor F (2017) Impact of the determination of the sorption-desorption curves on the prediction of the hemp concrete hygrothermal behavior. Constr Build Mater 157:108–116. https://doi.org/10.1016/j.conbuildmat.2017.09.077
Galbraith GH, Mclean RC, Guo J (1997) Moisture permeability data presented as a mathematical function applicable to heat and moisture transport models. In: 5th international building performance simulation association conference, Pragues, Czech Republic
Gourlay E, Glé P, Marceau S, Foy C, Moscardelli S (2017) Effect of water content on the acoustical and thermal properties of hemp concretes. Constr Build Mater 139:513–523. https://doi.org/10.1016/j.conbuildmat.2016.11.018
Guggenheim EA (1966) Application of statistical mechanics. Oxford University Press, Chapter 11
Gustafsson SE (1991) Transient plane source techniques for thermal conductivity and thermal diffusivity measurements of solid materials. Rev Sci Instrum 62:797–804. https://doi.org/10.1063/1.1142087
Hadamard J (1902) Sur les problèmes aux dérivés partielles et leur signification physique. Princeton Univ Bull 13:49–52
Hansen N, Ostermeier A (2001) Completely derandomized self-adaptation in evolution strategies. Evol Comput 9:159–195
Hens H (2007) Building physics – heat, air and moisture, fundamentals and engineering methods with examples and exercises. Ernst & Sohn A Wiley. ISBN 978-3-433-01841-5
Huang HC, Tan YC, Liu CW, Chen CH (2005) A novel hysteresis model in unsaturated soil. Hydrol Process 19:1653–1665. https://doi.org/10.1002/hyp.5594
Hundt J, Kantelberg H (1978) Sorptionsuntersuchungen an zemestein, zementmörtel und beton [In German]. Deutscher Ausschuss für Stahlbeton Heft 297:25–39

Ishida T, Maekawa K, Kishi T (2007) Enhanced modeling of moisture equilibrium and transport in cementitious materials under arbitrary temperature and relative humidity history. Cem Concr Res 37:565–578. https://doi.org/10.1016/j.cemconres.2006.11.015

Jamali A, Kouhila M, Ait Mohamed L, Jaouhari JT, Idlimam A, Abdenouri N (2006) Sorption isotherms of Chenopodium ambrosioides leaves at three temperatures. J Food Eng 72:77–84. https://doi.org/10.1016/j.jfoodeng.2004.11.021

Janssen H (2011) Thermal diffusion of water vapour transport in porous materials: fact or fiction? Int J Heat Mass Transf 54:1548–1562. https://doi.org/10.1016/j.ijheatmasstransfer.2010.11.034

Janssen H, Scheffler GA, Plagge R (2016) Experimental study of dynamic effects in moisture transfer in building materials. Int J Heat Mass Transf 98:141–149. https://doi.org/10.1016/j.ijheatmasstransfer.2016.03.031

Kool JB, Parker JC (1987) Development and evaluation of closed form expressions for hysteretic soil hydraulic properties. Water Resour Res 23:105–114. https://doi.org/10.1029/WR023i001p00105

Korjenic A, Petránek V, Zach J, Hroudova J (2011) Development and performance evaluation of natural thermal-insulation materials. Energ Buildings 43:2518–2523. https://doi.org/10.1016/j.enbuild.2011.06.012

Krus M (1996) Moisture transport and storage coefficients of porous mineral building materials. Theoretical principles and new test methods. PhD thesis. Fraunhaufer IRB Verlag ISBN 3-8167-4535-0, Stuttgart, Germany

Künzel H (1995) Simultaneous heat and moisture transport in building components. PhD thesis. Fraunhofer IRB Verlag ISBN 3-8167-4103-7, Fraunhofer IBP, Stuttgart, Germany

Kwiatkowski J, Woloszyn M, Roux JJ (2009) Modelling of hysteresis influence on mass transfer in building materials. Build Environ 44:633–642. https://doi.org/10.1016/j.buildenv.2008.05.006

Langmuir L (1918) The adsorption of gases on plane surfaces of glass, mica and platinum. J Am Chem Soc 40:1361–1403. https://doi.org/10.1021/ja02242a004

Latif E, Ciupala MA, Wijeyesekera DC (2014) The comparative in situ hygrothermal performance of hemp and stone wool insulations in vapour open timber frame wall panels. Constr Build Mater 73:205–213. https://doi.org/10.1016/j.conbuildmat.2014.09.060

Lelievre D, Colinart T, Glouannec P (2014) Hygrothermal behavior of bio-based building materials including hysteresis effects: experimental and numerical analyses. Energ Buildings 84:617–627. https://doi.org/10.1016/j.enbuild.2014.09.013

Levenberg K (1944) A method for the solution of certain problems in least squares. Q App Math 2:164–168. https://doi.org/10.1090/qam/10666

Lewis WK (1922) The evaporation of liquid into a gas. Trans Am Soc Mech Eng 1849:325–340

Luikov AV (1975) System of differential equation of heat and mass transfer in capillary porous bodies. Int J Heat Mass Transf 18:1–14. https://doi.org/10.1016/0017-9310(75)90002-2

Maalouf C, Tran Le AD, Umurigirwaa SB, Lachi M, Douzane O (2014) Study of hygrothermal behaviour of a hemp concrete building envelope under summer conditions in France. Energ Buildings 77:48–57. https://doi.org/10.1016/j.enbuild.2014.03.040

Marquardt D (1963) An algorithm for least-squares estimation of nonlinear parameters. J Appl Math 11:431–444. https://doi.org/10.1137/0111030

Mazhoud B, Collet F, Pretot S, Lanos C (2017) Mechanical properties of hemp-clay and hemp stabilized clay composites. Constr Build Mater 155:1126–1137. https://doi.org/10.1016/j.conbuildmat.2017.08.121

McLaughlin CP, Magee TRA (1998) The determination of sorption isotherm and the isosteric heats of sorption for potatoes. J Food Eng 35:267–280. https://doi.org/10.1016/S0260-8774(98)00025-9

McMinn WAM, Magee TRA (2003) Thermodynamic properties of moisture sorption of potato. J Food Eng 60:157–165. https://doi.org/10.1016/S0260-8774(03)00036-0

Miller SA (2018) Natural fiber textile reinforced bio-based composites: mechanical properties, creep, and environmental impacts. J Clean Prod 198:612–623. https://doi.org/10.1016/j.jclepro.2018.07.038

Milly PCD (1984) A linear analysis of thermal effects on evaporation from soil. Water Resour Res 20:1075–1085. https://doi.org/10.1029/WR020i008p01075
Moujalled B, Aït Oumeziane Y, Samri D, Stephan E, Moissette S, Bart M, Lanos C (2015) Experimental and numerical evaluation of the hygrothermal performance of a hemp-lime building. In: First international conference on bio-based building materials, Clermont-Ferrand, France
Moujalled B, Aït Oumeziane Y, Moissette S, Bart M, Lanos C, Samri D (2018) Experimental and numerical evaluation of the hygrothermal performance of a hemp lime concrete building: a long-term case study. Build Environ 136:11–27. https://doi.org/10.1016/j.buildenv.2018.03.025
Mualem Y (1974) A conceptual model of hysteresis. Water Resour Res 10:514–520. https://doi.org/10.1029/WR010i003p00514
Mualem Y (1977) Extension of the similarity hypothesis used for modeling the soil water characteristics. Water Resour Res 13:773–780. https://doi.org/10.1029/WR013i004p00773
Mualem Y (1984) A modified dependent-domain theory of hysteresis. Soil Sci 137:283–291. https://doi.org/10.1097/00010694-198405000-00001
Mualem Y (2009) General scaling rules of the hysteretic water retention function based on Mualem's domain theory. Eur J Soil Sci 60:652–661. https://doi.org/10.1111/j.1365-2389.2009.01130.x
Néel L (1942) Théorie des lois d'aimantation de Lord Rayleigh. 1ère partie: les déplacements d'une paroi isolée. [In French]. Cah Phys 12:1–20
Néel L (1943) Théorie des lois d'aimantation de Lord Rayleigh. 2ère partie: multiples domaines et champ coercitif. [In French]. Cah Phys 13:18–30
Niyigena C, Amziane S, Chateauneuf A, Arnaud L, Bessette L, Collet F, Lanos C, Escadeillas G, Lawrence M, Magniont C, Marceau S, Pavia S, Peter U, Picandet V, Sonebi M, Walker P (2016) Variability of mechanical properties of hemp concrete. Mat Today Comm 7:122–133. https://doi.org/10.1016/j.mtcomm.2016.03.003
Parker JC, Lenhard RJ (1987) A model for hysteretic constitutive relations governing multiphase flow: saturation—pressure relations. Water Resour Res 23:2187–2196. https://doi.org/10.1029/WR023i012p02187
Pedersen CR (1990) Transient calculation on moisture migration using a simplified description of hysteresis in sorption isotherms. In: Proceedings of 2nd Nordic symposium on building physics, Trondheim, Norway
Philip JR, De Vries DA (1957) Moisture movement in porous material under temperature gradients. Eos Trans AGU 38:222–232. https://doi.org/10.1029/TR038i002p00222
Pierre T, Colinart T, Glouannec P (2014) Measurement of thermal properties of biosourced building materials. Int J Thermophys 35:1832–1852. https://doi.org/10.1007/s10765-013-1477-0
Pittau F, Krause F, Lumia G, Habert G (2018) Fast-growing bio-based materials as an opportunity for storing carbon in exterior walls. Build Environ 129:117–129. https://doi.org/10.1016/j.buildenv.2017.12.006
Poulovassilis A (1962) Hysteresis of pore water, an application of the concept of independent domains. Soil Sci 93:405–412. https://doi.org/10.1007/978-3-642-65523-4_17
Powers TC, Brownyard TL (1948) Studies of the physical properties of the hardened cement paste. Portland Cement Association Bulletin n°22
Poyet S (2009) Experimental investigation of the effect of temperature on the first desorption isotherm of concrete. Cem Concr Res 39:1052–1059. https://doi.org/10.1016/j.cemconres.2009.06.019
Poyet S, Charles S (2009) Temperature dependence of the sorption isotherms of cemented based materials: heat of sorption and Clausius-Clapeyron formula. Cem Concr Res 39:1060–1067. https://doi.org/10.1016/j.cemconres.2009.07.018
Preisach P (1935) Über die magnetische Nachwirkung. [In German]. FZ Physik 94:277–302. https://doi.org/10.1007/BF01349418

Pretot S, Collet F, Garnier C (2014) Life cycle assessment of a hemp concrete wall: impact of thickness and coating. Build Environ 72:223–231. https://doi.org/10.1016/j.buildenv.2013.11.010

Promis G, Douzane O, Tran Le AD, Langlet T (2018) Moisture hysteresis influence on mass transfer through bio-based building materials in dynamic state. Energ Buildings 166:450–459. https://doi.org/10.1016/j.enbuild.2018.01.067

Promis G, Freitas Dutra L, Douzane O, Tran Le AD, Langlet T (2019) Temperature-dependent sorption models for mass transfer throughout bio-based building materials. Constr Build Mater 197:513–525. https://doi.org/10.1016/j.conbuildmat.2018.11.212

Radjy F, Richards CW (1973) Effect of curing and heat treatment history on the dynamic mechanical response and the pore structure of hardened cement paste. Cem Concr Res 3:7–21. https://doi.org/10.1016/0008-8846(73)90057-4

Radjy F, Sellevold EJ, Hansen KK (2003) Isosteric vapor pressure–temperature data for water sorption in hardened cement paste: enthalpy, entropy and sorption isotherms at different temperatures. Report BYG-DTU R057. Technical University of Denmark (DTU)

Rahim M, Douzane O, Tran Le AD, Promis G, Laidoudi B, Crigny A, Dupre B, Langlet T (2015) Characterization of flax lime and hemp lime concretes: hygric properties and moisture buffer capacity. Energ Buildings 88:91–99. https://doi.org/10.1016/j.enbuild.2014.11.043

Rahim M, Tran Le AD, Douzane O, Promis G, Langlet T (2016a) Characterization and comparison of hygric properties of rape straw concrete and hemp concrete. Constr Build Mater 102:679–687. https://doi.org/10.1016/j.conbuildmat.2015.11.021

Rahim M, Douzane O, Tran Le AD, Langlet T (2016b) Effect of moisture and temperature on thermal properties of three bio-based materials. Constr Build Mater 111:119–127. https://doi.org/10.1016/j.conbuildmat.2016.02.061

Reuge N, Moissette S, Bart M, Collet F, Lanos C (2019) Water transport in bio-based porous materials: a model of local kinetics of sorption – application to three hemp concretes. Trans Porous Media. https://doi.org/10.1007/s11242-019-01272-4

Rode C (2005) Moisture buffering of building materials. Report BYG·DTU R-126, Technical University of Denmark

Rode C, Hansen KK (2011) Hysteresis and temperature dependency of moisture sorption – new measurements. In: 9th Nordic symposium on building physics, 2

Rouchier S, Woloszyn M, Kedowide Y, Bejat T (2015) Identification of the hygrothermal proprieties of a building envelope material by the covariance matrix adaptation evolution strategy. J Build Perform Simul 9:1–14. https://doi.org/10.1080/19401493.2014.996608

Rouchier S (2018) Solving inverse problems in building physics: an overview of guidelines for a careful and optimal use of data. Energ Buildings 166:178–195. https://doi.org/10.1016/j.enbuild.2018.02.009

Samri D (2008) Analyse physique et caractérisation hygrothermique des matériaux de construction: approche expérimentale et modélisation numérique. PhD thesis [In French]. ENTPE Lyon, France

Scott PS, Farquhar GJ, Kouwen N (1983) Hysteretic effects on net infiltration. Adv Infiltration 11–83:163–170

Seng B, Magniont C, Lorente S (2018a) Characterization of a precast hemp concrete. Part I: physical and thermal properties. J Build Eng. https://doi.org/10.1016/j.jobe.2018.07.016

Seng B, Magniont C, Lorente S (2018b) Characterization of a precast hemp concrete block. Part II: hygric properties. J Build Eng. https://doi.org/10.1016/j.jobe.2018.09.007

Shea A, Lawrence M, Walker P (2012) Hygrothermal performance of an experimental hemp–lime building. Constr Build Mater 36:270–275. https://doi.org/10.1016/j.conbuildmat.2012.04.123

Sing KSW, Everett DH, Haul RAW, Moscou L, Pierotti RA, Rouqerol J, Siemieniewska T (1986) Reporting physisorption data for gas/solid systems with special reference to the determination of surface area and porosity. Pure Appl Chem 57:603–619. https://doi.org/10.1351/pac198557040603

Somé CS, Ben Fraj A, Pavoine A, Chehade MH (2018) Modeling and experimental characterization of effective transverse thermal properties of hemp insulation concrete. Constr Build Mater 189:384–396. https://doi.org/10.1016/j.conbuildmat.2018.08.210

Spierling S, Knüpffer E, Behnsen H, Muderbasch M, Krieg H, Springer S, Albrecht S, Herrmann C, Endres HJ (2018) Bio-based plastics – a review of environmental, social and economic impact assessments. J Clean Prod 185:476–491. https://doi.org/10.1016/j.jclepro.2018.03.014

Standberg-De Bruijn P, Johansson P (2014) Moisture transport properties of lime hemp concrete determined over the complete moisture range. Biosyst Eng 122:31–41. https://doi.org/10.1016/j.biosystemseng.2014.03.001

Staudt PB, Kechinski C, Tessaro IC, Marczak LDF, Soares RDP, Cardozo NSM (2013a) A new method for predicting sorption isotherms at different temperatures using the BET model. J Food Eng 114:139–145. https://doi.org/10.1016/j.jfoodeng.2013.04.013

Staudt PB, Tessaro IC, Marczak LDF, Soares RP, Cardozo NSM (2013b) A new method for predicting sorption isotherms at different temperatures: extension to the GAB model. J Food Eng 118:247–255. https://doi.org/10.1016/j.jfoodeng.2013.04.013

Steeman HJ, Van Belleghem M, Janssens A, De Paepe M (2009) Coupled simulation of heat and moisture transport in air and porous materials for the assessment of moisture related damage. Build Environ 44:2176–2184. https://doi.org/10.1016/j.buildenv.2009.03.016

Tariku F, Kumaran K, Fazio P (2010) Transient model for coupled heat, air and moisture transfer through multilayered porous media. Int J Heat Mass Transf 53:3035–3044. https://doi.org/10.1016/j.ijheatmasstransfer.2010.03.024

Taylor SA, Cary JW (1964) Linear equations for the simultaneous flux of matter and energy in a continuous soil system. Soil Sci Soc Am Proc 28:167–172

Topp GC, Miller EE (1966) Hysteretic moisture characteristics and hydraulic conductivities for glass-bead media. Soil Sci Soc Am Proc 30:156–162. https://doi.org/10.2136/sssaj1966.03615995003000020008x

Tran Le AD, Maalouf C, Mai TH, Wurtz E, Collet F (2010) Transient hygrothermal behaviour of a hemp concrete building envelope. Energ Buildings 42:1797–1806. https://doi.org/10.1016/j.enbuild.2010.05.016

Tran Le AD (2011) Etude des transferts hygrothermiques dans le béton de chanvre et leur application a bâtiment. PhD thesis [In French]. Université de Reims Champagne Ardennes, France

Tran Le AD, Nguyen ST, Langlet T (2019) A novel anisotropic analytical model for effective thermal conductivity tensor of dry lime-hemp concrete with preferred spatial distributions. Energ Buildings 182:75–87. https://doi.org/10.1016/j.enbuild.2018.09.043

Trong LN, Asamoto S, Matsui K (2018) Sorption isotherm and length change behaviour of autoclaved aerated concrete. Cem Concr Compos 94:136–144. https://doi.org/10.1016/j.cemconcomp.2018.09.003

Van Belleghem M, Steeman HJ, Steeman M, Janssens A, De Paepe M (2010) Sensitivity analysis of CFD coupled non-isothermal heat and moisture modelling. Build Environ 45:2485–2496. https://doi.org/10.1016/j.buildenv.2010.05.011

Van Genuchten MT (1980) A closed-form equation for predicting the hydraulic conductivity of unsatured soils. Soil Sci Soc Am J 4:892–898

Walker R, Pavía S (2014) Moisture transfer and thermal properties of hemp–lime concretes. Constr Build Mater 64:270–276. https://doi.org/10.1016/j.conbuildmat.2014.04.081

Whitaker S (1977) Simultaneous heat, mass and momentum transfer in porous media: a theory of drying. Adv Heat Tran 13:119–203. https://doi.org/10.1016/S0065-2717(08)70223-5

Whitaker S (1986a) Flow in porous media I: a theoretical derivation of Darcy's law. Trans Porous Media 1:3–25. https://doi.org/10.1007/BF01036523

Whitaker S (1986b) Flow in porous media II: the governing equations for immiscible, two-phase flow. Trans Porous Media 1:105–125. https://doi.org/10.1007/BF00714688

Williams J, Lawrence M, Walker P (2017) The influence of the casting process on the internal structure and physical properties of hemp-lime. Mater Struct:50–108. https://doi.org/10.1617/s11527-0-016-0976-4

Xie Y, Hill CAS, Jalaludin Z, Curling SF, Anandjiwala RD, Norton AJ, Newman G (2011) The dynamic water sorption behaviour of natural fibres and kinetic analysis using the parallel exponential kinetics model. J Mater Sci 46:479–489. https://doi.org/10.1007/s10853-010-4935-0

Yan Z, Sousa-Gallagher MJ, Oliveira FAR (2008) Sorption isotherms and moisture sorption hysteresis of intermediate moisture content banana. J Food Eng 86:342–348. https://doi.org/10.1016/j.jfoodeng.2007.10.009

Zhang X, Zillig W, Künzel HM, Mitterer C, Zhang X (2016) Combined effects of sorption hysteresis and its temperature dependency on wood materials and building enclosures-part II: hygrothermal modeling. Build Environ 106:181–195. https://doi.org/10.1016/j.buildenv.2016.06.033

Chapter 7
Hempseed Protein: Processing and Functional Properties

Anne Pihlanto, Markus Nurmi, and Sari Mäkinen

Abstract Increasing the utilization of plant proteins is needed to support the production of protein-rich foods that could replace animal proteins in the human diet so to reduce the strain that intensive animal husbandry poses to the environment. The seeds of non-drug *Cannabis sativa* L. commonly referred to as hemp, are an important source of nutrition. Hempseed typically contains over 30% oil and about 25% protein, with considerable amounts of dietary fiber, vitamins, minerals, and specific phenolic compounds. However, the utilization of hempseed in food products is still at the very beginning. Previously, few research groups have reported trials on extracting and purifying proteins from hempseed. However, most of these methods have focused on protein isolates using different precipitation techniques, but these methods may not be suitable for commercial scale production, due to their intensive costs. Also, precipitation techniques may adversely affect the functional properties of hempseed proteins. This review aims to provide an overview of the current knowledge of the hempseed protein extractions and the functional properties of the enriched protein fractions.

Keywords Hemp seed protein · Extraction · Isolation · Functional properties · Digestibility

Abbreviations

EA	emulsifying activity
EC	emulsifying capacity
ES	emulsion stability
HPC	Hemp protein concentrate
HPI	Hemp protein isolate
HSM	Hempseed meal or cake
PDCAAS	protein digestibility corrected amino acid score

A. Pihlanto (✉) · M. Nurmi · S. Mäkinen
Natural Resources Institute Finland, Jokioinen, Finland
e-mail: anne.pihlanto@luke.fi; markus.nurmi@luke.fi; sari.makinen@luke.fi

G. Crini, E. Lichtfouse (eds.), *Sustainable Agriculture Reviews 42*, Sustainable Agriculture Reviews 42, https://doi.org/10.1007/978-3-030-41384-2_7

7.1 Introduction

There is an accruing body of evidence showing the need to shift toward a more plant-based diet for both environmental and health reasons. The global population is rapidly growing and simultaneously the demand for dietary protein, mainly of animal origin, is projected to increase by more than 50% by 2030 compared to the year 2000. The traditional Western dietary pattern focuses predominantly on animal-based products to satisfy protein requirements. This diet is environmentally detrimental, as it relies on intensive livestock farming, which contributes to the depletion of natural resources. Besides, the high intake of meat, especially red and processed meat, is associated with a higher incidence of coronary heart diseases, type 2 diabetes mellitus and several forms of cancer (Wolk 2017).

The development and market of the new meat and dairy analogue has accelerated in recent years, with some of the most promising alternatives based on plant sources such as soybeans and peas. Plant protein-based meat and dairy substitutes can deliver high nutritional and sensory quality, and at the same time fulfill the World's priority challenge of reducing greenhouse gas emissions and limiting the destruction of forestland (Dijkstra et al. 2003).

The production and consumption of food that includes high-protein crops may contribute toward achieving a sustainable diet. As reviewed by Callaway (2004) hemp has been an important source of food, fiber, and medicine for thousands of years. Hemp grows well in a variety of climates and soil types and has a short maturity period of 3–4 months. It also absorbs carbon dioxide up to five times more efficiently (thus slow global warming) than the same acreage of forest trees.

Hemp is a multipurpose crop due to its considerable amounts of over 30% oil and about 25% of easily digested protein containing significant amounts of all essential amino acids (Callaway 2004; Kriese et al. 2004; Teh and Birch 2013). After the oil is extracted from the food-grade hempseeds by cold pressing technology, the remaining hempseed meal contains high protein, fibre and carbohydrate amount as illustrated in Table 7.1. Milled hempseed meal is nowadays offered commercially as a source of vegetable protein and dietary fibre in the form of different products such as protein powders, flours, shake drinks, snacks, etc. Over the past few years, the availability of non-drug varieties with low δ-9-tetrahydrocannabinol (THC) contents has increased its industrial utilization for food product manufacture.

This review describes the properties of hemp seed as a food source and the current knowledge on producing functional protein concentrates from hemp seed.

7.2 Hempseed as Food

According to House et al. (2010), the crude protein concentration ranged between 21.3% and 27.5% in fresh whole hempseed products, 30.3–38.7% in dehulled hempseeds and 31–53.3% in hemp seed meal. The protein is concentrated in the

Table 7.1 Typical nutritional content (%) of hempseed

	Whole seed[a]	Seed meal[a]	Seed cake[b]	Whole seed[c]
Oil (%)	35.5	11.1	14.0	34.6
Protein	24.8	33.5	33.5	25.6
Carbohydrates	27.6	42.6	22.1	34.4
Moisture	6.5	5.6	7.1	6.7
Ash	5.6	7.2	5.9	5.4
Energy (kJ/100 g)	2200	1700	ND	2301
Total dietary fiber (%)	27.6	42.6	14.4	33.8
Soluble fiber	5.4	16.4	ND	2.9
Non-soluble fiber	22.2	26.2	ND	30.9

[a]Callaway 2004; [b]Teh et al. 2014; [c]Mattila et al. 2018 ND not determined

inner parts of the seed because the hulls contain lower levels of protein, 8.8–16.3% (Mattila et al. 2018). The crude protein content of commercial hemp seed flour was also analyzed by Multari et al. (2016) and their result was 38.6%. There was only a minor variation in the protein content between the ten industrial cultivars grown in southern Quebec (23.8–28.0% fresh weight) (Vonapartis et al. 2015). The quality of the hempseed protein is high and comparable to high-quality proteins sources such as egg white and soybean (Callaway 2004).

Besides the protein content, also the total carbohydrates content of whole hempseed (cv Finola) and cold-pressed hempseed meal has been reported to be high, 27.6% and 42.6%, respectively (Table 7.1). Hempseed is also a rich source of dietary fibre, especially non-digestible.

The mean results for oil content by House et al. (2010) for 11 hempseed products, by Vonapartis et al. (2015) for 10 industrial hemp cultivars and by Galasso et al. (2016) for seeds of 20 hemp traits and cultivars were 30.4%, 29.2%, and 31.9%, respectively. The hemp seed oil has a unique fatty acid composition. The seed oil typically contains over 90% unsaturated fatty acids which have large amounts of essential fatty acids (EFAs) which cannot be synthesized by humans.

7.2.1 *Hempseed Proteins*

Hempseed protein consists mainly of globulin (edestin) and albumin. Edestin accounts for approximately 60–80% of the total protein content, while albumin constitutes majority of the rest (House et al. 2010).

7.2.1.1 Globulins

The globular edestin is located inside the aleurone grains as large crystalloidal substructures and is composed of six identical subunits and each subunit consists of acidic and basic subunits linked by one disulphide bond (Patel et al. 1994). Hemp seed legumin consists of mainly the 11S and 7S protein types, which can be separated using the pH shifting technique as described by Wang et al. (2008). The molecular weight (MW) of edestin is estimated to be approximately 300 kDa. The acid subunit is approximately 34.0 kDa and is relatively homogeneous, while basic subunit consists mainly of two subunits of about 20.0 and 18.0 kDa (Wang et al. 2008). Kim and Lee (2011) isolated and characterized the edestin protein from Korean variety. The first seven and six amino acid residues of the acid subunit had a sequence of Ile-SerArg-Ser-Ala-Val-Tyr in the N-terminus, while two constituents of basic subunit showed an identical N-terminus of Gly-Leu-Glu-Glu-Thr-Phe. Wang et al. (2008) isolated the 7S and 11S fractions using a similar extraction method than for 7S and 11S fractions of soy protein isolate (SPI). The main component in hemp 11S is edestin, basic subunit and a subunit of about 4.8 kDa, makes up the 7S. Further analysis showed that the 7S polypeptide has no thermal transition, while the 11S protein exhibits a similar denaturation temperature as hemp protein isolate (HPI) at 91.9 °C, indicating that the HPI thermal property was due mainly to the 11S component (Wang et al. 2008).

7.2.1.2 Albumins

The albumin fraction constitutes about 25% of hempseed storage protein. Malomo and Aluko (2015a) found that the albumin fraction contains few disulfide-bonded proteins and hence a less compact structure with greater flexibility than the globulin fraction. This was further confirmed by intrinsic fluorescence and circular dichroism analysis which illustrated greater exposures of tyrosine residues when compared with globulin. On the other hand, albumins had highly ordered secondary structure and very little tertiary conformation at pH 3.0 but the tertiary conformation increased at higher pH values. The high degree of flexibility and ordered secondary structure are probably structural factors that contribute to the high solubility and foaming capacity of albumin in comparison to the more compact or aggregated globulin.

7.2.1.3 Other Proteins

Additionally, a methionine- and cysteine-rich seed protein (10 kDa protein, 2S albumin) has been isolated from hempseed. The protein consisted of two polypeptide chains (small and large) with 27 and 61 amino acid residues, respectively (Odani and Odani 1998). The two polypeptide chains contain 18% by weight of sulfur-containing amino acids (cysteine and methionine) and are held together by two disulfide bonds. This protein had no trypsin inhibitory activity and could serve as a rich thiol source to improve the nutritional quality of plant-based foods since various plant food proteins, especially legumin proteins from soybean, pea, and beans, are deficient in sulfur. The gene families encoding the precursor polypeptides of 2S albumin have recently been identified by Ponzoni et al. (2018), and two genomic isoforms for 2S albumin were obtained, namely, Cs2S-1 and Cs2S-2. The alignment of the deduced gene with the mature 2S protein sequence published in the literature (Odani and Odani 1998) showed that Cs2S is 97% identical to the mature 2S protein.

There is not much information available concerning protease inhibitors in hemp. Pojić et al. (2014) and Mattila et al. (2018) were able to measure trypsin inhibitory activity from HSM, seed hull and whole hemp seeds.

7.2.1.4 Protein Quality for Food

In human nutrition, the quality of a protein is defined by (1) the relative contribution that the amino acids contained in the protein make to an individual's amino acid requirement and (2) the digestibility of the protein. Hempseed and products derived from it contain all essential amino acids required by humans. The respective amino acid scores are presented in Fig. 7.1. The amino acid score of a protein reflects the extent to which a dietary protein meets the needs of an individual for a particular amino acid. The essential amino acids of hempseed are comparable to other high-quality proteins, such as casein and soy protein (Tang et al. 2006), and are sufficient for the Food and Agriculture Organization (FAO)/World Health Organization (WHO) suggested requirements for 2- to 5-year-old children. Hemp protein contains an exceptionally high amount of arginine and glutamine (Lu et al. 2010). Arginine accounts for approximately 12% of hempseed protein when compared with less than 7% for most other food proteins, including the proteins from potato, wheat, maize, rice, soy, rapeseed, egg white, and whey (Callaway 2004). Furthermore, whole hempseed and hemp protein products contain excellent amounts of the sulfur

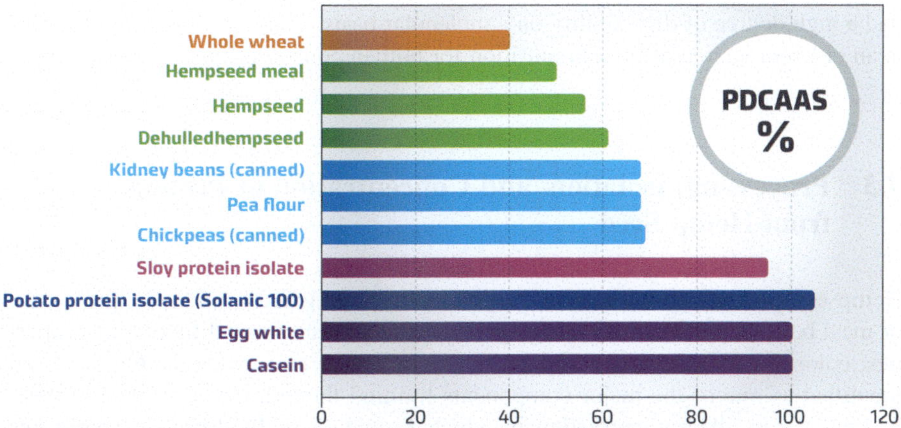

Fig. 7.1 Protein digestibility-corrected amino acid scores (PDCAAS) of hempseed protein sources in comparison to other food proteins. The values are from FAO/WHO (2011), House et al. 2010 and https://www.avebe.com/nutritional-value-of-potato-protein/

amino acids cysteine and methionine (Callaway 2004; Russo and Reggiani 2013; Mattila et al. 2018). Total sulfur-containing amino acids are in the range of 3.5–5.9%, which is close to the reference protein profiles established by United Nations University (UNU) as requirements for infants and preschool children 2- to 5-year-old. House et al. (2010) calculated the respective amino acid scores of hemp protein, identifying that lysine (score 0.5–0.62) was the first-limiting amino acid in hemp protein, followed by leucine and tryptophan.

Digestibility of dietary proteins affects the bioavailability of amino acids and thus, is a critical factor for the nutritional quality of the proteins. The digestion of dietary proteins depends on enzyme accessibility, which is affected by the molecular structure as well as other components associated with proteins. House et al. (2010) measured the protein digestibility and corrected amino acid score (PDCAAS) of whole hempseed, dehulled hempseed and HSM using a rat bioassay for protein digestibility and the FAO/WHO amino acid requirement of children (2–5 years of age) as reference. The protein digestibility of dehulled hempseed, depending on the sources, was 90.8–97.5%, almost comparable to 97.6% for casein. The PDCAAS values for hemp protein sources varied between 0.48 and 0.61. These values are within the range of major pulse proteins and are above the values of cereal grain products. Lysine as the main limiting amino acid in hempseeds and rather low level of tryptophan presumably contribute to the relatively low PDCAAS score.

Wang et al. (2008) compared the digestibility of hempseed protein isolates (7S, and 11S) and soy protein isolate using an *in vitro* digestion model. During the pepsin digestion, edestin was rapidly degraded similarly to the digestion of the soy protein isolates, and oligo-peptides with MWs less than 10 kDa were released. In addition, the total digestibility (pepsin plus trypsin digestion) of hempseed protein isolates (88–91%) was distinctly higher than that of soy protein isolate (71%). The most recent studies by Mamone et al. (2019) showed that hemp flour and protein isolate had a high degree of digestibility on a molecular basis. These findings hint that HPI is an efficient source of protein nutrition for human consumption.

7.3 Processing, Isolation, and Concentration of Protein from Hemp Seed

Hempseed proteins must be extracted or otherwise enriched from seeds, press cake or meal before application as food ingredient due to their limited functional properties, color and antinutritive compounds present in the raw materials. Oil has been identified as one of the major components limiting the extraction of plant protein, particularly in alkaline extraction in which formation of lipid-protein complexes hinder the recovery of protein (Manamperi et al. 2011). Processing options for producing hempseed protein products are illustrated in Fig. 7.2.

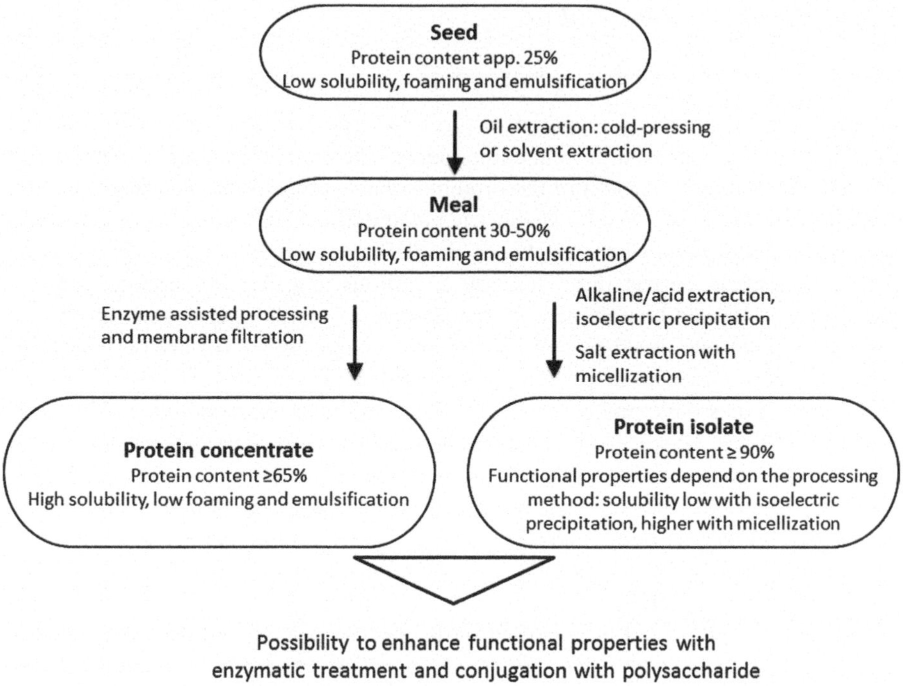

Fig. 7.2 Processing options to produce hempseed protein concentrates and isolates from hempseed

7.3.1 Hempseed Meal

The oil extraction by-product of crushed hempseeds is commonly referred to as hempseed meal or cake. The protein content in hempseed meal ranges from 30% to 50% in the dry matter depending on the used hempseed variety, the oil extraction method (cold-pressing or solvent extraction) and efficiency (Malomo et al. 2014).

7.3.2 Hemp Protein Concentrate

Hemp protein concentrate is prepared from dehulled and defatted hempseed or hempseed meal by removing most of the water-soluble, nonprotein constituents. The protein concentrates contains at least 65% protein (N $\times$ 6.25) on a dry weight basis. Malomo and Aluko (2015b) obtained protein concentrate by enzymatic digestion of fiber using carbohydrase and phytase coupled with membrane ultrafiltration that enriched protein content up to 70%. Protein digestibility of the protein concentrate was significantly higher than that of meal and traditional isoelectric protein isolate.

7.3.3 *Hemp Protein Isolate*

The most purified and enriched form of the commercial protein product, protein isolate (>90% protein), is prepared to meet food processing needs with minimal influence of unwanted non-protein components. The applied extraction method will affect the final protein content of the protein isolate, composition, and functionality. Alkaline extraction followed by isoelectric precipitation is the most common method to prepare hemp protein isolates (Lu et al. 2010; Malomo et al. 2014; Wang et al. 2008). Depending on specific extraction conditions (e.g pH, temperature, time), a purity up to 94% can be obtained. At the alkaline extraction, pH is generally 9–10, since the native hempseed proteins are tightly compacted, and may be integrated with other components, like phenolic compounds and phytic acid.

Several studies have reported that elevated extraction temperatures can improve protein solubility. However, an adverse chemical reaction such as the formation of lysinoalanine compounds from cysteine and serine residues can occur at highly alkaline conditions during heating. For example, protein isolates extracted at pH 10 and room temperature had a low level of lysinoalanine (0.8 mg/100 g protein) but at pH 12 at 40 °C for 5 min, the lysinoalanine content increased to 4 mg/100 g protein (Wang et al. 2018).

Teh et al. (2014) used acid extraction to prepare protein isolates. The yield of protein extracted at acidic pH was lower than that extracted at alkaline pH. In addition, a method, known as "salt extraction with micellization", has been described by Dapčević-Hadnađev et al. (2018). Protein isolate obtained by this method has a very high purity (98.9% protein, on a dry basis).

However, the limited functional properties, especially protein solubility, reduce the application of this protein in food formulations (Tang et al. 2006). The poor functional properties have been linked to the formation of covalent disulfide bonds between individual proteins and subsequent aggregation at neutral or acidic pH, due to its high free sulfhydryl content from sulfur-containing amino acids. The effects of limited or extensive enzymatic protein hydrolysis as a means of improving functional properties of hemp seed proteins have been reported (Yin et al. 2008).

7.4 Functional Properties of Hemp Seed Products

Besides nutritional value, the quality of food proteins is often determined by their techno-functional properties. The value and usefulness of a protein ingredient depend on its functionality, namely, its behavior and performance in food systems during preparation, processing, storage, and consumption. The functional properties of proteins depend on the nature and extent of interactions among protein molecules and with the presence of other components (for example, water and oil) in the food system. Since many formulated foods exist as beverages, emulsions, foams or solids, the ability of proteins to bind water or fat and their solubility, foaming, gelation, emulsification and film formation properties are essential to the quality of food products. Although the functional properties of hemp seed proteins have not yet been extensively studied, some properties have been reported. Below is summarized the most studied functional properties (solubility, foaming, and emulsification) of hempseed products.

7.4.1 Solubility

Solubility and dispersion stability of proteins are important factors that influence several other functional properties. Proteins are generally most soluble at their native state, whereas denaturation by heat, shear or chemicals may promote aggregation due to changed surface properties of proteins in the unfolded state. Hemp seed protein products have shown a typical U-shaped pH-solubility profile, with the minimum between pH 4.0 and 7.0 (Malomo et al. 2014; Tang et al. 2006; Dapčević-Hadnađev et al. 2018). In general, hemp seed protein exhibits low solubility in comparison to other plant proteins, such as soy. The low solubility is attributed to edestin aggregation at pH below 7 (Tang et al. 2006; Malomo et al. 2014). Albumin fraction has minimum solubility at pH 3.0 and globulin fraction at pH 5.0 (Malomo and Aluko 2015a, b). Albumin fraction has a higher solubility than globulin fraction due to the reduced level of aromatic and hydrophobic amino acids or a high isoelectric point when compared to the globulin fraction. Protein isolate made by isoelectric precipitation showed low solubility across the pH 3 to 9 (Malomo et al. 2014; Malomo and Aluko 2015a) or pH 4 to 7 (Dapčević-Hadnađev et al. 2018). The hemp seed protein product made by micellization had higher solubility across the pH range from 3.0 to 5.0 when compared to protein isolates made by isoelectric precipitation. In general, micelle protein isolates were found to have significantly higher solubility compared to those prepared by isoelectric precipitation. This was ascribed to the more preserved native state of protein isolates extracted using the micellization technique, while the isoelectric precipitation method led to protein denaturation and thereby resulted in hydrophobic interactions among protein molecules and formation of insoluble aggregates. HSM proteins are also insoluble, mainly because crosslinking by phytate and high levels of insoluble fiber that reduce the protein-water interactions. On the other hand, protein concentrates have shown higher solubility, indicating that less denaturation has occurred during processing. Especially using membrane ultrafiltration to prepare protein concentrates, up to 74% protein solubility was seen at pH 4.0–5.0 (Malomo and Aluko 2015a, b). The results indicated differences in polypeptide types, especially the absence of protein aggregation.

7.4.2 Emulsification

Emulsifying properties of proteins are related to many factors, for example, the rate of protein adsorption at the oil-water interface, the amount of protein adsorbed (loading), the conformational rearrangement at the interface, the extent of interfacial tension reduction, and the rheology of the cohesive film. A number of quality indexes, such as emulsifying activity (EA), emulsifying capacity (EC), emulsion stability (ES), and droplet size, are commonly used to evaluate the emulsifying properties of proteins. Even when proteins are not fully soluble, they may form stable dispersions eg. gels, foams and emulsions.

Reports on the EA of hemp seed protein products are highly variable (Tang et al. 2006; Malomo and Aluko 2015a, b), most probably due to different protein extraction procedures causing variation in the edestin and albumin ration and state of the proteins. Malomo and Aluko (2015b) showed that protein concentrates had generally a low EC, and the size of oil droplets was typically 6–15 μm while protein isolates formed emulsions with oil droplet sizes less than 1 μm. The results show, that the technique used to prepare the product influences on the emulsifying properties. Protein isolates preserved the native state in the isolation process and thus, were able to form emulsions with small droplets with sufficient droplet-droplet static repulsion properties. The protein isolates stabilized emulsions possessed low viscosity which enabled fast droplet movement and led to increased creaming and coalescence at lower protein concentrations (0.25–0.75% w/w). The isolation technique favored pH-induced structural unfolding of protein molecules and exposure of hydrophobic sites and sulfhydryl groups. Subsequently, protein connected droplet aggregates were formed during emulsification.

7.4.3 Foaming Properties

Foaming properties of protein isolates are mainly determined by their molecular flexibility and ability to reduce surface tension. Factors such as protein concentration, solubility, and hydrophobicity/hydrophilicity ratio can also influence on the foaming properties (Damodaran 2006; Malomo et al. 2014). The differences in structural conformation of hemp seed protein products at different pH values were reflected as different foaming capacities. Products made from protein isolates had better foaming capacity at pH 3.0 when compared to pH 5.0, 7.0 and 9.0 (Malomo et al. 2014). Albumin fraction had significantly higher foaming capacity than globulin fraction at pH range from 3.0 to 9.0 (Malomo and Aluko 2015a). The higher foaming capacity of the albumin is consistent with the observed higher solubility and suggests that greater interactions with the aqueous phase enhance the ability of the protein molecules to encapsulate air particles. This is because interactions with the hydrophilic aqueous phase will enhance protein unfolding and hence increase the foam forming ability (Sai-Ut et al. 2009). In hemp seed meals, foaming capacity improved as the environment changed from pH 3.0 to 9.0, which suggests increased interaction with water and protein unfolding at neutral and alkaline pH values. The foaming capacity and stability of the meal were inferior when compared to the purer protein isolates (Malomo and Aluko 2015a). Interference from nonprotein materials may have reduced the ability of meal proteins to form and stabilize foams.

7.5 Processing to Improve Functional Properties

As described above, due to the compact structure, hemp seed protein generally has limited functionality and industrial application when compared to many other food proteins. Therefore, structural modifications to improve the functionality are needed. Several physical, chemical, and enzymatic treatments have been applied to hemp protein, of which heat treatment, enzymatic hydrolysis, acylation, and pH shift have demonstrated promising efficacies.

Raikos et al. (2015) heated protein isolates from 40 to 100 °C for 10 min to modify the protein structure. Heating at 80 °C or higher temperatures produced insoluble large molecular aggregates via covalent linkages. Approximately 60–80% of the native protein was converted to an aggregated state. A similar observation was reported by Yin et al. (2008), who heated protein isolates at 95 °C for 10 min, noting that the thermally treated isolates had a significantly lower protein solubility than untreated isolates at pH 3.0 to 10.0. However, Wang et al. (2018) found that heating at 80 °C up to 60 min actually slightly improved the solubility of protein isolate. This might be due to different testing methods, as well as the protein composition. Heat treatment increased EA of protein isolates at pH values away from the isoelectric point, water holding capacity and the foam stability but did not affect the foaming capacity (Yin et al. 2008).

Several studies have been published on applying enzymes to improve the functionality of protein isolates (Tang et al. 2009; Wang et al. 2009; Yin et al. 2008). The studies are mainly focused on enhancing the antioxidant properties with commercial proteases, such as alcalase, flavourzyme, neutrase, protamex, pepsin, and trypsin (Tang et al. 2009; Wang et al. 2009), however, trypsin has also been applied to modify the protein solubility, water holding, emulsifying and foaming properties (Yin et al. 2008). Trypsin treatment to hydrolysis degree of 2–6% increased remarkably protein solubility but decreased emulsifying properties, water holding and fat absorption capacity (Yin et al. 2008). The results suggest, that mild protease treatment to enhance protein solubility could be combined to further treatment with transglutaminase or Maillard reaction with polysaccharide to improve the other functional properties (e.g. Martineza et al. 2005). This approach might be useful on producing hemp protein ingredients with diverse functional properties.

For plant proteins in general, conjugation with polysaccharides is suggested to be a feasible option for improving functional properties, reviewed recently by Akhtar and Ding (2017). Linking of proteins with polysaccharides has been reported to enhance the functional properties of native proteins in different circumstances, such as low and neutral pH conditions and with coloring agents (Liu et al. 2012). Alternative processing technologies, such as a spinning disc reactor, have also been applied for producing the protein-polysaccharide conjugates. In the spinning disc reactor, there is no need for additives and freeze drying, the benefits include decreased processing time and lower energy need (Burns and Jachuck 2003). Although the benefits of conjugating proteins with polysaccharides have been reported widely, more research is needed to achieve a better understanding on the conjugate structure-function relationship and also, the possible effects of the conjugates on digestion and gut health.

7.6 Conclusions

Due to hemp's well-recognized nutritional value, food manufacturers have developed a wide range of retail products from hemp, such as nuts, oil, protein flour, energy bars, granola, hemp nut butter, pasta, and ice cream (Leson 2006). A recent emphasis has been on hemp seed protein, which is used not only as a nutritive additive but also as a functional ingredient in formulated foods to enhance the product quality attributes. The low allergenicity of hemp protein when compared with most of the other plant proteins also permits it as a substitute for other proteins in some food product. The use of hemp seed protein products as an alternative to the commonly used casein, whey, wheat, and soy protein is on a rise. For instance, some studies have shown that hemp seed protein products can be used as value-added ingredients in the production of bread with increased protein and macro- and microelement contents, and lower baking loss and baking time (Korus et al. 2017; Lukin and Bitiutskikh 2017; Pojić et al. 2015). The key to keeping hemp protein competitive in the plant protein market is to assure its nutritional value, functionality, safety, and acceptable sensory characteristics.

Hemp seeds, an emerging protein-rich plant material, is becoming an important alternative protein source in the food due to consumers worldwide interest and demand of plant proteins are expected to grow rapidly. Although research has made progress in recent years in understanding the chemical composition, nutritional and health benefits, processing properties, and functional behavior of hemp seed proteins in food processing, much remains unknown about it. Therefore, it is clear that more systematic research is required to explore the structure–functionality relationship of hemp protein, technologies to modify functionalities must be vigorously explored through scientific research to convert hemp protein into a more suitable form. Additional research is also needed to investigate the health benefits. The research is essential to develop this valuable protein source and broadening its market potential in the food industry.

Acknowledgements We are grateful for funding from the Strategic Research Program (SRC) Academy of Finland (grant numbers 293045 and 314243 ScenoProt, Novel protein Sources of Food Security).

References

Akhtar M, Ding R (2017) Covalently cross-linked proteins & polysaccharides: formation, characterisation and potential applications. Curr Opin Colloid Interface Sci 28:31–36. https://doi.org/10.1016/j.cocis.2017.01.002

Burns JR, Jachuck RJJ (2003) Recent advances on spinning disc reactor technology: better processes for bigger profits. 5th international conference on process intensification. BHR Group, Maastricht

Callaway J (2004) Hemp seed as a nutritional resource: an overview. Euphytica 140:65–72. https://doi.org/10.1007/s10681-004-4811-6

Damodaran S (2006) Protein stabilization of emulsions and foams. J Food Sci 70:R54–R66. https://doi.org/10.1111/j.1365-2621.2005.tb07150.x

Dapčević-Hadnađev T, Hadnađev M, Lazaridour A, Moschakis T, Biliaderis CG (2018) Hemp seed meal protein isolates prepared by different isolation techniques. Part II. Gelation properties at different ionic strengths. Food Hydrocoll 81:481–489. https://doi.org/10.1016/j.foodhyd.2018.03.022

Dijkstra DS, Linnemann AR, van Boekel TA (2003) Towards sustainable production of protein-rich foods: appraisal of eight crops for western Europe. PART II: analysis of the technological aspects of the production chain. Crit Rev Food Sci Nutr 43:481–506. https://doi.org/10.1080/10408690390246332

FAO/WHO Consultation, FAO Expert (2011) Dietary protein quality evaluation in human nutrition. FAO Food Nutr Pap 92:1–66

Galasso I, Russo R, Mapelli S, Ponzoni E, Brambilla IM, Battelli G, Reggiani R (2016) Variability in seed traits in a collection of *Cannabis sativa L.* genotypes. Front Plant Sci 7:688. https://doi.org/10.3389/fpls.2016.00688

House JD, Neufeld J, Leson G (2010) Evaluating the quality of protein from hemp seed (*Cannabis sativa L.*) products through the use of the protein digestibility-corrected amino acid score method. J Agric Food Chem 58:11801–11807. https://doi.org/10.1021/jf102636b

Kim JJ, Lee MY (2011) Isolation and characterization of edestin from Cheungsam hemp seed. J Appl Biol Chem 54:84–88. https://doi.org/10.3839/jabc.2011.015

Korus J, Witczak M, Ziobro R, Juszczak L (2017) Hemp (*Cannabis sativa* subsp. *sativa*) flour and protein preparation as natural nutrients and structure forming agents in starch based gluten-free bread. LWT-Food Sci Technol 84:143–150. https://doi.org/10.1016/j.lwt.2017.05.046

Kriese U, Schumann E, Weber W, Beyer M, Brühl L (2004) Oil content, tocopherol composition and fatty acid patterns of the seeds of 51 *Cannabis sativa L.* genotypes. Euphytica 137:339–351. https://doi.org/10.1023/b:euph.0000040473.23941.76

Leson G (2006) Hemp foods in North America: status and joint industry research. J Ind Hemp 11:87–93. https://doi.org/10.1300/j237v11n01_08

Liu J, Ru Q, Ding Y (2012) Glycation a promising method for food protein modification: physicochemical properties and structure, a review. Food Res Int 49:170–183. https://doi.org/10.1016/j.foodres.2012.07.034

Lu R-R, Qian P, Sun Z, Zhou X-H, Chen T-P, He J-F, Zhang H, Wu J (2010) Hemp seed protein derived antioxidative peptides: purification, identification and protection from hydrogen peroxide-induced apoptosis in PC12 cells. Food Chem 123:1210–1218. https://doi.org/10.1016/j.foodchem.2010.05.089

Lukin A, Bitiutskikh K (2017) On potential use of hemp flour in bread production. Bull Transilvania University of Brasov For Wood Ind Agric Food Eng Series II 10:113–118

Malomo SA, Aluko RE (2015a) A comparative study of the structural and functional properties of isolated hemp seed (*Cannabis sativa L.*) albumin and globulin fractions. Food Hydrocoll 43:743–752. https://doi.org/10.1016/j.foodhyd.2014.08.001

Malomo SA, Aluko RE (2015b) Conversion of a low protein hemp seed meal into a functional protein concentrate through enzymatic digestion of fibre coupled with membrane ultrafiltration. Innov Food Sci Emerg Technol 31:151–159. https://doi.org/10.1016/j.ifset.2015.08.004

Malomo SA, He R, Aluko RE (2014) Structural and functional properties of hemp seed protein products. J Food Sci 79:C1512–C1521. https://doi.org/10.1111/1750-3841.12537

Mamone G, Picariello G, Ramondo A, Nicolai MA, Ferranti P (2019) Production, digestibility and allergenicity of hemp (*Cannabis sativa L.*) protein isolates. Food Res Int 115:562–571. https://doi.org/10.1016/j.foodres.2018.09.017

Manamperi WA, Wiesenborn DP, Chang SK, Pryor SW (2011) Effects of protein separation conditions on the functional and thermal properties of canola protein isolates. J Food Sci 76: E266–E273. https://doi.org/10.1111/j.1750-3841.2011.02087.x

Martineza KD, Baeza RI, Millán F, Pilosof AMR (2005) Effect of limited hydrolysis of sunflower protein on the interactions with polysaccharides in foams. Food Hydrocoll 19:361–369. https://doi.org/10.1016/j.foodhyd.2004.10.002

Mattila P, Mäkinen S, Eurola M, Jalava T, Pihlava JM, Hellström J, Pihlanto A (2018) Nutritional value of commercial protein-rich plant products. Plant Food Hum Nutr 73:108–115. https://doi.org/10.1007/s11130-018-0660-7

Multari S, Neacsu M, Scobbie L, Cantlay L, Duncan G, Vaughan N, Stewart D, Russell WR (2016) Nutritional and phytochemical content of high-protein crops. J Agric Food Chem 64:7800–7811. https://doi.org/10.1021/acs.jafc.6b00926

Odani S, Odani S (1998) Isolation and primary structure of a methionine-and cystine-rich seed protein of Cannabis sativa. Biosci Biotechnol Biochem 62:650–654. https://doi.org/10.1271/bbb.62.650

Patel S, Cudney R, McPherson A (1994) Crystallographic characterization and molecular symmetry of edestin, a legumin from hemp. J Mol Biol 235:361–363. https://doi.org/10.1016/s0022-2836(05)80040-3

Pojić M, Mišan A, Sakač M, Dapčević Hadnađev T, Šarić B, Milovanović I, Hadnađev M (2014) Characterization of byproducts originating from hemp oil processing. J Agric Food Chem 62:12436–12442. https://doi.org/10.1021/jf5044426

Pojić M, Dapcevic Hadnađev T, Hadnađev M, Rakita S, Brlek T (2015) Bread supplementation with hemp seed cake: a by-product of hemp oil processing. J Food Qual 38:431–440. https://doi.org/10.1111/jfq.12159

Ponzoni E, Brambilla I, Galasso I (2018) Genome-wide identification and organization of seed storage protein genes of Cannabis sativa. Biol Plant 62:693–702. https://doi.org/10.1007/s10535-018-0810-7

Raikos V, Duthie G, Ranawana V (2015) Denaturation and oxidative stability of hemp seed (*Cannabis sativa L.*) protein isolate as affected by heat treatment. Plant Food Hum Nutr 70:304–309. https://doi.org/10.1007/s11130-015-0494-5

Russo R, Reggiani R (2013) Variability in antinutritional compounds in hemp seed meal of Italian and French varieties. Plant 1:25–29. https://doi.org/10.11648/j.plant.20130102.13

Sai-Ut S, Ketnawa S, Chaiwut P, Rawdkuen S (2009) Biochemical and functional properties of proteins from red kidney, navy and adzuki beans. Asian J Food Agro-Ind 2:493–504

Tang C, Ten Z, Wang X, Yang X (2006) Physicochemical and functional properties of hemp (*Cannabis sativa L.*) protein isolate. J Agric Food Chem 54:8945–8950. https://doi.org/10.1021/jf0619176

Tang CH, Wang X-S, Yang X-Q (2009) Enzymatic hydrolysis of hemp (*Cannabis sativa L.*) protein isolate by various proteases and antioxidant properties of the resulting hydrolysates. Food Chem 114:1484–1490. https://doi.org/10.1016/j.foodchem.2008.11.049

Teh S, Birch J (2013) Physicochemical and quality characteristics of cold-pressed hemp, flax and canola seed oils. J Food Compos Anal 30:26–31. https://doi.org/10.1016/j.jfca.2013.01.004

Teh SS, Bekhit ADE, Carne A, Birch J (2014) Effect of the defatting process, acid and alkali extraction on the physicochemical and functional properties of hemp, flax and canola seed cake protein isolates. J Food Meas Charact 8:92–104. https://doi.org/10.1007/s11694-013-9168-x

Vonapartis E, Aubin M, Seguin P, Mustafa AF, Charron J (2015) Seed composition of ten industrial hemp cultivars approved for production in Canada. J Food Compos Anal 39:8–12. https://doi.org/10.1016/j.jfca.2014.11.004

Wang XS, Tang CH, Yang XQ, Gao WR (2008) Characterization, amino acid composition and in vitro digestibility of hemp (*Cannabis sativa L.*) proteins. Food Chem 107:11–18. https://doi.org/10.1016/j.foodchem.2007.06.064

Wang XS, Tang CH, Chen L, Yang X-Q (2009) Characterization and antioxidant properties of hemp protein hydrolysates obtained with Neutrase®. Food Technol Biotechnol 31:428–434

Wang Q, Jin Y, Xiong YL (2018) Heating-aided pH shifting modifies hemp seed protein structure, cross-linking, and emulsifying properties. J Agric Food Chem 66:10827–10834. https://doi.org/10.1021/acs.jafc.8b03901

Wolk A (2017) Potential health hazards of eating red meat. J Intern Med 281:106–122. https://doi.org/10.1111/joim.12543

Yin S, Tang C, Cao J, Hu E, Wen Q, Yang X (2008) Effects of limited enzymatic hydrolysis with trypsin on the functional properties of hemp (*Cannabis sativa L.*) protein isolate. Food Chem 106:1004–1013. https://doi.org/10.1016/j.foodchem.2007.07.030

Chapter 8
Functional and Bioactive Properties of Hemp Proteins

Tamara Dapčević-Hadnađev, Miroslav Hadnađev, Manda Dizdar, and Nataša Jovanović Lješković

Abstract Hemp seeds primarily represent a source of edible oil, which comprises over 30% of the whole seed. However, the role of the hemp seed, as a valuable material for protein extraction, could not be neglected. While the hemp seed is characterized with the protein content of 25%, after oil extraction, the protein content in a hempseed cake, a by-product of the oil extraction process, can increase up to 50%. Hempseed protein mainly consists of a legumin type protein, edestin, which accounts for 60–80% of the total protein content, followed by albumin. Recently, the ability of hemp protein to act as a techno-functional agent in different food applications has been investigated. The role of hemp protein as emulsifiers, foaming agent, gel-forming and biodegradable film-forming material was studied pointing the possibility to replace synthetic agents with the natural ones. Moreover, a large number of studies have revealed a bio-functionality of hemp proteins, i.e. application of enzymatic hydrolysis for the production of bioactive peptides. Bioactivity was mostly investigated by determining antioxidant properties and antihypertensive effects of enzymatic hemp seed protein hydrolysates and their peptide fractions. The hydrolysis was achieved by employing a range of proteases as well as different degrees of hydrolysis, which resulted in significant differences in the antioxidant properties of obtained hemp protein hydrolysates. The present chapter is a review of recent information on hemp protein extraction techniques, with the special emphases to its techno- and bio-functionality.

Keywords Hemp protein · Isolation technique · Hydration · Gelling · Interfacial properties · Bioactive peptides

---

T. Dapčević-Hadnađev (✉) · M. Hadnađev
University of Novi Sad, Institute of Food Technology, Novi Sad, Serbia

University Business Academy in Novi Sad, Faculty of Pharmacy, Novi Sad, Serbia
e-mail: tamara.dapcevic@fins.uns.ac.rs; miroslav.hadnadjev@fins.uns.ac.rs

M. Dizdar · N. J. Lješković
University Business Academy in Novi Sad, Faculty of Pharmacy, Novi Sad, Serbia
e-mail: manda.dizdar@faculty-pharmacy.com; natasa.ljeskovic@faculty-pharmacy.com

G. Crini, E. Lichtfouse (eds.), *Sustainable Agriculture Reviews 42*, Sustainable Agriculture Reviews 42, https://doi.org/10.1007/978-3-030-41384-2_8

Abbreviations

ACE	angiotensin-I-converting enzyme
DPPH	2,2-diphenyl-1-picrylhydrazyl
DPPH•	2,2-diphenyl-1-picrylhydrazyl radical
DSC	differential scanning calorimetry
HPH	hemp protein hydrolysate
HPI	hemp protein isolate
IEP	alkaline extraction-isoelectric precipitation
MW	molecular weight
OHC	oil holding properties
pI	isoelectric point
S	Svedberg unit
WHC	water holding properties

8.1 Introduction

Investigations on different alternative sources of proteins have experienced a boom in the last decade, due to the growing interest in the economical valorisation of agro-food by-products, along with concerns about food allergies, animal welfare, higher demands for dietary protein due to population increase, as well as the negative impact on the environment associated with the animal-derived proteins (Söderberg 2013; Hadnađev et al. 2017; Pihlanto et al. 2017). However, plant protein sources are still underutilized compared to animal proteins because of their lower nutritional values due to their amino acid profile and the presence of anti-nutrients, which can impair protein digestion and increase consumption toxicity (Pihlanto et al. 2017). Nevertheless, the processed forms of plant-based food (protein concentrates or isolates) are found to have lower levels of anti-nutritional factors such as trypsin inhibitors, alkaloids, lectins, tannins, phytic acid and hemagglutinins than their corresponding raw materials (Arntfield et al. 1985; Rodríguez-Ambriz et al. 2005; Mondor et al. 2009). Moreover, in comparison to animal-derived proteins, plant proteins exhibit poorer physico-chemical and sensory properties such as solubility, bitterness, off-flavour, dark colour which may hinder their maximal utilization (Hadnađev et al. 2017; Pihlanto et al. 2017). Therefore, there has been considerable effort to improve both nutritional and functional properties of plant proteins through the exploitation of innovative processing conditions as well as potential novel applications (Gómez-Guillén et al. 2011; Pihlanto et al. 2017).

Hemp seeds, which are primarily grown for oil production, are rich in proteins. After oil extraction, the protein content in a hempseed cake or meal, a by-product of the extraction process, can increase up to 44%, making this product a valuable material for protein isolation (Pojić et al. 2014). Hempseed meal and obtained isolates can be incorporated into various food systems, such as bread (Pojić et al. 2015; Korus et al. 2017) to increase its nutritional value and/or to provide specific functional attributes, such as rheological, texture and sensory characteristics. However, most of the studies investigating hemp proteins focus on its functional characteristics and the antioxidant potential of its hydrolysates. In order to improve its functionality and bioactivity, hemp proteins are isolated from hempseed meal using various techniques (Malomo and Aluko 2015b; Hadnađev et al. 2018a) and/or modified by physical (Yin et al. 2008), enzymatic (Yin et al. 2008; Tang et al. 2009a) and chemical (Yin et al. 2009) processes.

This chapter gives an overview of the relationships between the hemp protein structure and physico-chemical properties and its functionality. Strategies to improve hemp protein techno-functional (e.g. gelling, foaming, emulsifying) and bio-functional (e.g. antioxidant activity, cholesterol-lowering) properties by innovative isolation and modification processes will be also discussed.

8.2 Techniques for Hemp Protein Extraction

Extraction of proteins is generally governed by many different parameters such as pH, temperature, duration of extraction, type of solvent/salt used, ionic strength, solid to solution ratio, particle size of starting raw material, etc. In order to optimize protein yield, different protein extraction procedures have been developed (Singhal et al. 2016). While dry processing (air classification) has been widely used to separate protein fraction from some legume crops, due to high fat content, hemp proteins are typically extracted using wet processing. The wet extraction processes which are being exploited in the preparation of protein-rich material (isolates or concentrates) include alkaline extraction-isoelectric precipitation, salt extraction (micellization) and ultrafiltration.

8.2.1 Alkaline Extraction-Isoelectric Precipitation

One of the most widely used protein extraction technique is alkaline extraction-isoelectric precipitation (IEP) procedure. This procedure involves several steps and has been commonly used for oilseed and legume sources. First step is alkaline extraction procedure under which proteins are dissolved. It is assumed that at pH higher than pI protein-water interactions are promoted and dispersibility and solubility of protein is improved (Singhal et al. 2016). In order to facilitate the hemp protein extraction procedure, Tang et al. (2006) proposed a previous defatting step of hemp seed meal. After alkaline extraction, insoluble materials such as insoluble fibres, carbohydrates and insoluble proteins are removed by centrifugation step. The pH value of the obtained supernatant is adjusted to pI value of extracted protein (cca 5.0 for hemp proteins). Around the pI value proteins tend to aggregate via van der

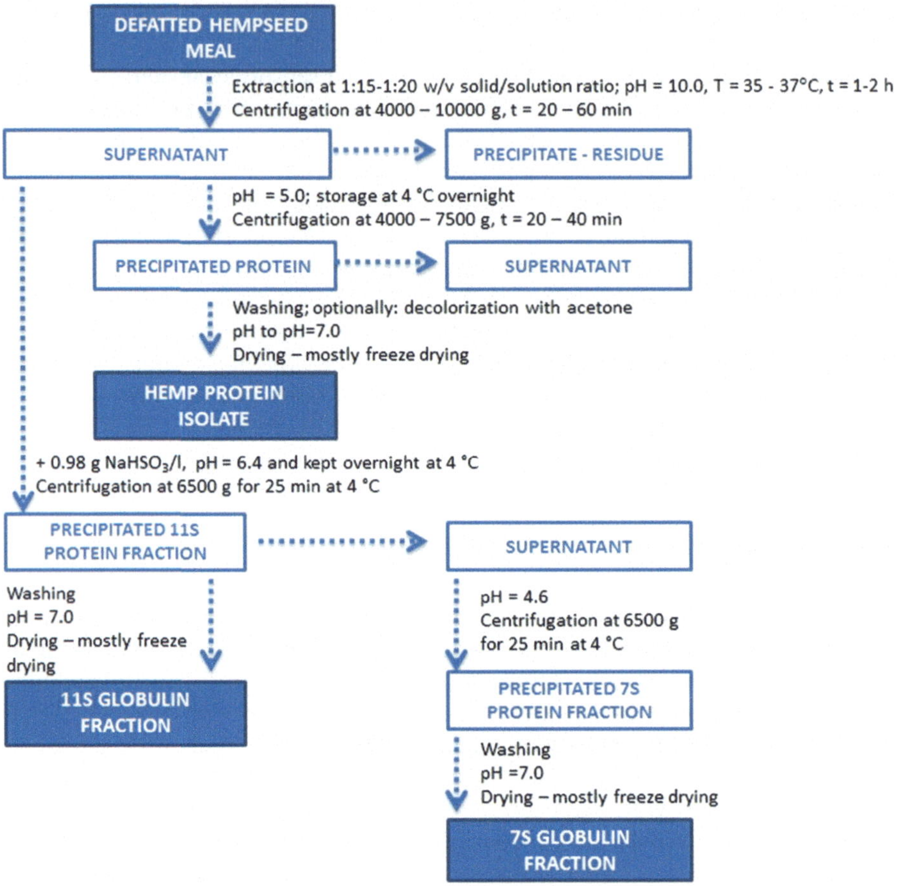

Fig. 8.1 Flow chart depicting the steps of hemp protein extraction by alkaline extraction-isoelectric precipitation technique and fractionation into 11S – legumin-type globulins and 7S – vicilin-type globulins. Starting material and obtained products are presented in text boxes, while the overview of the used procedures is listed along the arrows. *T* temperature, *t* time

Waals forces and hydrophobic interactions which are now more dominant than protein-water interaction resulting in protein precipitation from the solution (Singhal et al. 2016). After the precipitation, centrifugation procedure followed by several washing steps is commonly employed. Consequently, the obtained precipitate is optionally dispersed in distilled water and its pH value is adjusted to 7.0. The final step involves drying of obtained protein precipitate; most commonly by freeze drying technique (Fig. 8.1).

In order to fractionate hemp globulin protein, Wang et al. (2008) proposed the addition of $NaHSO_3$ in supernatant obtained after alkaline protein extraction and after centrifugation procedure (Fig. 8.1), followed by pH adjustment to 6.4 and precipitation. The obtained precipitate is referred to 11S protein fraction whereas the

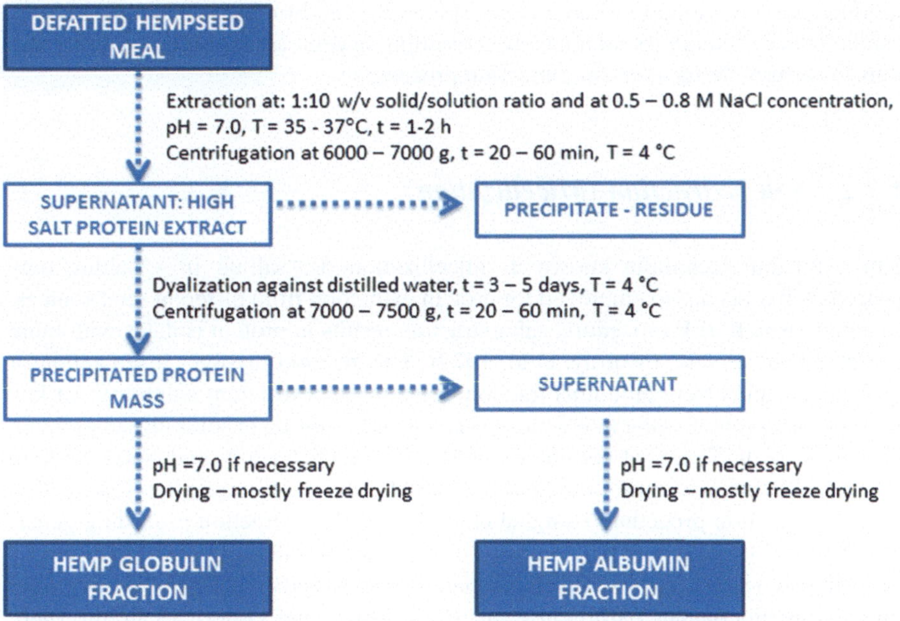

Fig. 8.2 Flow chart depicting the steps of hemp protein extraction by salt extraction (micellization) technique and fractionation into globulin and albumin fractions. Starting material and obtained products are presented in text boxes, while the overview of the used procedures is listed along the arrows. *T* temperature, *t* time

residual supernatant represents dissolved 7S protein fraction. The precipitation of 7S protein fraction is generally conducted by adjusting to pH = 4.6, followed by centrifugation step.

In order to optimize protein extraction procedure concerning the protein yield, purity, quality, etc., different approaches were discussed and different extraction conditions are summarised in Figs. 8.1 and 8.2. Zhang et al. (2008) found that optimum extraction conditions, obtained by response surface methodology, were T = 39.58 °C, extraction pH = 9.27, ratio of liquid to solid 24.22 and time 0.5 h, precipitation pH = 5.0. Under these conditions 96.4% of protein was precipitated from supernatant, the obtained protein yield was 52.3% and protein content was 83.8%. According to Hadnađev et al. (2018a) hemp proteins were extracted at pH = 10.0 at 1:20 w/v solid/solution ratio, the obtained supernatant was precipitated at pH = 5.0 and the obtained precipitate was washed several times in order to remove soluble molecules. The authors recovered 50.6% of proteins found in starting hemp meal and the obtained protein content was 91.4%. Karaca et al. (2011), who investigated the isolates prepared from legume sources, obtained about 85% of overall protein content using this extraction procedure. Moreover, Malomo et al. (2014) obtained 37.9% of hemp protein yield, whereas Tang et al. (2006) reported a recovery of 73% protein relative to total protein content in hemp meal which can be related to different shearing history of raw material during pressing/extracting of oil, milling, etc. According to Malomo and Aluko (2015a) different mechanical stresses can influence protein/protein interaction resulting in decreased protein solubility and thus lower protein recovery in extraction process.

8.2.2 *Salt Extraction (Micellization)*

Salt extraction technique known as micellization or "salting in – salting out" procedure has been also employed for protein extraction from different seed sources. In comparison to IEP procedure, salt extraction results in protein isolates with more preserved native state (Murray, et al. 1979). This method involves a separation of globulin proteins from albumins fraction based on their different solubility. At low level of salts protein-water interactions is promoted resulting in solubility of globulin protein fraction. The most commonly used salts are $(NH_4)_2SO_4$ and NaCl. Sodium chloride solution at a concentration of 0.3–0.5 M is the most frequently used solution for this extraction procedure (Singhal et al. 2016). Salt extraction procedure generally involves: (i) protein extraction using a salt solution (salting-in step), followed by centrifugation step to remove insoluble material and (ii) precipitation of the extracted proteins in supernatant (Arntfield et al. 1985; López and Ordorica-Falomir 1986; Hadnađev et al. 2018a), which can be conducted using two different approaches (salting-out step). First method is to dilute supernatant with water in order to reduce ionic strength and second is to use dialysis to remove salts. The dilution method involves the addition of large quantity of water which has to be removed after protein precipitation. Therefore, the advantage is given to dialysis method which is easier to perform. According to Malomo and Aluko (2015a) and Hadnađev et al. (2018a) hemp proteins have been extracted at 1:10 w/v (solid to solution ratio) and at 0.5 M NaCl and 0.8 M NaCl solution concentration, respectively, followed by dialysis using ultrafiltration cellulose membranes during 3–5 days at 4 °C. The precipitated proteins referring to globulin fraction were centrifuged and the obtained supernatant represents albumin protein fraction (Malomo and Aluko 2015a). Summarized conditions for salt extraction technique is presented in Fig. 8.2.

According to Alsohaimy et al. (2007) ammonium sulphate precipitation of legume proteins resulted in higher protein content in comparison to IEP procedure. However, study performed by Karaca et al. (2011) showed opposite trend according to which alkaline extraction procedure resulted in higher protein levels in comparison to salt extraction technique of legume proteins. Hadnađev et al. (2018a) prepared protein isolates from hempseed meal using IEP and salt extraction techniques. The latter method resulted in lower yield (19.24%) and higher protein content (98.87%) in comparison to the former method (yield – 24.24%, protein content – 91.44%). The lower yield can be related to higher selectivity of salt extraction technique towards a particular protein fraction (globulin) in comparison to more generic procedure of IEP. The yield of protein isolate relative to total protein in hemp meal obtained for salt extraction was 40.17%, which was lower in comparison to IEP (50.6%).

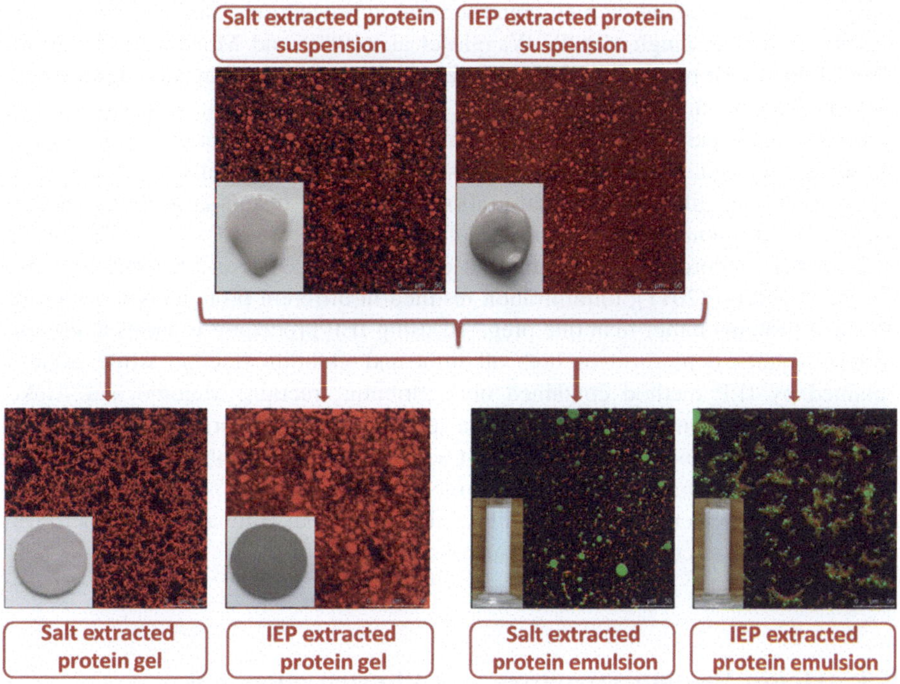

Fig. 8.3 Confocal laser scanning micrographs and photographs of 30% aqueous suspensions, 30% heat-induced gels and 10% sunflower oil-in-water emulsions (at pH = 3.0, emulsifier concentration = 1.5% calculated on continuous phase) prepared with alkali extracted-isoelectric precipitated (IEP) and salt extracted hemp proteins

Although both IEP and salt extraction procedure were characterized with high protein content, a higher purity for isolates obtained by micellization technique was observed by different authors (Mwasaru et al. 1999; Hadnađev et al. 2018a). Salt extraction procedure resulted in stronger protein/protein interactions resulting in exclusion of non-proteinaceous fractions. Moreover, the IEP procedure favoured co-extraction of phenolics from starting hemp meal material, resulting in dark green colour of obtained isolates and higher total phenolics in comparison to isolates prepared by salt extraction (Hadnađev et al. 2018a). Milder extraction conditions of micellization technique resulted in lower co-extraction of seed phenolics and obtained isolates were significantly lighter (Fig. 8.3).

8.2.3 Ultrafiltration

Ultrafiltration represents technique which involves implementation of different membranes and pressure as a driving force for protein separation (Singhal et al. 2016). Fredrikson et al. (2001) and Fuhrmeister and Meuser (2003) concluded that protein isolates obtained by membrane separation procedure exhibited higher functionality, whereas Singh (1988), Waggle et al. (1989) and Mondor et al. (2009) showed that those protein isolates had decreased level of anti-nutritional properties such as protease and amylase inhibitors, lectins, etc. The main principle of this method is that supernatant obtained after alkaline or acidic extraction is subjected to ultrafiltration or diafiltration in order to isolate proteins. Generally, extraction procedure involving ultrafiltration/diafiltration gave better results concerning protein content for different legume seed material in comparison to IEP (Vose 1980; Boye et al. 2010). According to Papalamprou et al. (2009, 2010) and Kiosseoglou and Paraskevopoulou (2011), ultrafiltration resulted in different protein composition of obtained proteins rather than one prepared using IEP procedure in terms that these protein isolates comprised of both albumin and globulin fraction while isolates obtained by IEP method contained only globulin fraction. Malomo and Aluko (2015b) treated hemp seed meal characterized with 37% protein content with different enzymes and the obtained digest was subjected to membrane ultrafiltration. As a result a membrane ultrafiltration protein concentrate had 74% protein content characterized by increased digestibility and solubility in comparison to isoelectric protein isolate and commercial protein concentrate.

8.3 Hemp Protein Structure and Physico-chemical Properties

Physico-chemical properties of proteins influence its quality and potential application and they are highly related to its structure. Structural features of hemp proteins have been studied by different techniques such as electrophoretic analysis (Tang et al. 2006; Wang et al. 2008; Yin et al. 2008; Malomo et al. 2014; Raikos et al. 2015; Malomo and Aluko 2015a, b; Hadnađev et al. 2018a), differential scanning calorimetry (DSC) (Tang et al. 2006; Wang et al. 2008; Yin et al. 2008; Hadnađev et al. 2018a), Fourier transform infrared spectroscopy (Hadnađev et al. 2018a), spectrofluorimetry (Malomo et al. 2014, 2015; Malomo and Aluko 2015a), circular dichroism (Malomo et al. 2014; Malomo and Aluko 2015a), size exclusion chromatography (Yin et al. 2008; Malomo et al. 2015), real-time qPCR (Ponzoni et al. 2018), mass spectrometry (Aiello et al. 2016), viscoelasticity (Dapčević-Hadnađev et al. 2019a), confocal microscopy (Dapčević-Hadnađev et al. 2019a).

It was shown that the majority of hemp proteins are globulin and albumin fractions, where globulins comprise up to 85% and albumins constitute about 15% of the total hemp protein (Tang et al. 2006; Wang et al. 2008). The salt soluble globulins can be further divided into 11S (S = Svedberg Unit) legumin-type globulins and 7S vicilin-type globulins (Singhal et al. 2016; Ponzoni et al. 2018). The 11S fraction of hemp proteins is called edestin. Ponzoni et al. (2018) reported detailed information on hemp storage protein genes and their genomic organization.

Their results demonstrate that two 2S albumin, one 7S vicilin-like, and six edestin genes are present in the hemp genome.

The 11S edestin is the main storage protein representing approximately 80% of the total seed protein (Tang et al. 2006; Wang et al. 2008). It is a hexamer (MW of ~300 kDa) composed of six identical subunits joined by non-covalent interactions. Each subunit pair consists of an acidic and basic chain linked by a disulfide bond (Patel et al. 1994; Kim and Lee 2011). The SDS-PAGE profiles, under non-reducing conditions, have indicated a major band of about 52 kDa corresponding to acidic-basic subunits. Upon adding 2-mercaptoethanol (reducing conditions), the disulphide bond between the acidic and basic subunits is disrupted. While acidic subunit (~34.0 kDa) is relatively homogenous, basic subunit (~18.0 and 20.0 kDa) consists of two bands (Wang et al. 2008; Raikos et al. 2015). According to amino acid composition Ponzoni et al. (2018) identified three edestin types (type1, −2 and −3). All edestin types showed typical 11S globulin features and were very rich in arginine and glutamic acid, but edestin type3 was the richest in cysteine and methionine.

Separation of the hemp protein in the 11S and 7S fractions using pH shift (pH = 6.4 and 4.6, respectively) and comparison to hemp protein isolate (HPI) precipitated at pH = 5.0 has indicated that 11S fraction showed similar SDS-PAGE profile to that of HPI. On the contrary, in the case of 7S fraction, the corresponding acidic-basic units (band of about 52 kDa) were almost absent (non-reducing conditions), and the relative content of the subunit with MW of ~48.0 kDa (reducing conditions) increased by almost fivefold relative to that of HPI (Wang et al. 2008). Ponzoni et al. (2018) indicated that Cs7S might be the gene that encodes the 7S minor polypeptide of approximately 48.0 kDa, which corresponds to the SDS-PAGE band reported by Wang et al. (2008).

Tang et al. (2006) and Wang et al. (2008) have found that water-soluble protein 2S albumin constitutes about 13% of the total hemp protein. According to Malomo et al. (2014), a minor band at SDS-PAGE profile of HPI with less than 14 kDa in size is likely to correspond to the albumin fraction. Gel electrophoresis, as well as intrinsic fluorescence and circular dichroism data, have indicated that the albumin is characterized with less disulfide bonds and hence a more open (flexible) structure in comparison to globulins (Malomo and Aluko 2015a).

According to the amino acid profiles of hemp seed products, hemp proteins contain a high amount of arginine (94–128 mg/g protein) (House et al. 2010), relative to other oilseeds, such as soybean (73.5 mg/g protein) (Wang et al. 2008). Hemp protein also represents a good source of the sulfur-containing amino acids methionine and cystine (15.5 mg/g protein, in comparison to 9.6 mg/g protein detected in soybean protein) (Callaway 2004; Wang et al. 2008). Compared to FAO/WHO reference protein for children 2–5 years of age, lysine is the first limiting amino acid in hemp protein, followed by leucine and tryptophan. The limitation in the lysine content of hemp protein results in its low amino acid scores (0.50–0.62) which are comparable to those of cereals (0.44 for whole wheat), but significantly lower relative to oilseeds (1.05 for soybean) (House et al. 2010). However, in vivo hemp protein digestibility, determined using casein (digestibility = 97.6%) as a reference standard, ranged between 85.2% and 86.7% (House et al. 2010), which was in a good agreement with in vitro digestibility (measured by nitrogen release) of hemp protein isolates (88–91%) (Wang et al. 2008). In general, hemp proteins are described as easily digested (Aiello et al. 2016).

It has been shown that separation of hemp protein into different fractions can significantly improve its amino acid profile. For example, the globulin fraction has a higher content of sulfur-containing amino acids, especially methionine, as well as aromatic amino acids and branched chain amino acids when compared to albumin fraction (Malomo and Aluko 2015a). Comparing two globulin fractions, edestin (11S globulin) is nutritionally superior protein in terms of essential and conditionally essential amino acids content than 7S globulin since it is characterized with higher content of methionine, arginine and tyrosine (Wang et al. 2008).

Differential scanning calorimetry (DSC) has shown that, unlike 7S protein fraction, edestin expressed ordered structure (Wang et al. 2008). DSC results were also highly impacted by isolation technique. Salt extracted isolates exhibited significantly higher structural order and thermal stability than IEP isolates which reflected the higher enthalpy and peak denaturation temperature of the former (Hadnađev et al. 2018a). While milder extraction conditions during micellization do not cause any irreversible changes in protein structure, hence allowing proteins to maintain their native conformation, highly alkaline conditions during IEP protein extraction led to extensive protein denaturation (Arntfield et al. 1985; Hadnađev et al. 2018a). In comparison to proteins isolated from other oilseeds, the enthalpy values of the hemp seed isolates (9.4–11.9 J/g protein) recovered by IEP technique (Tang et al. 2006; Wang et al. 2008; Yin et al. 2008; Hadnađev et al. 2018a) were close to that reported for canola (9.66 J/g protein) and soybean (14.23 J/g protein), but higher than that obtained for flaxseed (8.25 J/g), peanut (6.26 J/g) and pumpkin (5.1 J/g) protein preparations isolated with the same technique (Murray et al. 1985; Liu et al. 2011; Rezig et al. 2013; Kaushik et al. 2016). Hemp protein isolates obtained by the salt extraction also exhibited denaturation enthalpy values comparable to those found for canola (24.06 J/g protein) and soybean (24.56 J/g protein) micellar protein isolates, and almost two times higher than that reported for native peanut protein extracted by ammonium sulfate (12.4 J/g) and salt extracted pumpkin seed proteins (12.6 J/g) (Murray et al. 1985; Liu et al. 2011; Rezig et al. 2013).

Protein surface composition (presence of polar/non polar amino acids), together with different environmental conditions (pH, temperature, ionic strength and the type of ions present in the solution) greatly influence protein solubility. This physico-chemical property plays a key role for protein food applications as functional properties such as gelling, emulsification, foaming are closely related to protein solubility (Singhal et al. 2016). In general, proteins exhibit minimum solubility at their isoelectric point (pI) where they form aggregates and precipitate due to zero net surface charge. However, at pH values higher or lower that the pI, proteins possess negative or positive net charge which influence the formation of repulsive inter-molecule forces and increase in their solubility (Singhal et al. 2016). The solubility profiles of hemp protein isolates obtained by IEP and micellization were found to be the lowest between pH 5.0 and 6.0 (Tang et al. 2006; Yin et al. 2008;

Hadnađev et al. 2018a). The study by Hadnađev et al. (2018a) showed higher solubility at acidic conditions of hemp isolates prepared by the salt extraction method as compared to ones prepared by IEP method. This was ascribed to preparation conditions which in the case of IEP method favoured higher co-extraction of phytic acid and, consequently, formation of insoluble phytic acid-protein complexes at pH bellow pI (Hadnađev et al. 2018a). In general, hemp globulins exhibit low solubility in comparison to other vegetable proteins, such as soy protein (Tang et al. 2006), which is attributed to high free sulfhydryl content from sulfur-containing amino acids which favour the formation of covalent disulfide bonds between individual proteins and subsequent aggregation at neutral or acidic pH (Pihlanto et al. 2017). Due to reduced level of aromatic and hydrophobic amino acids and less rigid conformational structure, hemp albumins are characterized with higher solubility compared to the globulins (Malomo and Aluko 2015a). It was shown that hemp globulin solubility can be significantly increased by limited or extensive enzymatic hydrolysis (Yin et al. 2008) and succinylation or acetylation at low anhydride levels (Yin et al. 2009).

8.4 Techno-functionality of Hemp Protein

Apart from basic physicochemical properties, such as composition parameters, structural features, solubility, colour, taste, etc., the most important techno-functional properties of hemp proteins can be divided into: (i) properties associated with their gelling behaviour, i.e. water binding capacity, gel formation, thickening, (ii) properties related to their surface behaviour, which include emulsion and foam formation and stabilisation, and (iii) film-forming properties.

8.4.1 Gelling, Water and Oil Holding Properties

Water holding capacity represents the amount of water entrapped in the protein matrix, i.e. the ability of the protein to prevent water loss from its hydrated three-dimensional structure after applying centrifugal force (Ly et al. 1998). The water (WHC) and oil (OHC) holding properties of proteins are important in formulation of food products, its quality, shelf life and consumer acceptability since they influence product texture and mouth feel (Singhal et al. 2016). WHC and OHC values of hemp proteins have been determined by different authors (Tang et al. 2006; Teh et al. 2014; Malomo et al. 2014; Hadnađev et al. 2018a) and fall in the range of 0.80–12.01 g/g and 1.62–13.70 g/g, respectively, suggesting that both protein isolation procedure and method used to determine WHC influence values. Namely, as there are no standard methods for WHC and OHC determination, the results among studies are largely influenced by centrifugation force, time allowed for hydration, absorption, etc. Hadnađev et al. (2018a) reported that IEP protein isolates had higher WHCs than the ones prepared by micellization procedure. However, no significant differences in OHC were observed between the isolates produced by different techniques (Hadnađev et al. 2018a). Higher WHCs values of proteins isolated by IEP were ascribed to partial denaturation of the protein during extraction which led to extensive conformational changes and exposure of polar amino acid side chains and, consequently, increased water uptake through capillary mechanisms and formation of hydrogen bonding between the exposed amino acid residues and water molecules (Hadnađev et al. 2018a).

According to Tang et al. (2006) hemp protein isolates showed significantly lower water absorption capacity in comparison soy protein isolates. This could be explained by the fact that the polar groups of the proteins are found in the interior of the aggregates resulting in lower WHC values of hemp protein isolates in comparison to soy proteins. However, oil holding capacity was almost the same as of the soy protein isolates suggesting that surface hydrophobic groups can interact with lipids causing higher OHC values. Concerning the flax seed and canola proteins, Teh et al. (2014) revealed that hemp protein isolates demonstrated the highest WHC as well as OHC values especially in comparison to flax seed proteins. Moreover, hemp and canola protein isolates expressed higher WHC and lower OHC values in comparison to hemp and canola seed meal suggesting that alkaline or acid extraction resulted in conformational changes of hemp and canola proteins leading to more pronounced hydrophilic surface and WHC values than the starting raw material.

Formation of more or less ordered macroscopic structures, such as gels, is governed by the tendency of proteins to aggregate or to form intermolecular cross-links on heating and change of pH (Caetano da Silva Lannes and Natali Miquelim 2013). In order to measure the least gelation concentration of hemp products, Malomo et al. (2014) placed samples of different concentrations in water bath at 95 °C and left them in refrigerator for 14 h. The concentration of protein samples at which the gel was formed, i.e. it did not slip by inverting the tube was referred to the least gelation concentration. The authors revealed that gelation capacity of hemps seed meal was 12%, whereas for hemp protein isolate was 22%. This was most likely attributed to poor gel-forming ability of hemp protein isolates probably caused by significant protein aggregation in IEP procedure resulting in reduced protein flexibility needed for network structuration. However, hemp meal which was not treated by alkaline/acid was not affected by protein aggregation and consequently had lower least gelation concentration values (Malomo et al. 2014).

Dapčević-Hadnađev et al. (2019a) compared gelling properties of hemp protein isolates obtained by alkaline extraction-isoelectric precipitation and salt extraction procedure at 30% protein concentration. At lower protein concentrations, isolates obtained by salt extraction gave low viscous dispersions which did not built homogenous gels upon heating. Alkaline extraction procedure resulted in protein gels that upon heating had initially higher storage modulus values and less pronounced drop in system stiffness in comparison to salt extracted proteins. This could be attributed to lower dissociation of edestin subunits during heating since the part of dissociation had already taken place in more aggressive alkaline extraction procedure.

Subsequent heating, however, resulted in gels of similar strengths, i.e. they had similar values of elastic modulus. The gels of differently isolated proteins were also prepared by pouring protein dispersions in hermetically sealed disc shape vessel and immersing them in oil bath at 120 °C which enabled fast gel formation. According to Dapčević-Hadnađev et al. (2019a) slight frequency dependence of both moduli, which is typical for gel-like behaviour, was observed. Moreover, according to strain sweep measurements earlier decrease in yield stress values was observed for gels prepared by salt extracted proteins in comparison to gels obtained by alkaline extraction. This could be a result of weaker transient network structure of those gels. Generally, both gels were characterized by low tan δ values (<0.4) indicated that they built true gel structure. Confocal micrographs (Fig. 8.3) have shown that IEP isolated protein gel networks were characterized by fairly large globular entities, whereas salt extracted protein gels exhibited a finer network microstructure.

8.4.2 Interfacial Properties

One of important techno-functionalities of proteins is surface activity, i.e. ability to reduce the surface tension at interface between a solid, liquid or gas phase. Surface phenomena play a key role in food processing and knowledge of colloidal interactions at interface represents important prerequisite for controlling food physical properties, such as stability (Caetano da Silva Lannes and Natali Miquelim 2013).

Emulsions and foams as dispersed systems are often present in food. Proteins are widely used as functional ingredients for the formation and stabilisation of emulsions and foams, due to their propensity to concentrate at most interfaces. Protein surface properties are based on the presence of both polar or hydrophilic, and non-polar or lipophilic groups. During the emulsion and foam formation, they migrate toward surface, adsorb and form a film at the surface of the oil droplets or gas bubbles, thus protecting the system against destabilisation (Rouimi et al. 2005; Caetano da Silva Lannes and Natali Miquelim 2013).

8.4.2.1 Emulsifying Properties

Unlike small molecular weight emulsifiers, proteins have the ability to act as both emulsifier and stabilizer since they have high tendency to adhere to the oil–water interfaces and form viscoelastic films at the surface of the oil droplets providing electrostatic and steric stabilization (Rouimi et al. 2005; Tcholakova et al. 2006; Lam and Nickerson 2013).

The emulsifying properties of food proteins are mostly evaluated by measuring: (i) emulsion activity, i.e. the maximum interfacial area per unit mass of protein in a stabilized solution; (ii) emulsion capacity, i.e. the maximum amount of oil per unit mass of protein that can be emulsified prior to a phase inversion taking place, (iii) emulsion stability, i.e., the capacity of a protein to form an emulsion that remains unchanged for a certain time period at a specific conditions (Amarowicz 2010; Lam and Nickerson 2013).

According to the values of emulsion activity and stability indexes Tang et al. (2006) concluded that emulsifying activities of hemp protein isolate are poor, when compared to the once of soy protein. Both proteins exhibited the highest emulsifying activity indexes at pH = 3.0 which was attributed to their high solubility at that pH value. Studies concerning emulsifying properties of hemp protein prepared by IEP, as well as globulin and albumin fractions prepared by salt extraction revealed that all the fractions can form emulsions with small particle sizes ($d_{3.2}$ less than 0.6 μm at protein concentration 50 mg/ml) which are comparable to ones observed for whey protein isolate and lower in comparison to other plant proteins (Malomo et al. 2014; Malomo and Aluko 2015a). In the range of pH 3.0–9.0, IEP extracted isolate exhibited the lowest emulsification ability at pH 3.0 where higher solubility value was recorded (Malomo et al. 2014) which was in contrast to a report of Tang et al. (2006). On the contrary, hemp globulin fraction obtained by salt extraction did not express pH-dependant emulsion droplet size, while albumin stabilized emulsions had higher droplet sizes at pH 7.0 and 9.0 as measured at protein concentration of 50 mg/mL. All the studies reported that emulsion stabilities were nearly unchanged with pH. The differences between the studies (Tang et al. 2006; Malomo et al. 2014; Malomo and Aluko 2015a) could be ascribed to the differences in emulsion preparation. While Tang et al. (2006) prepared emulsions containing 1 part of oil in 3 parts of aqueous phase with 0.2% (w/v) proteins, emulsions investigated by Malomo et al. (2014) and Malomo and Aluko (2015a) had 1 part of oil in 5 parts of aqueous phase containing 1–5% (w/v) of proteins. Dapčević-Hadnađev et al. (2019b) studied the interfacial and emulsifying properties of hemp proteins extracted with IEP and micellization, in order to evaluate the differences in their ability to stabilize 10% oil-in-water emulsions at pH = 3.0. They have shown that emulsification mechanism of two hemp proteins was influenced by the differences in their denaturation degree which contributed to the alteration in their solubility, adsorption rate and conformational flexibility at the interface. While salt extracted protein formed emulsions with small droplets and enough droplet-droplet static repulsion, in the emulsions stabilized by IEP extracted protein, isolation technique favoured pH-induced structural unfolding of protein molecules, exposure of hydrophobic sites and sulfhydryl groups and subsequent droplets bridging flocculation (Fig. 8.3) (Dapčević-Hadnađev et al. 2019b).

Although enzymatic hydrolysis is considered a powerful tool in the modification of emulsifying characteristics of proteins (Amarowicz 2010), Yin et al. (2008) have shown that limited enzymatic hydrolysis with trypsin led to significant decline in emulsifying activity index at neutral and acidic pH values, relative to the hemp protein isolate. On the contrary, it was shown that hemp protein physical modification (thermal treatment) or chemical modification by acetylation or succinylation can led to increase in its emulsifying activity index (Yin et al. 2008, 2009). The improvement of emulsifying activity index by thermal treatment or acylation is attributed to unfolding of protein molecule and subsequent exposure of hydrophobic groups initially buried in the interior of protein molecules (Yin et al. 2008, 2009).

8.4.2.2 Foaming Properties

Proteins are considered to be good foaming agents since they can strongly adsorb at the gas-liquid interface, provide steric and electrostatic stability, form films with cohesive structure and a high module of rheological interface as a result of the interactions between the adsorbed molecules. Their adsorption at the gas-liquid interface is facilitated by the presence of hydrophobic regions due to non polar amino acids (Caetano da Silva Lannes and Natali Miquelim 2013). Foaming properties are mostly characterized by measuring: (i) foaming capacity, i.e. increase in volume of the protein dispersion upon mixing and (i) foam stability, i.e. the percentage of foam remaining after certain period of time. Higher foaming capacities of hemp proteins were recorded at lower pH values, which was attributed to their higher solubility at those pHs, as well as greater exposure of aromatic groups leading to enhanced protein–protein interactions and formation of cohesive viscoelastic interfacial membranes. However, the pH value did not influence foam stability (Malomo et al. 2014). It was also revealed that hemp albumins have higher foaming capacity than hemp globulin due to higher solubility of the former which enable their diffusion and unfolding at the air-water interface to encapsulate air bubble. However, globulins had higher foam stability than albumins which may be attributed to higher contents of hydrophobic amino acids in globulin fraction, which enhance protein-protein interactions and formation of a strong interfacial viscoelastic membrane (Malomo and Aluko 2015a). Hemp protein physical modification via heat treatment did not affect foaming capacity, while it influenced an increase in foam stability. On the contrary, enzymatic modification with trypsin led to decreases in both foaming capacity and foam stability indicating the role of the polypeptide chain length in foam stabilization (Yin et al. 2008).

8.4.3 Film-Forming Properties

Edible films either formed as food coatings or free-standing films are defined as a thin layer of material which can be consumed and provides a barrier to food from moisture, oxygen and solute migration (Bourtoom 2008). Different biological compounds such as polysaccharides, proteins, lipids and their derivatives can be used for the biopolymer films and coatings preparation (Yin et al. 2007). Proteins are characterized by the ability to form continuous network and wide range of different proteins (wheat gluten, maize zein, soy proteins, gelatine, collagen, egg proteins, pea proteins, milk proteins, etc.) have already been investigated for this purposes (Bourtoom 2008).

Yin et al. (2007) compared cast films obtained from hemp protein isolate and from soy protein isolate in terms of moisture content, total soluble mass, tensile strength and elongation at break as well as surface hydrophobicity. The amount of glycerol which served as plasticizer and its influence on abovementioned parameters were also monitored. At certain glycerol amounts, films prepared with hemp protein isolates expressed similar moisture content, but much less total soluble mass and elongation at break, higher surface hydrophobicity as well as higher tensile strength in comparison to films prepared by soy protein isolates. Moreover, surface hydrophobicity increased with plasticizer amount increase. According to protein films solubility in different solvents as well as to free sulfhydryl group content it was estimated that disulfide bonds were the most dominant force in the formation of films prepared from hemp protein isolates, whereas in the formation of soy protein films hydrogen bonds and hydrophobic interactions were also employed. The obtained findings showed that hemp protein isolates might have good properties for film preparations characterized by low solubility and high surface hydrophobicity.

8.5 Bio-Functionality of Hemp Protein Hydrolysates

Biopeptides are natural chemical compounds found in plants and animals that have the ability to improve certain health functions when found in the human organism (Hernandez-Ledesma et al. 2007). They represent short amino acid sequences (2–20 amino acid residues) found in the primary protein structure. Biopeptides are derived from food proteins by the action of enzymes and when they enter the body they can act as modulators of certain physiological processes, similar to endogenous peptides with hormonal activity (Tang et al. 2009b). Potential physiological effects of biopeptides are related to: regulation of elevated blood pressure (inhibition of angiotensin and converting enzyme), antioxidant activity, prevention of platelet aggregation, modulation of the immune system, lowering of cholesterol and triglycerides levels in plasma, stimulation of the nervous system, antimicrobial effect, improvement of the transport and absorption of minerals (Korhonen and Pihlanto 2006; Hartmann and Meisel 2007).

Besides abovementioned bio-functionalities, it was shown that protein hydrolysates can exhibit a variety of techno-functionalities (Fig. 8.4) (Amarowicz 2010).

8.5.1 Antioxidant Properties

Antioxidants are important to the human body as they may provide a defence against Reactive Oxygen Species (Ryan et al. 2011). The supplementation with the antioxidants may provide additional support to endogenous antioxidants in the defence against oxidative stress (Kunwar and Priyadarsini 2011). The antioxidant activities of enzymatic hydrolysates from animal and plant food proteins, including bovine caseins and whey proteins (Pihlanto 2006), soy proteins (Moure et al. 2006), wheat protein (Zhu et al. 2006), porcine haemoglobin, collagen and myofibrillar protein (Chang et al. 2007), and fish proteins (Rajapakse et al. 2005; Kim et al. 2007), have been widely investigated by use of many *in vitro* antioxidant evaluation systems

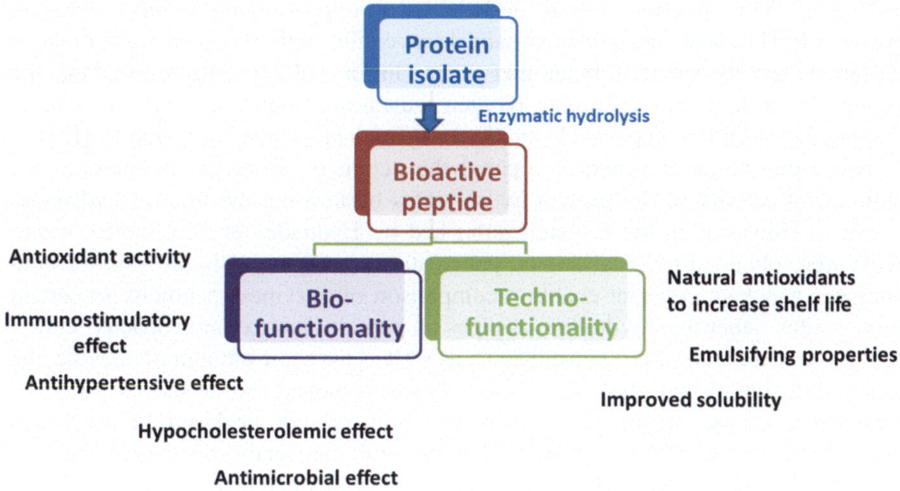

Fig. 8.4 Versatile functionalities of protein hydrolysates obtained by hydrolysis of protein isolate. List of investigated bio- and techno-functionalities

(water-soluble and oil-soluble). The antioxidant property of these hydrolysates was shown to be dependent on protease specificity, degree of hydrolysis and nature of released peptides (e.g. molecular weight and amino acid composition) and it has been attributed to cooperative or combined effects of a number of properties, including their ability to scavenge free radicals, to act as chelating agents of metal ions, or act as hydrogen donator.

The initial HPI extracts have limited bioactive properties as measured by 2,2-diphenyl-1-picrylhydrazyl (DPPH) radical scavenging, oxygen radical absorbance capacity and angiotensin-I-converting enzyme (ACE)-inhibitory activity. The bioactivities of the HPI extracts can be increased upon hydrolysis by the proteases. The degree of hemp protein hydrolysates (HPH) bioactivity depends on the hydrolysis time, as well as the type of protease and substrate (Teh et al. 2016).

To examine the antioxidant properties of hemp seed peptides, protein hydrolysates are prepared using different enzymes, and different hydrolysis times (Tang et al. 2009a). Tang et al. (2009a) conducted hemp protein isolate hydrolysis with alcalase, flavourzyme, neutrase, protamex, pepsin and trypsin for 2 and 4 h, separately, followed by testing for antioxidant activities such as DPPH$^{\bullet}$ scavenging, iron (III) reduction, and iron chelation. They have found that the DPPH radical scavenging and Fe^{2++} chelating abilities of the hydrolysates were positively correlated with the yield of trichloroacetic acid-soluble peptides or surface hydrophobicity values. Similar results were obtained when HPH were produced using Neutrase® (Wang et al. 2009).

Girgih et al. (2011a) performed simulated gastrointestinal hydrolysis of hemp seed proteins using pepsin and pancreatin followed by membrane ultrafiltration fractionation to obtain fractions with peptide sizes less than 1, 1–3, 3–5, and

5–10 kDa. While fractionation of the HPH led to improvements in ferric reducing power, DPPH radical, and hydroxyl radical scavenging activities, a decrease in metal chelation capacity occurred. In another paper, Girgih et al. (2013) confirmed that the peptide fractions exhibited higher oxygen radical absorbance capacity as well as scavenging of DPPH, superoxide and hydroxyl radicals when compared to HPH.

Numerous research papers compared the action of different enzymes on the antioxidant activity of the protein hydrolysates by keeping the time of hydrolysis constant. However, in the research published by Hadnađev et al. (2018b), where HPH were obtained by alcalase and pancreatin, the degree of hydrolysis for both enzymes was kept constant enabling comparison of enzymes specificity to certain protein sites rather than hydrolysis degrees on peptides antioxidant capacity. Utilizing the same amount of protein substrate, as well as the same amount of enzyme, the required degree of hydrolysis (3, 6 and 9%) was achieved for the shorter period of time using alcalase enzyme than pancreatin. In the case of alcalase time range was between 1.8 and 13.5 min, and in the samples with pancreatin it was between 10.9 and 90.9 min (Hadnađev et al. 2018b). This can be explained by the fact that alcalase has a greater affinity for peptide cleavage. Therefore, alcalase could be observed as a more effective choice for the hemp proteins hydrolysis than pancreatin. In general, alkaline proteases, such as alcalase, show better activity in peptide hydrolysis in comparison to acid and neutral proteases (Klompong et al. 2007).

Literature data shows that the ability of the sample (Tang et al. 2009a) to scavenge DPPH$^•$ is closely related to protein hydrolysis degree. The strongest antioxidant activity was detected in samples characterized by the highest degree of hydrolysis (9%) regardless of the enzyme employed. Comparing the IC50 values (the concentration of a sample needed to quench 50% of the initial amount of DPPH$^•$), it can be concluded that hydrolysates obtained by pancreatin showed a stronger antioxidant activity than hydrolysates derived by alcalase (Hadnađev et al. 2018b). Previous studies showed that the ability of protein isolates to scavenge DPPH$^•$ was associated with the high hydrophobicity of the proteins themselves. It was also indicated that smaller protein fractions possessed a higher proportion of hydrophobic amino acids (Pownall et al. 2010). Moreover, it is well known that the antioxidant activity of the proteins is related to amino acid composition (Chen et al. 1995). Since pancreatin represents a mixture of enzymes with different specificities, results of scavenging activity on DPPH$^•$ indicated that in the case of pancreatin hydrolysis, bioactive peptides with a higher proportion of hydrophobic amino acids have been released. The results confirmed that protein hydrolysates exhibited greater scavenging activity on DPPH$^•$ than the isolate (HPI). Moreover, it can be assumed that certain DPPH antiradical activity of HPI was probably due to polyphenol compounds and pigments that have been co-extracted with the isolate, as indicated by the green colour of the isolate itself (Hadnađev et al. 2018b). Namely, phenolic compounds are the reason of the dark colour and undesirable taste of protein isolates. According to Xu and Diosady (2002), phenolic compounds bind to proteins in aqueous solution through the following mechanisms: hydrogen bonds, covalent bonds, hydrophobic interactions and ionic bonds, and their removal represents a major challenge.

IC50 values for reducing power of the protein hydrolysates were in the range of 1.98–4.70 mg/mL, depending on the enzyme used, as well as the hydrolysis time or degree (Hadnađev et al. 2018b). These results are in agreement with the results obtained by Tang et al. (2009a) who investigated the effect of various proteases on the antioxidant properties of the resulting hemp protein hydrolysates. Glutathione exhibited the greatest ability in reducing Fe^{3+} compared to hemp protein isolates and hydrolysates. It is a tripeptide containing cysteine – amino acid with sulfhydryl group. Sulfhydryl groups are considered to be responsible for the expression of the reducing activity (Battin and Brumaghim 2009). The hydrolysates obtained by using pancreatin showed a higher reducing power relative to those obtained by alcalase treatment. Battin and Brumaghim (2009) stated that the type of amino acids, as well as peptide type plays an important role in reducing power activity. Since hydrolysates obtained using pancreatin and alcalase treatment had a lower proportion of sulfhydryl groups in their composition than glutathione, they exhibited poorer ability to reduce Fe^{3+} (Isinguzo 2011).

By fractionation of HPH, Girgih et al. (2014) identified Trp-Val-Tyr-Tyr (WVYY) and Pro-Ser-Leu-Pro-Ala (PSLPA) as the most active antioxidant peptides with 67% and 58% $DPPH^{\bullet}$ scavenging and metal chelation activity of 94% and 96%, respectively.

8.5.2 Antihypertensive/ACE Inhibitory Activity

Elevated blood pressure is one of the major risk factors for cardiovascular disease, a leading cause of death in the developed world (Erdmann et al. 2008). Angiotensin I-converting enzyme (ACE) plays a crucial role in the regulation of blood pressure as it promotes the conversion of angiotensin I to the vasoconstrictor angiotensin II as well as inactivates the vasodilator bradykinin. ACE is an exopeptidase that cleaves dipeptides from the C-terminal of peptide substrate. Binding to ACE is mainly affected by the C-terminal tripeptides sequence of the substrate. Though the mechanism is not well known, previous studies showed that the presence of aromatic, hydrophobic and positively charged amino acid residues at the C-terminus has a major influence in the ACE inhibitory potency of the peptides. Hydrophobic residues tryptophan, tyrosine, phenylalanine and proline were shown to be the most potent ACE inhibitory peptides (Nagpal et al. 2011). Synthetic ACE inhibitors have been used for decades as the antihypertensive agents. In recent years, some food proteins have been identified as sources of ACE inhibitory peptides and are currently the best known class of bioactive peptides (FitzGerald et al. 2004). Although additional investigation is needed, the ACE inhibitory peptides have shown to have potential to be used in initial treatment in mildly hypertensive individuals or as supplemental treatment. They would also represent a less costly alternative in treatment of hypertension. Another advantage is that these peptides have not been associated with the some side effects reported for synthetic ACE inhibitors such as dry cough, skin rashes and angioedema, probably due to the lower ACE inhibitory activity determined in vitro (FitzGerald and Meisel 2000).

Short-term (24 h) oral administration (200 mg/kg body weight) of hemp seed protein hydrolysate (HPH) to spontaneously hypertensive rats was shown to reduce blood pressure (−30 mmHg after 8 h) and was positively correlated with the in vitro ACE and renin inhibitions (Girgih et al. 2011b). Spontaneously hypertensive rats are considered one of the best experimental models for evaluating antihypertensive drugs or food-based inhibitors (Quiñones et al. 2010).

Long term effect of hemp protein products was also determined by Girgih et al. (2014) who investigated the ability of diet containing HPH and HPI to attenuate elevated blood pressure development in spontaneously hypertensive rats. They have showed that after 8 weeks HPH more significantly reduced blood pressure than HPI, demonstrated that a simulated gastrointestinal digested hemp seed protein hydrolysate inhibited renin (IC_{50} of 0.81 mg/mL) and ACE (IC_{50} of 0.67 mg/mL) activities in vitro. Weaker antihypertensive effect of the unhydrolysed protein (HPI) reflects the superior absorption properties of predigested peptides. The higher potency of the HPH diets may also be due to the higher contents of aromatic and hydrophobic amino acids when compared to the HPI. Those results are similar to the once reported for soybean protein hydrolysate (Yang et al. 2004).

The results confirm the potential of HPH as an ingredient that can be used to formulate functional foods for the prevention and for the support in therapy of hypertension.

8.6 Conclusion

Although having a range of techno-functionalities (gelling, emulsifying, foaming, film-forming agents), utilization of hemp proteins in different food matrices is limited since their behaviour highly depend on their structure and composition, environmental factors (pH, ionic strength, type of solvent/salt used, temperature) and isolation technique and conditions. Therefore, incorporation of hemp proteins into food products represents a technological challenge. At the same time, knowledge about hemp proteins limitation to adapt their functionality to environmental and isolation conditions represents a technological advantage which is reflected in target application of the one type of protein for different techno-functionalities in food systems. It was shown that poor solubility of hemp proteins (an important prerequisite for a protein to be an effective techno-functional agent) can be altered by changing the isolation procedure or by application of enzymatic or chemical modification. Enzymatic hydrolysis of hemp protein proved as a powerful tool in providing versatility in its techno-functionalities, as well as in enabling simultaneous expression of techno- and bio-functionality in the same food system since protein hydrolysates can also act as antioxidants in foams, emulsions, edible film, etc.

Acknowledgements This work was financially supported by the Ministry of Education, Science and Technological Development of the Republic of Serbia under the Agreement on the Implementation and Financing of Research of the Institute of Food Technology (No. 451-03-68/2020-14/200222) and the Provincial Secretariat for Higher Education and Scientific Research, Republic of Serbia [grant number 142-451-2138/2019-01/02] as a part of the project entitled "Techno-functionality of proteins from alternative plant sources from Vojvodina region".

References

Aiello G, Fasoli E, Boschin G, Lammi C, Zanoni C, Citterio A, Arnoldi A (2016) Proteomic characterization of hempseed (*Cannabis sativa* L.). J Proteome 147:187–196. https://doi.org/10.1016/j.jprot.2016.05.033

Alsohaimy SA, Sitohy MZ, El-Masry RA (2007) Isolation and partial characterization of chickpea, lupine and lentil seed proteins. World J Agric Sci 3:123–129. ISSN:1817-3047

Amarowicz R (2010) Modification of emulsifying properties of food proteins by enzymatic hydrolysis. Eur J Lipid Sci Technol 112:695–696. https://doi.org/10.1002/ejlt.201000382

Arntfield SD, Ismond MA, Murray ED (1985) The fate of antinutritional factors during the preparation of a fababean protein isolate using a micellization technique. Can Inst Food Sci Technol J 18:137–143. https://doi.org/10.1016/S0315-5463(85)71771-3

Battin EE, Brumaghim JL (2009) Antioxidant activity of sulfur and selenium: a review of reactive oxygen species scavenging, glutathione peroxidase, and metal-binding antioxidant mechanisms. Cell Biochem Biophys 55:1–23. https://doi.org/10.1007/s12013-009-9054-7

Bourtoom T (2008) Edible films and coatings: characteristics and properties. Int Food Res J 15:237–248

Boye JI, Akshay S, Ribéreau S, Mondor M, Farnworth E, Rajamohamed SH (2010) Comparison of the functional properties of pea, chickpea and lentil protein concentrates processed using ultrafiltration and isoelectric precipitation techniques. Food Res Int 43:537–546. https://doi.org/10.1016/j.foodres.2009.07.021

Caetano da Silva Lannes S, Natali Miquelim J (2013) Interfacial behavior of food proteins. Curr Nutr Food Sci 9:10–14. https://doi.org/10.2174/157340113804810914

Callaway JC (2004) Hempseed as a nutritional resource: an overview. Euphytica 140:65–72. https://doi.org/10.1007/s10681-004-4811-6

Chang CY, Wu KC, Chiang SH (2007) Antioxidant properties and protein compositions of porcine haemoglobin hydrolysates. Food Chem 100:1537–1543. https://doi.org/10.1016/j.foodchem.2005.12.019

Chen HM, Muramoto K, Yamauchi F (1995) Structural analysis of antioxidative peptides from soybean β-Conglycinin. J Agric Food Chem 43:574–578. https://doi.org/10.1021/jf00051a004

Dapčević-Hadnađev T, Hadnađev M, Lazaridou A, Moschakis T, Biliaderis CG (2019a) Hempseed meal protein isolates prepared by different isolation techniques. Part II. Gelation properties at different ionic strengths. Food Hydrocoll 81:481–489. https://doi.org/10.1016/j.foodhyd.2018.03.022

Dapčević-Hadnađev T, Dizdar M, Pojić M, Krstonošić V, Zychowski LM, Hadnađev M (2019b) Emulsifying properties of hemp proteins: effect of isolation technique. Food Hydrocoll 89:912–920. https://doi.org/10.1016/j.foodhyd.2018.12.002

Erdmann K, Cheung BWY, Schroder H (2008) The possible roles of food derived bioactive peptides in reducing the risk of cardiovascular disease. J Nutr Biochem 19:643–654. https://doi.org/10.1016/j.jnutbio.2007.11.010

FitzGerald RJ, Meisel H (2000) Milk protein-derived peptide inhibitors of angiotensin-I-converting enzyme. Br J Nutr 8:33–37. https://doi.org/10.1017/S0007114500002221

FitzGerald RJ, Murray BA, Walsh DJ (2004) Hypotensive peptides from milk proteins. J Nutr 134:980–988. https://doi.org/10.1093/jn/134.4.980S

Fredrikson M, Biot P, Alminger ML, Carlsson NG, Sandberg AS (2001) Production process for high quality pea protein isolate with low content of oligosaccharides and phytate. J Agric Food Chem 49:1208–1212. https://doi.org/10.1021/jf000708x

Fuhrmeister H, Meuser F (2003) Impact of processing on functional properties of protein products from wrinkled peas. J Food Eng 56:119–129. https://doi.org/10.1016/S0260-8774(02)00241-8

Girgih AT, Udenigwe CC, Aluko RE (2011a) In vitro antioxidant properties of hemp seed (*Cannabis sativa* L.) protein hydrolysate fractions. J Am Oil Chem Soc 88:381–389. https://doi.org/10.1007/s11746-010-1686-7

Girgih AT, Udenigwe CC, Li H, Adebiyi AP, Aluko RE (2011b) Kinetics of enzyme inhibition and antihypertensive effects of hemp seed (*Cannabis sativa* L.) protein hydrolysates. J Am Oil Chem Soc 88:1767–1774. https://doi.org/10.1007/s11746-011-1841-9

Girgih AT, Udenigwe CC, Aluko RE (2013) Reverse-phase HPLC separation of hemp seed (*Cannabis sativa L.*) protein hydrolysate produced peptide fractions with enhanced antioxidant capacity. Plant Foods Hum Nutr 68:39–46. https://doi.org/10.1007/s11130-013-0340-6

Girgih AT, Alashi AM, He R, Malomo SA, Aluko RE (2014) Preventive and treatment effects of hemp seed (*Cannabis sativa* L.) meal protein hydrolysate against high blood pressure in spontaneously hypertensive rats. Eur J Nutr 53:1237–1246. https://doi.org/10.1007/s00394-013-0625-4

Gómez-Guillén MC, Giménez B, López-Caballero MA, Montero MP (2011) Functional and bioactive properties of collagen and gelatin from alternative sources: a review. Food Hydrocoll 25:1813–1827. https://doi.org/10.1016/j.foodhyd.2011.02.007

Hadnađev M, Hadnađev-Dapčević T, Pojić M, Šarić B, Mišan A, Jovanov P, Sakač M (2017) Progress in vegetable proteins isolation techniques: a review. Food Feed Res 44:11–21. https://doi.org/10.5937/FFR1701011H

Hadnađev M, Dapčević-Hadnađev T, Lazaridou A, Moschakis T, Michaelidou AM, Popović S, Biliaderis CG (2018a) Hempseed meal protein isolates prepared by different isolation techniques. Part I. Physicochemical properties. Food Hydrocoll 79:526–533. https://doi.org/10.1016/j.foodhyd.2017.12.015

Hadnađev M, Dizdar M, Dapčević Hadnađev T, Jovanov P, Mišan A, Sakač M (2018b) Hydrolyzed hemp seed proteins as bioactive peptides. J Process Energy Agric 22:90–94. https://doi.org/10.5937/JPEA1802090H

Hartmann R, Meisel H (2007) Food-derived peptides with biological activity: from research to food applications. Curr Opin Biotechnol 18:163–169. https://doi.org/10.1016/j.copbio.2007.01.013

Hernandez-Ledesma B, Quiros A, Amigo L, Recio I (2007) Identification of bioactive peptides after digestion of human milk and infant formula with pepsin and pancreatin. Int Dairy J 17:42–49. https://doi.org/10.1016/j.idairyj.2005.12.012

House JD, Neufeld J, Leson G (2010) Evaluating the quality of protein from hemp seed (*Cannabis sativa* L.) products through the use of the protein digestibility-corrected amino acid score method. J Agric Food Chem 58:11801–11807. https://doi.org/10.1021/jf102636b

Isinguzo G (2011) Physicochemical, functional and *in vitro* bioactive properties of hempseed (*Cannabis sativa*) protein isolates and hydrolysates. MSc theses, Faculty of Graduate Studies of the University of Manitoba, Canada

Karaca AC, Low N, Nickerson M (2011) Emulsifying properties of chickpea, faba bean, lentil and pea proteins produced by isoelectric precipitation and salt extraction. Food Res Int 44:2742–2750. https://doi.org/10.1016/j.foodres.2011.06.012

Kaushik P, Dowling K, McKnight S, Barrow CJ, Wang B, Adhikari B (2016) Preparation, characterization and functional properties of flax seed protein isolate. Food Chem 197:212–220. https://doi.org/10.1016/j.foodchem.2015.09.106

Kim JJ, Lee MY (2011) Isolation and characterization of edestin from Cheungsam hempseed. J Appl Biol Chem 54:84–88. https://doi.org/10.3839/jabc.2011.015

Kim SY, Je JY, Kim SK (2007) Purification and characterization of antioxidant peptide from hoki (*Johnius belengerii*) frame protein by gastrointestinal digestion. J Nutr Biochem 18:31–38. https://doi.org/10.1016/j.jnutbio.2006.02.006

Kiosseoglou V, Paraskevopoulou A (2011) Chapter 3: Functional and physicochemical properties of pulse proteins. In: Tiwari BK, Gowen A, Mckenna B (eds) Pulse foods processing: quality and nutritional applications. Academic, San Diego, pp 57–89

Klompong V, Benjakul S, Kantachote D, Shahidi F (2007) Antioxidative activity and functional properties of protein hydrolysate of yellow stripe trevally (*Selaroidesleptolepis*) as influenced by the degree of hydrolysis and enzyme type. Food Chem 102:1317–1327. https://doi.org/10.1016/j.foodchem.2006.07.016

Korhonen H, Pihlanto A (2006) Bioactive peptides: production and functionality. Int Dairy J 16:945–960. https://doi.org/10.1016/j.idairyj.2005.10.012

Korus J, Witczak M, Ziobro R, Juszczak L (2017) Hemp (*Cannabis sativa subsp. sativa*) flour and protein preparation as natural nutrients and structure forming agents in starch based gluten-free bread. Lebensm Wiss Technol 84:143–150. https://doi.org/10.1016/j.lwt.2017.05.046

Kunwar A, Priyadarsini KI (2011) Free radicals, oxidative stress and importance of antioxidants in human health. J Med Allied Sci 1:53–60. ISSN:2231-1696

Lam RS, Nickerson MT (2013) Food proteins: a review on their emulsifying properties using a structure–function approach. Food Chem 141:975–984. https://doi.org/10.1016/j.foodchem.2013.04.038

Liu Y, Zhao G, Ren J, Zhao M, Yang B (2011) Effect of denaturation during extraction on the conformational and functional properties of peanut protein isolate. Innov Food Sci Emerg Technol 12:375–380. https://doi.org/10.1016/j.ifset.2011.01.012

López OP, Ordorica-Falomir C (1986) Production of safflower protein isolates: composition, yield and protein quality. J Sci Food Agric 37:1097–1103. https://doi.org/10.1002/jsfa.2740371107

Ly YP, Johnson LA, Jane J (1998) Chapter 6: Soy protein as biopolymer. In: Kaplan D (ed) Biopolymers from renewable resources. Springer, Berlin/Heidelberg, pp 144–176

Malomo SA, Aluko RE (2015a) A comparative study of the structural and functional properties of isolated hemp seed (*Cannabis sativa* L.) albumin and globulin fractions. Food Hydrocoll 43:743–752. https://doi.org/10.1016/j.foodhyd.2014.08.001

Malomo SA, Aluko RE (2015b) Conversion of a low protein hemp seed meal into a functional protein concentrate through enzymatic digestion of fibre coupled with membrane ultrafiltration. Innov Food Sci Emerg Technol 31:151–159. https://doi.org/10.1016/j.ifset.2015.08.004

Malomo SA, He R, Aluko RE (2014) Structural and functional properties of hemp seed protein products. J Food Sci 79:C1512–C1521. https://doi.org/10.1111/1750-3841.12537

Malomo S, Onuh J, Girgih A, Aluko R (2015) Structural and antihypertensive properties of enzymatic hemp seed protein hydrolysates. Nutrients 7:7616–7632. https://doi.org/10.3390/nu7095358

Mondor M, Aksay S, Drolet H, Roufik S, Farnworth E, Boye JI (2009) Influence of processing on composition and antinutritional factors of chickpea protein concentrates produced by isoelectric precipitation and ultrafiltration. Innov Food Sci Emerg Technol 10:342–347. https://doi.org/10.1016/j.ifset.2009.01.007

Moure A, Domínguez H, Parajó JC (2006) Antioxidant properties of ultrafiltration-recovered soy protein fractions from industrial effluents and their hydrolysates. Process Biochem 41:447–456. https://doi.org/10.1016/j.procbio.2005.05.014

Murray ED, Myers CD, Barker LD (1979) U.S. Patent No. 4,169,090. U.S. Patent and Trademark Office, Washington, DC

Murray ED, Arntfield SD, Ismond MA (1985) The influence of processing parameters on food protein functionality II. Factors affecting thermal properties as analyzed by differential scanning calorimetry. Can Inst Food Sci Technol J 18:158–162. https://doi.org/10.1016/S0315-5463(85)71774-9

Mwasaru MA, Muhammad K, Bakar J, Man YBC (1999) Effects of isolation technique and conditions on the extractability, physicochemical and functional properties of pigeonpea

(*Cajanus cajan*) and cowpea (*Vigna unguiculata*) protein isolates. I. Physicochemical properties. Food Chem 67:435–443. https://doi.org/10.1016/S0308-8146(99)00150-8

Nagpal R, Behare P, Rana R, Kumar A, Kumar M, Arora S, Morotta F, Jain S, Yadav H (2011) Bioactive peptides derived from milk proteins and their health beneficial potentials: an update. Food Funct 2:18–27. https://doi.org/10.1039/c0fo00016g

Papalamprou EM, Doxastakis GI, Biliaderis CG, Kiosseoglou V (2009) Influence on preparation methods on physicochemical and gelation properties of chickpea protein isolates. Food Hydrocoll 23:337–343. https://doi.org/10.1016/j.foodhyd.2008.03.006

Papalamprou EM, Doxastakis GI, Kiosseoglou V (2010) Chickpea protein isolates obtained by wet extraction as emulsifying agents. J Sci Food Agric 90:304–313. https://doi.org/10.1002/jsfa.3816

Patel S, Cudney R, McPherson A (1994) Crystallographic characterization and molecular symmetry of edestin, a legumin from hemp. J Mol Biol 235:361–363. https://doi.org/10.1016/S0022-2836(05)80040-3

Pihlanto A (2006) Antioxidative peptides derived from milk proteins. Int Dairy J 16:1306–1314. https://doi.org/10.1016/j.idairyj.2006.06.005

Pihlanto A, Mattila P, Mäkinen S, Pajari AM (2017) Bioactivities of alternative protein sources and their potential health benefits. Food Funct 8:3443–3458. https://doi.org/10.1039/C7FO00302A

Pojić M, Mišan A, Sakač M, Dapčević Hadnađev T, Šarić B, Milovanović I, Hadnađev M (2014) Characterization of byproducts originating from hemp oil processing. J Agric Food Chem 62:12436–12442. https://doi.org/10.1021/jf5044426

Pojić M, Dapčević Hadnađev T, Hadnađev M, Rakita S, Brlek T (2015) Bread supplementation with hemp seed cake: a by-product of hemp oil processing. J Food Qual 38(6):431–440. https://doi.org/10.1111/jfq.12159

Ponzoni E, Brambilla IM, Galasso I (2018) Genome-wide identification and organization of seed storage protein genes of *Cannabis sativa*. Biol Plant 62:693–702. https://doi.org/10.1007/s10535-018-0810-7

Pownall TL, Udenigwe CC, Aluko RE (2010) Amino acid composition and antioxidant properties of pea seed (*Pisum sativum* L.) enzymatic protein hydrolysate fractions. J Agric Food Chem 58:4712–4718. https://doi.org/10.1021/jf904456r

Quiñones M, Sánchez D, Muguerza B, Moulay L, Laghi S, Miguel M, Aleixandre A (2010) Long-term intake of CocoanOX attenuates the development of hypertension in spontaneously hypertensive rats. Food Chem 122:1013–1019. https://doi.org/10.1016/j.foodchem.2010.03.059

Raikos V, Duthie G, Ranawana V (2015) Denaturation and oxidative stability of hemp seed (*Cannabis sativa L.*) protein isolate as affected by heat treatment. Plant Foods Hum Nutr 70:304–309. https://doi.org/10.1007/s11130-015-0494-5

Rajapakse N, Mendis E, Byun HG, Kim SK (2005) Purification and in vitro antioxidative effects of giant squid muscle peptides on free radical-mediated oxidative systems. J Nutr Biochem 16:562–569. https://doi.org/10.1016/j.jnutbio.2005.02.005

Rezig L, Chibani F, Chouaibi M, Dalgalarrondo M, Hessini K, Guéguen J, Hamdi S (2013) Pumpkin (*Cucurbita maxima*) seed proteins: sequential extraction processing and fraction characterization. J Agric Food Chem 61:7715–7721. https://doi.org/10.1021/jf402323u

Rodríguez-Ambriz SL, Martínez-Ayala AL, Millán F, Davila-Ortiz G (2005) Composition and functional properties of Lupinus campestris protein isolates. Plant Foods Hum Nutr 60:99–107. https://doi.org/10.1007/s11130-005-6835-z

Rouimi S, Schorsch C, Valentini C, Vaslin S (2005) Foam stability and interfacial properties of milk protein–surfactant systems. Food Hydrocoll 19:467–478. https://doi.org/10.1016/j.foodhyd.2004.10.032

Ryan JT, Ross RP, Bolton D, Fitzgerald GF, Stanton C (2011) Bioactive peptides from muscle sources: meat and fish. Nutrients 3:765–791. https://doi.org/10.3390/nu3090765

Singh U (1988) Antinutritional factors of chickpea and pigeonpea and their removal by processing. Plant Foods Hum Nutr 38:251–261. https://doi.org/10.1007/BF01092864

Singhal A, Karaca AC, Tyler R, Nickerson M (2016) Chapter 3: Pulse proteins: from processing to structure-function relationships. In: Goyal AK (ed) Grain legumes. IntechOpen, Rijeka, pp 55–78

Söderberg J (2013) Functional properties of legume proteins compared to egg proteins and their potential as egg replacers in vegan food. Swedish University of Agricultural Sciences

Tang CH, Ten Z, Wang XS, Yang XQ (2006) Physicochemical and functional properties of hemp (*Cannabis sativa* L.) protein isolate. J Agric Food Chem 54:8945–8950. https://doi.org/10.1021/jf0619176

Tang CH, Wang XS, Yang XQ (2009a) Enzymatic hydrolysis of hemp (*Cannabis sativa* L.) protein isolate by various proteases and antioxidant properties of the resulting hydrolysates. Food Chem 114:1484–1490. https://doi.org/10.1016/j.foodchem.2008.11.049

Tang CH, Peng J, Zhen DW, Chen Z (2009b) Physicochemical and antioxidant properties of buckwheat (*Fagopyrumesculentum Moench*) protein hydrolysates. Food Chem 115:672–678. https://doi.org/10.1016/j.foodchem.2008.12.068

Tcholakova S, Denkov ND, Ivanov IB, Campbell B (2006) Coalescence stability of emulsions containing globular milk proteins. Adv Colloid Interf Sci 123:259–293. https://doi.org/10.1016/j.cis.2006.05.021

Teh SS, Bekhit AE, Carne A, Birch J (2014) Effect of the defatting process, acid and alkali extraction on the physicochemical and functional properties of hemp, flax and canola seed cake protein isolates. J Food Meas Charact 8:92–104. https://doi.org/10.1007/s11694-013-9168-x

Teh SS, Bekhit AE, Carne A, Birch J (2016) Antioxidant and ACE-inhibitory activities of hemp (*Cannabis sativa* L.) protein hydrolysates produced by the proteases AFP, HT, Pro-G, actinidin and zingibain. Food Chem 203:199–206. https://doi.org/10.1016/j.foodchem.2016.02.057

Vose JR (1980) Production and functionality of starches and protein isolates from legume seeds (field peas and horse beans). Cereal Chem 57:406–410

Waggle DH, Steinke FH, Shen JL (1989) Isolated soy proteins. In: Matthews RH (ed) Legumes: chemistry, technology and human nutrition. Marcel Dekker, New York, pp 99–138

Wang XS, Tang CH, Yang XQ, Gao WR (2008) Characterization, amino acid composition and in vitro digestibility of hemp (*Cannabis sativa* L.) proteins. Food Chem 107:11–18. https://doi.org/10.1016/j.foodchem.2007.06.064

Wang XS, Tang CH, Chen L, Yang XQ (2009) Characterization and antioxidant properties of hemp protein hydrolysates obtained with Neutrase®. Food Technol Biotechnol 47:428–434. ISSN:1330-9862

Xu L, Diosady LL (2002) Removal of phenolic compounds in the production of high-quality canola protein isolates. Food Res Int 35:23–30. https://doi.org/10.1016/S0963-9969(00)00159-9

Yang HY, Yang SC, Chen JR, Tzeng YH, Han BC (2004) Soyabean protein hydrolysate prevents the development of hypertension in spontaneously hypertensive rats. Br J Nutr 92:507–512. https://doi.org/10.1079/BJN20041218

Yin SW, Tang CH, Wen QB, Yang XQ (2007) Properties of cast films from hemp (*Cannabis sativa L.*) and soy protein isolates. A comparative study. J Agric Food Chem 55:7399–7404. https://doi.org/10.1021/jf071117a

Yin SW, Tang CH, Cao JS, Hu EK, Wen QB, Yang XQ (2008) Effects of limited enzymatic hydrolysis with trypsin on the functional properties of hemp (*Cannabis sativa L.*) protein isolate. Food Chem 106:1004–1013. https://doi.org/10.1016/j.foodchem.2007.07.030

Yin SW, Tang CH, Wen QB, Yang XQ (2009) Functional and structural properties and in vitro digestibility of acylated hemp (*Cannabis sativa* L.) protein isolates. Int J Food Sci Technol 44:2653–2661. https://doi.org/10.1111/j.1365-2621.2009.02098.x

Zhang W, Yang YJ, Wang JJ (2008) Study on preparation of industrial hemp protein isolates [Chinese]. Sci Technol Food Ind 03. TS201.21

Zhu KX, Zhou HM, Qian HF (2006) Antioxidant and free radical-scavenging activities of wheat germ protein hydrolysates (WGPH) prepared with alcalase. Process Biochem 41:1296–1302. https://doi.org/10.1016/j.procbio.2005.12.029

Chapter 9
Hemp Seed as a Source of Food Proteins

François Potin and Rémi Saurel

Abstract The galloping population growth is leading to a significant increase in the protein demand. Conventional protein sources, particularly those of animal origin, will no longer be sufficient to meet this demand. The use of plant proteins, less costly in terms of resources and with a much lower environmental impact, is an interesting alternative to meet future societal and environmental challenges. Hemp seed is an undervalued co-product resulting from the cultivation of industrial hemp. This plant resource has significant contents in protein and oil of nutritional value, about 26% and 36% respectively, that may help to meet the challenges of sustainable food. The specific valorization of hemp proteins for human consumption is a recent issue. Through this chapter, we showed the current state of knowledge on hemp proteins in terms of composition, nutritional aspect, extraction, and physicochemical, functional and biological properties. Different extraction routes have been proposed to recover the main hemp protein fractions from oil press cakes in general. Extraction yields generally vary from 34% to 51%, and the protein purity of the resulting protein isolates from 87% to 94%. The proteins, usually extractible from hemp meal, belong to the globulin and albumin families. Most of hemp protein isolates contain predominantly globulins, mainly the 11S edestins. The hemp proteins have an interesting essential amino acid profile and have a high digestibility of about 90%. Many authors have highlighted that hemp protein hydrolysates possess a wide range of health biological activities such as antioxidant properties, metal chelation, antihypertensive, hypoglycemic properties... Hemp proteins also have techno-functional properties such as gelling, emulsifying and foaming properties adapted to the development of new plant-based foods. Different treatments of hemp proteins can improve their functional properties, such as enzymatic and chemical modifications or pH- and heat-induced denaturation. Despite limited solubility, hemp protein ingredients represent an alternative to current cereal and legume protein materials in human diet in the future.

F. Potin · R. Saurel (✉)
Procédés Alimentaires et Microbiologiques, UMR A 02.102, University of Bourgogne Franche-Comté, AgroSup Dijon, Dijon, France
e-mail: remi.saurel@agrosupdijon.fr

G. Crini, E. Lichtfouse (eds.), *Sustainable Agriculture Reviews 42*,
Sustainable Agriculture Reviews 42, https://doi.org/10.1007/978-3-030-41384-2_9

Keywords Hemp · Protein · Extraction · Techno-functional properties · Biological activity

Abbreviations

HEC	Hemp Enzyme Concentrate, hemp protein concentrate obtained by enzyme-assisted extraction
HMI	Hemp Micellized Isolate, hemp protein isolate obtained by micellization
HPH	Hemp Protein Hydrolysate
HPI	Hemp Precipitated Isolate, hemp protein isolate obtained by globulin isoelectric point precipitation
SPI	Soy Precipitated Isolate, soy protein isolate obtained by globulin isoelectric point precipitation

9.1 Introduction

According to the forecasts of the Food and Agriculture Organization of the United Nations (FAO), the world's population is expected to reach 9.1 billion by 2050, requiring a 70% increase in world food production. In 2017, 11% of the world's population was undernourished and 13% of adults were obese (FAO 2018). Feeding the planet by fighting against global warming are the major issues that will have to be faced, especially as global warming is leading to a decrease in crop yields each year (FAO). Ovovegetarian food is known to require less energy, land and water resources than meat food (Pimentel and Pimentel 2003). The consumption of more plant proteins is a solution to face these problems. Global protein demand was expected to increase by 40% between 2010 and 2030. This increase was expected to be 33% and 43% for animal and plant protein demands respectively (Ozanne 2014).

Cereals, legumes and oilseeds are the main sources of plant proteins. It is estimated that only 35% of the total proteins produced by the agricultural sector is used for human consumption. The rest is used for livestock feed, non-food applications, or simply treated as a waste. Soy proteins or corn and wheat flours, are already widely used. In contrast, "conventional" sources of protein should no longer be sufficient to meet the global demand for protein in the coming years. Finding alternative sources of protein is strongly recommended (Sari et al. 2015).

Hemp is an annual herbaceous plant, highly harvested worldwide, mainly for its fibers in stems (Ash 1948). Asia is the biggest producer of hemp, and cultivates nearly 75% of the world's hemp. In Asia, production is mainly shared between China, which grows 50% of the world's hemp, and North Korea. Europe is the second largest hemp growing continent. In 2004, it produced 14% of the world's hemp, but it has seen a renewed interest in hemp cultivation since 2011. The area under cultivation has increased from 8,000 ha in 2011 to 33,000 ha in 2016. France is the European leader and produces nearly half of European production, placing it in third position worldwide behind North Korea (Bouloc 2006; Carus et al. 2013; Amaducci et al. 2015).

Hemp is an ecologically and economically interesting plant for many reasons. It has a strong and rapid growth since it is harvested 4 months after seeding and produces a significant amount of biomass (Ash 1948). It does not need phytosanitary products because it has the ability to eradicate weeds, and is not affected by any pests or diseases. It is one of the few agricultural products to be grown without phytosanitary products in non-organic farms (Carus et al. 2013). It is used in rotation culture and increases the yield of the next culture by 10–15%. It has the ability to remove large quantities of heavy metals from the soil, and its long root system helps to prevent soil erosion. It also has low water and input requirements (Ranalli and Venturi 2004).

One hectare of hemp produces about 5 tons of straw and 1 ton of seed. All parts of hemp are recoverable. Fibers which represent about 30% of the straw are mainly used for pulp and paper, as well as for insulation. The woody core of the stems, that account for about 50% of the straw is mainly used as horse bedding or as garden mulch. The dusts, corresponding to about 15% of the straw are pelletized for incineration or used as compost. Flowers and leaves can also be used in pharmaceuticals or food supplements. Seeds are still undervalued, since they are mainly used for livestock feed, or consumed as such as human food (Carus et al. 2013). The large quantities of oils and proteins, of nutritional interests, contained in hemp seeds, are not yet valued at their true value in human food.

The purpose of this chapter is to synthesize scientific knowledge on hemp proteins in terms of composition, nutritional aspect, extraction, and physicochemical, functional and biological properties.

9.2 Hemp Seed Composition

The composition of hemp seed and its derivatives is reported Table 9.1 (Callaway 2004; Tang et al. 2006; House et al. 2010; Kim and Lee 2011b; Aluko 2017; Sarv 2017; Mattila et al. 2018; Balentić et al. 2019; Mamone et al. 2019). Hemp seeds have high contents in protein and fat. Whole seeds consist of approximately 26.2% protein and 36% lipid (all percentages in w/w dry basis). Hulled seeds have around 35.7% and 41.7% contents for proteins and lipids respectively. The seed hull contains about 69% carbohydrates, richer than the whole seeds with about 32% content, which are themselves richer in carbohydrates than the hulled seeds with about 16% content. Hemp carbohydrates are mostly fibers. Fibers account for about 93.2% of carbohydrates for the various hemp seed derivatives. Hemp seeds are usually cold-pressed to extract oil. The residual cakes or meals thus obtained have only around 11% lipids. The proteins are then concentrated at ~36%. Further delipidation by using organic solvent can be applied to the meal to remove almost all the lipids, increasing the protein content to ~43%. The minerals are mostly present in the seed pulp. The ash content of shelled seeds is indeed 6.7% against

Table 9.1 Composition of hemp seed and derived products

	Whole seeds				Hulled seeds		Seed hulls		Cold pressed hemp meal				Delipidated hemp meal			
g/100 g wet basis	Mean	±	SD	n	Mean	n	Mean	n	Mean	±	SD	n	Mean	±	SD	n
Proteins	24.5	±	0.4	5	35.9	1	12.7	1	33.4	±	5.2	4	39.2	±	9.7	3
					32.0	1	13.1	1								
Lipids	33.7	±	4.3	5	46.7	1	10.3	1	10.6	±	0.4	4	1.0			1
					32.5	1	11.5	1								
Moisture	6.4	±	0.4	4	4.9	1	5.1	1	6.4	±	1.7	4	8.6	±	1.6	3
							8.0	1								
Ashes	5.1	±	0.4	3	6.4	1	3.9	1	6.7	±	0.4	4	9.7			1
							3.2	1					3.2			1
Carbohydrates	27.6			1	15.1	1	64.1	1	45.3	±	3.5	3	57.3			1
	32.1			1												
Total fibers	21.6			1					42.6			1	45.4			1
	31.5			1					49.2			1				
Energy (kJ/100 g)	2248.9	±	133.5	3	2770.0	1	2020.0	1	2040.0			1				
							1738.8	1	1700.0			1				
References	Callaway (2004), House et al. (2010), Aluko (2017), (Sarv 2017), and Mattila et al. (2018)				House et al. (2010) and Kim and Lee (2011b)		House et al. (2010) and Mattila et al. (2018)		Callaway (2004), House et al. (2010), and Mamone et al. (2019)				Tang et al. (2006), Sarv (2017), and Balentić et al. (2019)			

SD: standard deviation, n: number of references

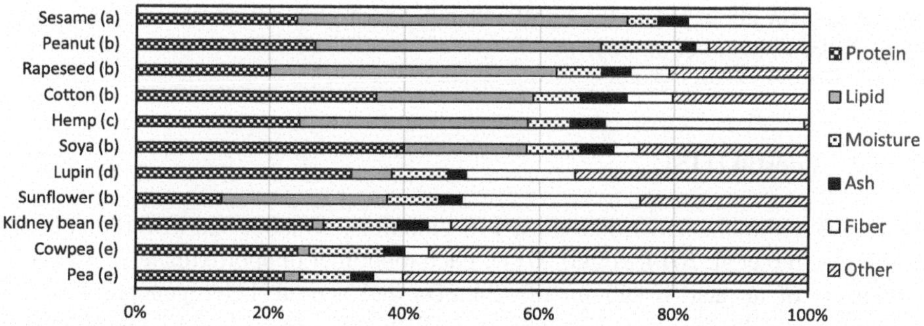

Fig. 9.1 Composition of different oleaginous and proteaginous seeds. a: Elleuch et al. (2007). b: Rodrigues et al. (2012). c: Callaway (2004), House et al. (2010), Aluko (2017), Sarv (2017), Mattila et al. (2018). d: Erbaş et al. (2005). e: Khattab et al. (2009)

3.8% for the seed hull. The energy value of hemp seed products ranges from 1997.9 kJ/100 g for meal to 2912.7 kJ/100 g for hulled seed.

The comparison of hemp seeds with other oilseeds and protein crops is presented Fig. 9.1 in terms of composition (Callaway 2004; Erbaş et al. 2005; Elleuch et al. 2007; Khattab et al. 2009; House et al. 2010; Rodrigues et al. 2012; Aluko 2017; Sarv 2017; Mattila et al. 2018). Hemp seeds have protein content in the range of other seeds and pulses. Their composition is similar to that of peas, cowpeas, red beans, peanuts, or sesame seeds. Their protein content is significantly higher compared to sunflower seeds or canola seeds, but still less rich than for their soybean, cotton and lupine counterparts. The amount of lipids contained in hemp seeds is higher than the average of seeds. Thus, hemp seeds contain more fat than peas, cowpeas, red beans, lupine seeds, sunflower seeds, soya beans and cotton seeds. They contain less fat than rapeseeds, sesame seeds and peanuts. Regarding protein and lipid contents, hemp seeds are then intermediate between oilseeds and proteaginous pulses.

9.3 Hemp Proteins

Hemp seed proteins are predominantly composed of about 65–80% of globulins. The remaining fraction corresponds to the albumins, amounting to about 20–30% (Tang et al. 2006; Park et al. 2012; Aluko 2017; Ponzoni et al. 2018).

9.3.1 Globulins

Hemp globulins are represented predominantly by 11S edestins representing about 65% of total proteins and 93% of globulins, and in a minority by 7S globulins accounting for about 5% of total proteins and 7% of globulins (Osburn 1992; Tang et al. 2006; Ponzoni et al. 2018).

9.3.1.1 Edestin (11S)

Edestin is a homohexamer, like most 11S globulins, of about 320 kDa (Patel et al. 1994; Docimo et al. 2014; Aluko 2017). Each monomer of approximately 54 kDa is composed of an acid α subunit of ~34 kDa and a more heterogeneous basic β subunit, generally forming 2 characteristic bands at ~18 and ~21 kDa upon gel electrophoresis (Tang et al. 2009; Kim and Lee 2011a; Pihlanto et al. 2017). The acid α and basic β subunits are linked together by a disulfide bridge. No disulfide bridge is involved between the monomers, and hydrophobic bonds seem to be preferentially at the origin of the hexameric structure (Tang et al. 2006; Wang et al. 2008, 2018; Kim and Lee 2011b; Malomo and Aluko 2015a; Hadnađev et al. 2018; Mamone et al. 2019). Kim and Lee (2011b) detected edestin bands by gel electrophoresis after periodic acid-Schiff staining, suggesting that edestin could contain carbohydrate parts in its structure.

Three types of edestin have been identified: CsEde1, CsEde2 and CsEde3 (Docimo et al. 2014; Ponzoni et al. 2018). Each type is itself composed of two isoforms. There are then 6 genes encoding edestin. The 2 genes coding for type 1 edestin, CdEde1A and CdEde1B, and the 2 genes coding for type 3 edestin, CdEde3A and CdEde3B, are carried by a common DNA fragment of 16,071 bp while the 2 genes coding for type 2 edestin, CdEde2A and CdEde2B, are carried by another fragment of 8,232 bp. An edestin gene, composed of 4 exons and 3 introns, codes jointly for the acid α and basic β subunits. A cleavage site is present between these 2 sequences to form the acid α and basic β subunits after post-translational modifications. The edestin forms coded by the 6 genes have a common N-terminal sequence of 23 amino acids and contain 4 cysteine residues; 3 of these residues locate on the acid α subunit and the last one on the basic β subunit. These cysteine residues allow the inter-chain disulfide bond between the α and β subunits and an intra-chain disulfide bond within the acid α subunit.

9.3.1.2 7S Globulin

7S globulin consists of a subunit of about 48 kDa (Tang et al. 2006, 2009; Ponzoni et al. 2018; Mamone et al. 2019). A single gene encoding 7S globulin has been identified on a 1893 bp DNA fragment. This gene has 5 exons and 4 introns and encodes a 53.5 kDa precursor polypeptide. The 7S polypeptide has a N-terminal sequence of 21 amino acids, 4 N-glycosylation sites and contains 3 cysteine residues. Wang et al. (2008) observed that the 48 kDa subunit of 7S globulin appeared to be partially associated with the β basic subunit of edestin through disulphide bond.

9.3.2 Albumins

According to Malomo and Aluko (2015a), the albumins consist mainly of 7 polypeptides of 6–35 kDa, unaffected by the addition of reducing agent, thus meaning no inter-chain disulfide bond is formed in their structure. Odani and Odani (1998) and Ponzoni et al. (2018) have identified specifically a 2S albumin of 10 kDa. It consists of 2 subunits of 7 and 3 kDa respectively, linked by 2 disulfide bridges in contradiction with the previous authors. Two genes encode this protein: Cs2S-1 and Cs2S-2. These 2 genes are present on the same 13,738 bp DNA fragment, and encode the same 142 amino acid precursor polypeptide. The 2 subunits are coded by the same gene. The precursor polypeptide contains a N-terminal sequence of 23 amino acids and a cleavage site, which leads to the dissociation of the 2 subunits by post-translational modifications. Albumin 2S is very rich in sulfur amino acids representing 18% of the total amino acids of this protein. The small subunit contains 2 cysteine and 3 methionine residues, while the large subunit contains 6 cysteine and 5 methionine residues. The largest subunit forms 2 intra-chain disulfide bonds.

9.3.3 Amino Acid Composition and Nutritional Value

9.3.3.1 Amino Acids

The amino acid composition of hemp seed and its derivatives is presented Table 9.2. As usual presentation, sufficient and limiting/deficient amounts of amino acids have been established based on the recommendations of the FAO/World Health Organization and United Nations University (FAO/WHO and UNU 2007) for feeding children aged from 1 to 2 years or for adults over 18 years old. Hemp proteins from whole and hulled seeds, seed hull, press-cake and hemp protein isolates, contain all essential amino acids in sufficient quantities for adult nutrition, except lysine, which is the only limiting amino acid. According to Malomo and Aluko (2015a), only the albumins contain lysine in sufficient quantity, but they appear limiting in isoleucine, leucine, tryptophan and in aromatic amino acids (Phenylalanine + Tyrosine) for both adult and children nutrition, and in valine only for children nutrition. These authors also determined the amount of amino acids in hemp globulin and albumin isolates. According to the combined amino acid data for albumins and globulins, total hemp proteins should be limiting in isoleucine, leucine, tryptophan and valine, which is not the case according to all other authors. The amino acid composition of hemp protein isolates resulting from selective isoelectric point precipitation (HPI) is representative of globulins (Tang et al. 2006; Girgih et al. 2010, 2013a; Malomo and Aluko 2015b; Ren et al. 2016; Hadnađev et al. 2018; Wang et al. 2018). Tryptophan is present in insufficient amount in the seed hull proteins for children nutrition. On the other hand, it is sufficient for adult nutrition (House et al. 2010; Mattila et al. 2018). Finally, the 7S globulin fraction and the 11S globulin fraction have rather similar amino acid compositions. However, the 7S

Table 9.2 Amino acid composition of proteins from hemp seed and derived products

g/100 g of tot AA	Whole seed		Hulled seed		Hull		Hemp presscake		HPI		HMI	Glb	Alb	7S	11S	FAO recommandations	
																1–2 Y/O	>18 Y/O
AA	Value ± SD	n	Value ± SD	n	Value ± SD	n	Value ± SD	n	Value ± SD	n	Value	Value	Value	Value	Value	Value	
Alanine	4.7 ± 0.8	3	4.6 ± 0.3	2	4.7 ± 0.1	2	4.4 ± 0.2	2	4.4 ± 0.3	8	4.5	2.8	4.1	5.1	4.9		
Arginine	11.9 ± 1.6	3	12.6 ± 0.8	2	10.3 ± 1.2	2	12.4 ± 2.8	2	12.9 ± 2.0	8	10.9	16.1	13.4	9.2	10.7		
Asx	11.1 ± 1.1	3	10.9 ± 0.5	2	10.8 ± 0.2	2	10.7 ± 1.4	2	11.0 ± 0.8	8	11.2	9.5	8.3	10.6	10.1		
Cysteine	1.7 ± 0.2	3	1.1 ± 1.1	2	1.9 ± 0.3	2	1.8 ± 0.2	2	0.9 ± 0.6	8	1.6	3.3	3.4	0.2	0.3		
Glycine	4.8 ± 0.3	3	4.7 ± 0.1	2	4.9 ± 0.0	2	4.8 ± 0.5	2	4.2 ± 0.2	8	4.1	4.1	8.7	4.6	4.4		
Glx	17.9 ± 1.9	3	19.8 ± 2.6	2	14.5 ± 0.5	2	18.1 ± 2.8	2	19.5 ± 2.2	8	19.1	21.5	21.4	19.1	19.1		
Histidine	2.7 ± 0.4	3	2.9 ± 0.2	2	2.6 ± 0.5	2	3.0 ± 0.7	2	2.9 ± 0.2	8	3.1	3.9	3.9	3.8	3.8	1.8	1.5
Isoleucine	3.9 ± 0.4	3	3.9 ± 0.2	2	4.3 ± 0.4	2	3.9 ± 0.0	2	4.0 ± 0.3	8	5.1	2.9	2.1	3.7	3.8	3.1	3.0
Leucine	6.9 ± 0.6	3	6.6 ± 0.6	2	7.7 ± 1.0	2	6.9 ± 0.8	2	6.8 ± 0.6	8	7.1	5.6	4.2	7.4	7.4	6.3	5.9
Lysine	4.0 ± 0.5	3	3.6 ± 0.0	2	3.9 ± 0.0	2	3.9 ± 0.5	2	3.5 ± 0.6	8	3.1	3.7	7.7	4.0	3.8	5.2	4.5
Methionine	2.5 ± 0.1	3	2.2 ± 0.8	2	2.4 ± 0.4	2	2.4 ± 0.1	2	1.8 ± 0.4	8	2.3	4.1	1.8	1.8	2.7		
Phenylalanine	4.7 ± 0.4	3	4.4 ± 0.4	2	5.7 ± 0.8	2	4.7 ± 0.5	2	4.6 ± 0.4	8	4.6	3.3	1.4	4.9	4.6		
Proline	4.3 ± 0.6	3	4.5 ± 0.3	2	6.5 ± 2.4	2	4.5 ± 0.4	2	4.2 ± 0.5	8	4.1	3.9	4.0	5.3	4.9		
Serine	5.3 ± 0.4	3	5.2 ± 0.4	2	4.9 ± 0.1	2	5.4 ± 1.1	2	5.1 ± 0.5	8	5.6	5.7	5.4	6.7	6.1		
Threonine	4.0 ± 0.6	3	3.8 ± 0.1	2	4.1 ± 0.3	2	3.8 ± 0.3	2	3.9 ± 0.6	8	3.6	2.6	4.9	4.8	4.6	2.7	2.3
Tryptophan	0.97 ± 0.1	2	1.10 ± 0.0	2	0.71 ± 0.0	1	1.10 ± 0.1	2	1.22 ± 0.2	3	ND	0.34	0.17	ND	ND	0.74	0.60
Tyrosine	3.4 ± 0.3	3	3.0 ± 0.9	2	3.9 ± 1.1	2	3.2 ± 0.2	2	3.7 ± 0.3	8	4.0	3.4	2.1	3.6	3.8		
Valine	5.2 ± 0.4	3	5.1 ± 0.1	2	6.3 ± 1.2	2	5.1 ± 0.0	2	5.3 ± 1.0	8	6.2	3.4	3.0	5.2	4.9	4.2	3.9
Met+Cys	4.2 ± 0.3	3	3.3 ± 1.8	2	4.3 ± 0.0	2	4.2 ± 0.1	2	2.6 ± 1.0	8	3.9	7.4	5.2	2.0	3.0	2.6	2.2
Phe+Tyr	8.1 ± 0.7	3	7.5 ± 0.5	2	9.6 ± 2.0	2	7.9 ± 0.7	2	8.3 ± 0.7	8	8.6	6.7	3.5	8.4	8.4	4.6	3.8
Hydrophilic	62.0 ± 7.0	3	63.0 ± 6.7	2	56.9 ± 4.3	2	62.3 ± 9.9	2	63.4 ± 7.7	8	62.1	69.7	70.4	61.9	62.3		
Hydrophobic	38.0 ± 3.8	3	37.0 ± 2.8	2	43.1 ± 6.4	2	37.7 ± 2.6	2	36.6 ± 3.9	8	37.9	30.3	29.6	38.1	37.7		
Phobic/Philic	0.6 ± 0.5	3	0.6 ± 0.4	2	0.8 ± 1.5	2	0.6 ± 0.3	2	0.6 ± 0.5	8	0.6	0.4	0.4	0.6	0.6		
Aromatic	9.1 ± 0.9	3	8.6 ± 1.4	2	10.4 ± 2.0	2	9.0 ± 0.8	2	9.6 ± 0.9	8	8.6	7.0	3.7	8.4	8.4		
References	Callaway (2004), House et al. (2010), and Mattila et al. (2018)		House et al. (2010), and a Kim and Lee (2011b)		House et al. (2010), and Mattila et al. (2018)		House et al. (2010), and Malomo and Aluko (2015b)		Tang et al. (2006), Wang et al. (2008), Girgih et al. (2010, 2013), Malomo and Aluko (2015b), Ren et al. (2016), Hadnađev et al. (2018), and Wang et al. (2018)		Hadnađev et al. (2018)	Malomo and Aluko (2015a)	Malomo and Aluko (2015a)	Wang et al. (2008)	Wang et al. (2008)	FAO/WHO and UNU (2007)	

Tot AA: total amino acids. Y/O: years old. HPI: hemp protein isolate obtained by isolelectric precipitation. HMI: hemp protein isolate obtained by micellization. Glb: globulins. Alb: albumins. Asx = aspartic acid + asparagine. Glx = glutamic acid + glutamine. Met: methionine. Cys: cysteine. Phe: phenylalanine. Tyr: tyrosine. Hydrophilic = arginie + Asx + cysteine + Glx + histidine + lysine + serine + threonine + tyrosine. Hydrophobic = alanine + glycine + isoleucine + leucine + methionine + phenylalanine + proline + tryptophan + valine. Phobic/Philic = hydrophobic/hydrophilic. Aromatic = phenylalanine + tryptophan + tyrosine. Grey background, limiting amino acids according to FAO recommendations for children aged from 1 to 2 Y/O. Underlined AA, essential amino acids globulin fraction is limiting in sulfur amino acids, unlike the 11S, explaining that the HPI, majorly represented by edestins are not deficient in sulfur amino acids (Callaway 2004; Wang et al. 2008; Kim and Lee 2011b).

9.3.3.2 Digestibility

Different in vitro and in vivo protein digestibility values available in the literature for hemp and other plant and animal protein sources are listed Table 9.3. The whole seed proteins of hemp and resulting press-cake have in vivo digestibility values of 85% and 87% respectively (House et al. 2010; Malomo and Aluko 2015b). Dehulling of hemp seeds increases in vivo protein digestibility by 10–95% (House et al. 2010). The hulled hemp seed proteins thus have digestibility values similar to those of animal proteins: 94% for meat and fish, 98% for beef, 94–95% for whole or skim milk, or 98% for eggs (Sarwar 1997; Schaafsma 2000; FAO/WHO and UNU 2007). In vivo digestibility values of hemp proteins are comparable to other vegetable proteins ranging from 71% for black beans to 99% for pea protein concentrates (PPC) (Sarwar 1997; Rutherfurd et al. 2015).

In vitro digestibility value of hemp meal proteins is close to that obtained in vivo: 85% vs. 87% respectively (House et al. 2010; Malomo and Aluko 2015b). This value is close to the in vitro protein digestibility of cottonseed meal measured at 81% (Hsu et al. 1977). Hemp protein isolate obtained by isoelectric point precipitation (HPI), has an in vitro digestibility of 88%, higher than that of soy protein isolate obtained by isoelectric point precipitation (SPI) evaluated at 80% (Hsu et al. 1977; FAO/WHO and UNU 2007; Wang et al. 2008; Malomo and Aluko 2015b).

9.4 Extraction of Hemp Proteins

In order to valorize all the components of the hemp seed, hemp proteins are generally extracted from cakes resulting from the mechanical pressing of whole seeds in hemp oil production.

Many authors have proposed the extraction of hemp proteins from hemp meal in order to characterize protein fraction properties and to produce hemp protein concentrates or isolates (Tang et al. 2006; Yin et al. 2007, 2008; Wang et al. 2008, 2009, 2018; Girgih et al. 2010, 2014; Kim and Lee 2011b; Malomo et al. 2014; Malomo and Aluko 2015b, 2016; Ren et al. 2016; Orio et al. 2017; Hadnađev et al. 2018; Mamone et al. 2019). The general extraction route of hemp proteins is presented Fig. 9.2. Hemp oil is extracted from whole or hulled seeds by cold mechanical pressing. The residual cake generally contains about 70% less oil than the original seeds. Further delipidation by solvent may be applied, generally with hexane, to substantially remove the lipids. The cakes are crushed and sometimes sifted before being hydrated. During extraction, the hydration is carried out typically by suspending 5% of cake (from 1% to 7.5%) in distilled water at pH 10 (from 8 to 10), 37 °C (from 20 °C to 37 °C), with 2 h stirring (from 1 h to 4 h). Successive

Table 9.3 In vivo and in vitro digestibility values of hemp proteins and other plant and animal proteins in percentage

Raw material		In vivo	In vitro
Hemp	Whole seed[a]	85	
	Hulled seed[a]	95	
	Meal[ab]	87	85
	HPI[bc]		88
	11S[c]		91
	7S[c]		88
Soy	Bean[de]	93	
	Flour[f]	86	
	Meal[g]	80	
	SPI[fghi]	95	80
Pea	Seed[ef]	86	
	PPC[i]	99	
Sunflower	Seed[e]	91	
	Flour[f]	90	
Cotton	Seed[f]	90	
	Meal[h]	85	81
Peanut[ef]		95	
Almond[j]			88
Faba bean[e]		88	
Black bean[g]		71	
Bean[f]		78	
Lentil[e]		84	
Rolled oat[e]		90	
Chickpea flour[k]		84	72
Rice PC[i]		88	
Wheat	Raw[d]	91	
	White flour[f]	96	
Milk	Whole[d]	95	
	Skim[g]	94	
	MPC[i]	98	
	Lactalbumin[g]	99	
Egg[d]		98	
Meat/fish[f]		94	
Beef[d]		98	

[a]House et al. (2010), [b]Malomo and Aluko (2015b), [c]Wang et al. (2008), [d]Schaafsma (2000), [e]Sarwar and Peace (1986), [f]FAO/WHO and UNU (2007), [g]Sarwar (1997), [h]Hsu et al. (1977), [i]Rutherfurd et al. (2015), [j]Ahrens et al. (2005), [k]Angulo-Bejarano et al. (2008). HPI and SPI, hemp and soy protein isolates respectively. PPC, rice PC and MPC, pea, rice and milk protein concentrates respectively

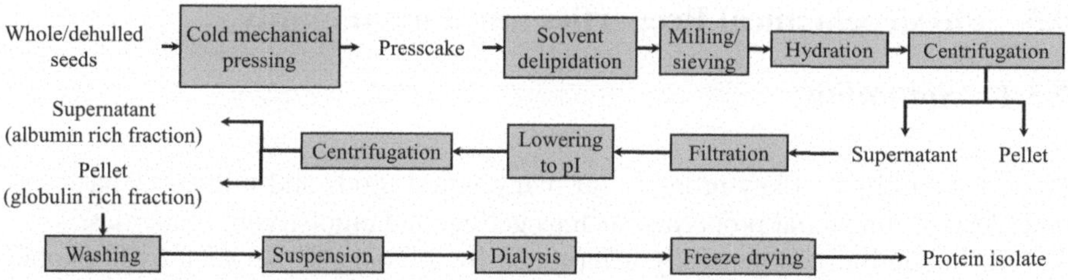

Fig. 9.2 General extraction procedure of proteins from hemp seed. pI: isoelectric point

centrifugation is carried out at 7000 g (from 6000 g to 10,000 g) at 20 °C (from 4 °C to 20 °C) during 30 min (from 15 min to 60 min). The pellet is discarded and the supernatant recovered. The pH of the supernatant is adjusted to pH 5 (from 4.5 to 5) and sometimes filtered on cheese-cloth or Whatman® paper. A new centrifugation step is applied at 8000 g (from 5000 g to 8000 g), at 20 °C for 10–60 min. The pellet is sometimes washed with distilled water and then resuspended at pH 7 (from pH 6.8 to 7). Dialysis is optionally applied, and the suspension is freeze-dried. The isolates obtained have protein content varying from 87% to 94% on a dry basis. The extraction yields are generally between 34 and 51% except in the work of Tang et al. (2006) where it reached 73%.

Alternative extraction methods based on micellization principle have been described in the literature. Malomo and Aluko (2015a) suspended the raw hemp meal at 10% in 0.5 M NaCl solution for 1 h at 24 °C. The supernatant obtained after centrifugation was filtered and dialyzed against distilled water. The content of the dialysis tubing was centrifuged. The supernatant was collected as the albumin fraction. After washing with distilled water and successive centrifugation, the resulting pellet was collected as the globulin fraction. After final freeze-drying, these authors obtained powders with protein contents (in w/w dry basis) and extraction yields of 80% and 6.33% for albumins and 100% and 6.07% for globulins, respectively. In a similar way, Hadnađev et al. (2018) proposed extraction by suspending 10% delipidated hemp cake in 0.8 M NaCl solution at pH 7 for 2 h at 35 °C. The supernatant obtained by centrifugation was diafiltered by ultrafiltration using a 12–14 kDa cut-off membrane with distilled water at pH 7. A new centrifugation step was performed on the retentate and the pellet was freeze-dried. The resulting protein isolate (HMI) had a high protein purity of ~99% with a protein yield of ~40%.

Otherwise, Malomo and Aluko (2015b) proposed an enzyme-assisted extraction method to facilitate the release of proteins from the plant material. A mixture of cellulase, hemicellulase, xylanase and phytase was applied to the aqueous suspension of hemp meal at pH 5. After incubation at 37 °C for 4 h, the solution was diafiltered by ultrafiltration at 10 kDa cut-off and then freeze-dried. The resulting powder (HEC) had at 74% protein content on a wet basis.

9.5 Physicochemical Properties and Functionality

9.5.1 *Solubility*

Protein solubility is a key-property for many liquid foods and influences basically other techno-functional properties such as gelling and emulsifying properties.

Protein solubility is generally determined by the balance between hydrophobic protein–protein and hydrophilic protein–water interactions. The common method to measure solubility evaluates the retention of proteins in the supernatant after centrifugation, which is influenced by the amount of aggregated proteins able to sediment. Protein solubility is sometimes referred to as "nitrogen solubility", since nitrogen from both protein and non-protein sources, such as nucleic acids, free amino acids, peptides, and phospholipids, is extracted in solubility tests. Nitrogen solubility can be defined as the ratio of solubilized nitrogen to total nitrogen in the sample, expressed as a percentage. Most of the authors who have investigated the solubility of hemp proteins have expressed the protein solubility, as relative nitrogen solubility. In this case, the amount of soluble protein, or soluble nitrogen, after hydration of the protein material in 0.1 M NaOH solution was considered to correspond to 100% solubility. The amount of soluble protein at the different pH of hydration is then divided by the amount of soluble protein in 0.1 M NaOH, to obtain the percentage of relative solubility. Thus, this method makes it possible to represent the solubility curve as a function of pH, but gives no indication of the actual solubility values of the protein material.

The extractability of hemp proteins from cake has been studied by Malomo et al. (2014) and Malomo and Aluko (2015b) by assessing their relative solubility from pH 3 to 9 (Fig. 9.3c). Relative solubility increased with increasing pH. From pH 3 to 5, it increased slightly from about 5 to 10%, respectively. Above pH 5, the increase in solubility is enhanced towards alkaline pH, reaching up about 25% at pH 9.

Many authors have studied the relative solubility of hemp protein isolates (HPI) produced by alkaline solubilization and isoelectric precipitation (Fig. 9.3c) (Tang et al. 2006; Kim and Lee 2011b; Malomo et al. 2014; Malomo and Aluko 2015b; Hadnađev et al. 2018; Wang et al. 2018; Dapčević-Hadnađev et al. 2019). Although the relative solubility values for a given pH varied considerably from one author to another, the profiles of solubility with pH were generally similar. The relative solubility is minimum at the isoelectric point of hemp globulins, around pH 5. It increases towards the alkaline pH to reach a maximum at pH 12. It seems that the maximum relative solubility is almost reached for pH 10. Starting from pH 5, relative solubility increases with decreasing pH up to pH 2. The maximum solubility encountered at acidic pH appears to be 2 times lower as the maximum solubility encountered at alkaline pH.

Malomo and Aluko (2015a) studied the relative solubility of hemp globulins and albumins separately from pH 3 to 9 (Fig. 9.3b). Hemp globulins presented the same solubility profile as HPI (Malomo and Aluko 2015b). This is consistent with the fact that HPI is obtained after selective isoelectric precipitation of globulins. The solubility profile of hemp albumins was totally different since their relative solubility

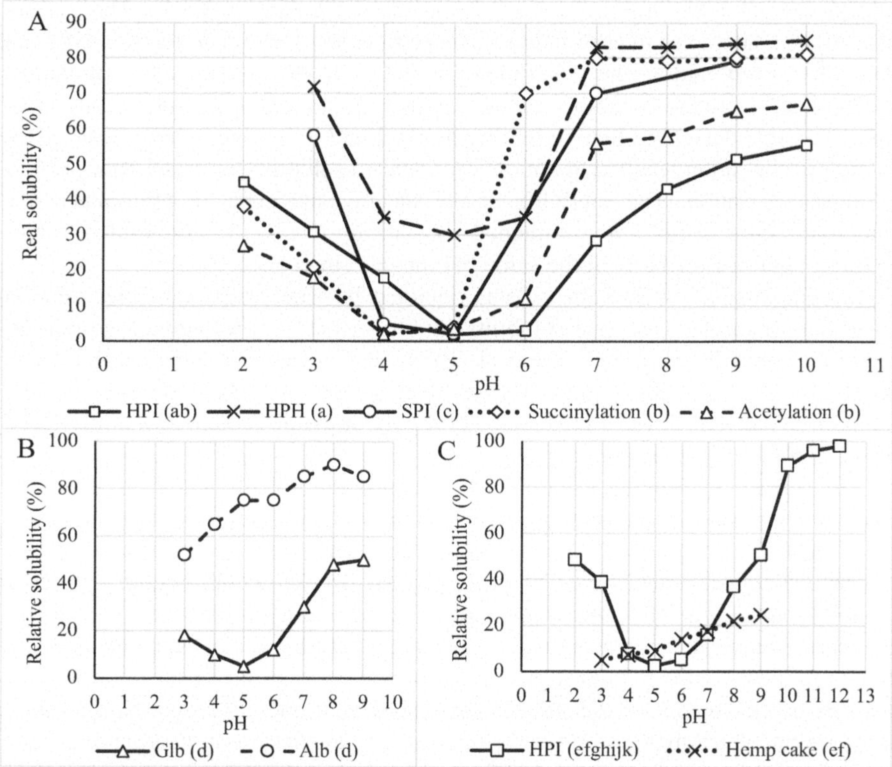

Fig. 9.3 Solubility of hemp proteins as a function of pH. (**a**) Real nitrogen solubility of hemp protein isolate obtained by isoelectric point precipitation (HPI), hemp protein hydrolysates (HPH), chemically modified HPI (succinylated or acetylated) and soy protein isolate obtained by isoelectric point precipitation (SPI). (**b**) Relative nitrogen solubility of isolated hemp globulins (Glb) and albumins (Alb). (**c**) Relative nitrogen solubility of hemp protein isolate obtained by isoelectric point precipitation (HPI) and hemp press cake. Results from a: Yin et al. (2008); b: Yin et al. (2009); c: Karki et al. (2009); d: Malomo and Aluko (2015a); e: Malomo et al. (2014); f: Malomo and Aluko (2015b); g: Kim and Lee (2011b); h: Hadnađev et al. (2018); i: Dapčević-Hadnađev et al. (2019); j: Tang et al. (2006); k: Wang et al. (2018)

tended to increase continuously with pH. The albumins showed significantly higher relative solubility values within the same pH range, varying from about 50 to 90%.

Yin et al. (2008, 2009) studied the real nitrogen solubility of HPI from pH 2 to 10 (Fig. 9.3a). They observed the same profile of solubility vs pH as the previous authors for relative solubility. Thus the solubility was minimal at pH 5 close to 0%, then increased with pH to reach the maximum value at pH 10 with about 60% of soluble protein. Solubility increased also from pH 5 to 2 up to about 45%. The effect of partial HPI hydrolysis by trypsin on the real protein solubility is shown in Fig. 9.3a (Yin et al. 2008). The hemp protein hydrolysates (HPH) having 2.3–6.7% degree of hydrolysis, generally showed the same solubility profile with pH as the initial HPI. On the other hand, the HPH solubility values were significantly higher within the entire pH range. The gain was 2–2.5 times at pH 3, more than

15 times at pH 5, about 3 times at pH 7 and about 1.5 times at pH 10. The authors also studied the impact of acetylation and succinylation of hemp proteins on HPI real solubility as reported in Fig. 9.3a (Yin et al. 2009). Both treatments did not change the profile of protein solubility curves. However, the chemical modifications of hemp proteins decreased their solubility at pH 2 and 3, but increased it from pH 6 to 10. Succinylation resulted in a more pronounced solubility increase from pH 6 to 10 than acetylation. The solubility of the succinylated hemp proteins reached a maximum value at pH 7 and remained constant up to pH 10, with similar values to those obtained for HPH in the same pH range (Yin et al. 2008).

In comparison, soy protein isolates obtained by isoelectric point precipitation (SPI) have a similar U-shaped solubility profile as that of HPI, with a minimum of solubility towards pH 4–5 (Karki et al. 2009) (Fig. 9.3a). Nevertheless, the actual solubility values are much higher for SPI than for HPI at acidic and basic pH. The solubility values of SPI and HPI are 60 and 30% at pH 3, 70 and 30% at pH 7 and 80 and 50% at pH 9, respectively.

The surface hydrophilic–hydrophobic balance of proteins and their ability to interact with water molecules are mainly related to their composition in amino acids and the net charge of these macromolecules. Therefore, the measure of surface hydrophobicity and net charge of proteins provide useful information regarding solubility behaviour of these compounds. The surface hydrophobicity of proteins can be measured through a fluorescent probe, mainly 8-anilino-1-naphthalene sulfonate or ANS, which fluorescence is amplified by binding to the hydrophobic pockets at the surface of the protein particle. The surface hydrophobicity values (Ho) are calculated by the slope of the fluorescence intensity as a function of protein concentration in mg/ml. Tang et al. (2009) and Wang et al. (2009, 2018) reported Ho values for hemp protein isolates obtained by isoelectric point precipitation (HPI) at pH 7 of 75, 89 and 179 respectively. These values were lower or close to the Ho values of 155 and 163 reported for soy protein isolates obtained by isoelectric point precipitation (SPI) by Sorgentini and Wagner (1999) and Jiang et al. (2009) respectively. Teh et al. (2016) measured the Ho values at neutral pH for acid soluble and alkali soluble hemp globulins respectively. Similar surface hydrophobicity values (Ho = 75) were obtained for these protein fractions. The same authors measured the zeta potential of hemp proteins at pH 7 from the same protein fractions. The 2 protein fractions had the same zeta potential value measured at −9.2 mV, significantly lower than those reported elsewhere for SPI at pH 7. Indeed Lam et al. (2007) and Liu and Tang (2015) reported zeta potential values of −25 mV and −33 mV for SPI, respectively. From these results, it can be assumed that the aqueous suspensions of hemp proteins are less stable than their soy counterparts at neutral pH.

9.5.2 Water Holding Capacity and Oil Holding Capacity

Water holding capacity (WHC) and oil holding capacity (OHC) are important properties for food products where soft texture is sought and limited water or lipid release during and after cooking are expected. Water or oil retention capacity of oilcakes are generally little affected by their delipidation. Hemp, flax and canola meals presented similar OHC values (Teh et al. 2013). Hemp cake has WHC values close to those of canola and higher than those of flax. In addition, hemp cake had slightly higher WHC value and slightly lower OHC value as those of hemp protein isolates obtained by isoelectric point precipitation (HPI) respectively (Malomo et al. 2014). The WHC and OHC values obtained for acid soluble hemp globulins and alkali soluble hemp globulins were slightly higher for the later ones (Teh et al. 2013). OHC and WHC values for the acid or alkali soluble globulins are close to those of canola proteins, and higher than those of flax ones.

The hemp protein isolates obtained by micellization (HMI) by Hadnađev et al. (2018) presented the same OHC value as those produced by isoelectric precipitation (HPI), whereas the WHC value for HMI was 2 times lower compared to HPI. In the previous work of Malomo and Aluko (2015b), the authors showed that hemp protein concentrates obtained by enzymatic digestion of hemp cake by hemicellulase, cellulase, xylanase and phytase (HEC) and successive ultrafiltration, led to an OHC value similar to that of HPI and a slightly higher WHC value. Yin et al. (2008) observed that the digestion of HPI by trypsin resulted in a 3 to 4-fold decrease in WHC. OHC was also decreased by protein hydrolysis, except at the highest hydrolysis degree close to 7% where it was similar to that of HPI. A heat treatment at 95 °C for 10 min applied by the same authors to HPI did not affect the WHC, but slightly decreased OHC of the isolates.

9.5.3 Gelation

A protein gel is a continuous macromolecular network entrapping an aqueous phase, formed under physical stress, mainly by heating, or by applying specific physico-chemical conditions from stable protein suspensions. Scientific data on the gelation properties of hemp proteins are scarce and only concern thermal gelation (Table 9.4).

Hemp protein isolates obtained by isoelectric point precipitation (HPI), mainly containing globulins, begin to denature at 86 °C (Ton: onset temperature) as observed by differential scanning calorimetry (DSC) and have a thermo-denaturation temperature (Td) peak at 94 °C (Table 9.5). The average enthalpy of denaturation (ΔH) of this protein fraction was estimated at 11 J/g of protein (Tang et al. 2006; Yin et al. 2008; Wang et al. 2008; Hadnađev et al. 2018). This denaturation could be predominantly ascribed to 11S globulins. In fact, 7S hemp proteins have been reported to present no thermal denaturation peak during calorimetry measurement (Wang et al. 2008). Tang et al. (2006) indicated that the degree of ordered structure of 11S edestin was related to both hydrophobic interactions and disulfide bridges as revealed by the decrease in ΔH in presence of sodium dodecyl sulfate and dithiothreitol respectively. The presence of disulfide bridges seems to play an effective role in the thermal stability of hemp 11S globulins since a slight decrease in Ton and Td values was also noted in presence of dithiothreitol.

Table 9.4 Summary of the functional properties studied in the scientific literature

Functionality	Protein material	Operating conditions	Performed analyses	References
Water or oil holding capacity	HPM + HPI		OHC, WHC	Malomo et al. (2014)
	HPM + HPI + HEC		OHC, WHC	Malomo and Aluko (2015b)
	Hemp, canola or flax: raw or delipidated cake, alkali or acid soluble globulins		OHC, WHC	Teh et al. (2013)
	HPI + SPI		OHC, WHC	Tang et al. (2006)
	HPI + HMI		OHC, WHC	Hadnađev et al. (2018)
	HPI + HPI heated 95 °C 10 min + HPH by trypsin		OHC, WHC	Yin et al. (2008)
Geling	HPM + HPI	pH 7; 2–20%; 95 °C 1 h	LGC	Malomo et al. (2014)
	HPI + HMI	pH 7; 30% protein suspension with 0, 50 or 300 mM NaCl; 120 °C 15 min	Confocal microscopy, rheology	Dapčević-Hadnađev et al. (2018)
Emulsifying	HPM + HPI	pH 3, 5, 7 or 9; 1, 2.5 or 5% protein suspension with 83/17 protein solution to oil ratio	Droplet size, ES	Malomo et al. (2014)
	HPM + HPI + HEC	pH 3, 5, 7 or 9; 1, 2.5 or 5% protein suspension with 83/17 protein solution to oil ratio	Droplet size, ES	Malomo and Aluko (2015b)
	Hemp, canola or flax: raw or delipidated cake, alkali or acid soluble globulins	pH 7; 2% protein suspension with 50/50 protein solution to oil ratio	EA, droplet size, ES, CS	Teh et al. (2013)
	HPI + SPI	pH 3 to 8; 0.2% protein suspension with 75/25 protein solution to oil ratio	EAI, ESI	Tang et al. (2006)
	Glb + Alb	pH 3, 5, 7 or 9; 1, 2.5 or 5% protein suspension with 83/17 protein	Droplet size, ES	Malomo and Aluko (2015a)

(continued)

Table 9.4 (continued)

Functionality	Protein material	Operating conditions	Performed analyses	References
		solution to oil ratio		
	HPI + HMI	pH 3; 0.25, 0.75 or 1.5% protein suspension with 90/10 protein solution to oil ratio	Droplet size, FI, CI, CS, confocal microscopy, rheology	Dapčević-Hadnađev et al. (2019)
	HPI + HPI shifted at pH 12 for 5 min at 20, 30, 40, 50 or 60 °C	pH 7; 2% protein suspension with 75/25 protein solution to oil ratio	EAI, ES, confocal microscopy	Wang et al. (2018)
	HPI + HPI heated 95 °C 10 min + HPH by trypsin	pH 3, 5, 7 or 9; 0.2% protein suspension with 75/25 protein solution to oil ratio	EAI, ESI	Yin et al. (2008)
Foaming	HPM + HPI	pH 3, 5, 7 or 9; 2, 4 or 6% protein suspension	FC, FS	Malomo et al. (2014)
	HPM + HPI + HEC	pH 3, 5, 7 or 9; 2, 4 or 6% protein suspension	FC, FS	Malomo and Aluko (2015b)
	Glb + Alb	pH 3, 5, 7 or 9; 2, 4 or 6% protein suspension	FC, FS	Malomo and Aluko (2015a)
	HPI + HPI heated 95 °C 10 min + HPH by trypsin	pH 7; 1% protein suspension	FC, FS	Yin et al. (2008)

HPM: Hemp protein meal. HPI: Hemp protein isolate obtained by isoelectric precipitation. HEC: Hemp protein concentrate obtained by enzymatic digestion of press-cake by cellulase, hemicellulase, xylanase and phytase. SPI: Soy protein isolate obtained by isoelectric precipitation. HMI: Hemp isolate obtained by micellization. HPH: Hemp protein hydrolysates. Glb: Hemp globulins. Alb: Hemp albumins. OHC: Oil holding capacity. WHC: Water holding capacity. LGC: Least gelation concentration. ES(I): Emulsifying stability (index). EA(I): Emulsifying activity (index). CS: Creaming stability. FI: Floculation index. CI: Coalescence index. FC: Foaming capacity. FS: Foaming stability

Hadnađev et al. (2018) compared DSC data between hemp protein isolates obtained by isoelectric precipitation (HPI) and micellization (HMI) respectively. These authors found that micellization extraction led to increased Td and ΔH values. This result revealed the production of HMI with higher thermal stability, suggesting higher level of structural organization and a more preserved native state of proteins compared to HPI.

Table 9.5 Differential scanning calorimetry data for hemp proteins and other plant protein sources

		Hemp				Sun-flower[e]	Rape-seed[f]	Peanut[g]	Soy[h]	Pea[i]	Chick-pea[j]
		HPI[abcd]	HMI[c]	11S[b]	7S[b]	pI protein isolates					
Ton	(°C)	86		86	None						
Td		94	96	92	None	99	98	104	90	82	88
ΔH	(J/g)	11	22	7	None	8	10	12	6	7	4

Td: denaturation temperature, ΔH: denaturation enthalpy, Ton: onset temperature, HPI: hemp protein isolate obtained by isoelectric precipitation, HMI: hemp protein isolate obtained by micellization

[a]Tang et al. (2006), [b]Wang et al. (2008), [c]Hadnađev et al. (2018), [d]Yin et al. (2008), [e]Salgado et al. (2011), [f]He et al. (2013), [g]Colombo et al. (2010), [h]Shen and Tang (2012), [i]Liang and Tang (2013), [j]Paredes-López et al. (1991)

Otherwise, Yin et al. (2008) indicated that the denaturation enthalpy values for HPI decreased in proportion to their degree of hydrolysis by trypsin, traducing a decrease in the extent of ordered structures within hemp polypeptides.

In addition, the HPI had thermo-denaturation temperature and denaturation enthalpy of the same order of magnitude as those of other oilseed protein isolates, prepared from sunflower seed, rapeseed or peanut, having Td and ΔH values ranging from 98 to 104 °C and 8 to 12 J/g of protein respectively (Colombo et al. 2010; Salgado et al. 2011; He et al. 2013). Legume proteins extracted from soybeans, peas or chickpeas, had slightly lower Td and ΔH values than those of HPI, the former ranging from 82 to 99 °C and between 4 and 7 J/g of protein respectively (Paredes-López et al. 1991; Shen and Tang 2012; Liang and Tang 2013).

According to Malomo et al. (2014), hemp meal had better gelling properties than hemp protein isolates obtained by isoelectric point precipitation (HPI), probably due to the contribution of sugars and polysaccharides. The minimum gelation concentrations of hemp meal and HPI were measured at 12 and 22% (w/w) respectively. Dapčević-Hadnađev et al. (2018) investigated also the gelling properties of hemp protein isolates obtained by alkaline extraction/isoelectric precipitation (HPI) or by micellization (HMI). The gels were formed from 30% (w/w) protein suspension at pH 7, heated at 120 °C for 15 min. For the 2 types of isolates, true gels were formed with very close gel forces evaluated by dynamic rheology. However, in comparison to HMI gels, the HPI gels displayed a slightly higher elastic modulus and a higher breakdown limit during amplitude sweep test, revealing the formation of stronger and more cohesive HPI-based gels. The microstructures of the 2 gel types were very different. From confocal microscopic observations, the HPI gels showed extended interconnections between large protein granules forming a low porous network, while HMI formed a filamentous protein network with large pores. The addition of NaCl up to 300 mM had little impact on the structure of the HPI gels, while it resulted in a pore size increase for HMI gels. Finally, the two types of gel had a yellowish-red color probably due to the co-extraction and polymerization of phenolic compounds.

9.5.4 Emulsion

Proteins are commonly used to stabilize oil in water emulsion in food products. Emulsions are thermodynamically unstable because they increase the interfacial surface and consequently the free energy of the system. The presence of proteins as emulsifiers, able to adsorb at the oil/water interfaces, helps to reduce interfacial tension and contributes to the stability of the emulsion. The ability of protein to form emulsion is then generally evaluated through particle size measurement or emulsifying activity index (EAI) determination (Table 9.4). Over time, oil droplets in emulsion are subject to flocculation, coalescence and creaming phenomena, since the system tends to minimize its free energy. Flocculation is the reversible or irreversible aggregation of droplets. Coalescence occurs when the interfacial film is broken, resulting in the irreversible fusion of dispersed droplets into larger ones. Finally, creaming is the migration of oily particles to the surface against gravity due to the difference in density between the two phases. The creaming kinetics is slowed down by the droplet size reduction and the viscosity increase of the continuous phase. Particle size change or phase separation in emulsion are evaluated as flocculation index (FI), coalescence index (CI) creaming stability (CS) and emulsion stability index (ESI).

In the literature, emulsifying properties of hemp proteins have been evaluated from hemp meals, hemp protein isolates and their separated fractions, i.e. globulins and albumins (Table 9.4).

9.5.4.1 Hemp Meal

Teh et al. (2013) have studied the emulsifying properties of hemp cakes at pH 7 compared to those obtained from rapeseed with or without delipidation. According to these authors, raw hemp meal had a lower emulsifying activity than canola meal. On the other hand, the delipidation of hemp cake increased its emulsifying activity, becoming higher than that of delipidated canola meal. Emulsions prepared from raw hemp cake were found more stable than those obtained from delipidated meal. Emulsion stability was higher for raw hemp meal than for raw canola meal. However, more stable emulsions were observed for delipidated canola meal than for their hemp counterpart.

9.5.4.2 Hemp Protein Isolate

Malomo et al. (2014) compared the emulsifying properties of hemp cake and hemp protein isolates obtained by isoelectric point precipitation (HPI) in the range of pH 3 to 9. Emulsions were prepared from 1% to 5% protein suspensions with a 83/17 protein solution to oil ratio. The authors found that the emulsifying properties of hemp meal were not significantly affected by pH and meal concentration. The droplet size measured by granulometry as the Sauter mean diameter (d_{32}) in the HPI emulsions was impacted by the pH, with some differences regarding the protein concentration used. In most cases, d_{32} values for HPI-based emulsions were higher than those prepared from hemp cake, except at 5% protein at pH 5, 7 and 9, where droplet size was slightly lower for the former emulsions. No difference in stability was observed between hemp cake-based and HPI-based emulsions.

Tang et al. (2006) investigated the emulsifying properties of HPI and soy protein isolate obtained by isoelectric point precipitation (SPI) from pH 3 to 8. Whatever the pH, the emulsifying activity (EAI) and emulsion stability (ESI) indexes were lower for HPI-based emulsions than for SPI-based emulsions. The EAI of both types of emulsion were maximum at pH 3, minimum around the isoelectric point, and relatively constant from pH 5 to 8. The ESI measured for the SPI-based emulsions followed the same profile as EAI with pH, whereas the ESI has been relatively constant for HPI-based emulsions in the same pH range.

Some authors have applied different pretreatments to HPI to improve their emulsifying properties. Yin et al. (2008) produced hydrolysates from HPI digestion by trypsin. The hydrolysates led to lower EAI than HPI at pH 3, 5, 7 and 9. The ESI was increased by the hydrolysis treatment for emulsions at pH 3, 5 and 9, but decreased for emulsions at pH 7. EAI and ESI of HPI-based emulsions were the highest at pH 9 and pH 7 respectively and then decreased in the following order: pH 7 > pH 3 > pH 5 and pH 9 > pH 3 > pH 5 respectively. The same authors have demonstrated that a heat treatment at 95 °C for 10 min applied to HPI allowed to improve the emulsifying activity at all pH except at pH 5. The heat treatment only led to improve emulsion stability at pH 7 and impaired it at pH 5.

Recently Wang et al. (2018) showed that mild heat treatment (30–60 °C) applied to HPI suspensions in strong alkaline conditions at pH 12 enhanced the emulsifying properties of hemp proteins at pH 7, by decreasing droplet size and improving emulsions stability. The efficiency of the pretreatment was reinforced with the increase in temperature.

Yin et al. (2009) observed that both preliminary acetylation and succinylation applied to HPI improved the emulsifying activity of the modified hemp proteins at pH 7. In addition, the ESI for HPI appeared slightly increased after acetylation and decreased after succinylation.

Other authors have compared the emulsifying properties of hemp protein isolates obtained by different extraction processes. Recently, Dapčević-Hadnađev et al. (2019) compared the properties of emulsions prepared at pH 3 from hemp protein isolates obtained by isoelectric precipitation (HPI) or by micellization (HMI). The emulsions with a 90/10 protein solution to oil ratio presented the lowest oil droplet size for HMI regardless protein concentration varying from 0.25% to 1.25%. Droplet size decreased as a function of protein concentration in the case of HPI-based emulsions. The flocculation index (FI) was the lowest for HMI and unaffected by protein concentration. The coalescence index (CI) decreased with increased protein concentration for both types of hemp protein isolate. CI values were relatively close between HPI- and HMI-based emulsions. Earlier, Malomo and Aluko (2015b) have compared the emulsifying properties of HPI-type proteins with hemp protein isolates obtained after enzymatic digestion of cake by cellulase, hemicellulase, phytase and xylanase, and successive ultrafiltration (HEC). The emulsions were prepared at pH 3, 5, 7 or 9 from 2, 4 or 6% protein suspension with a 83/17 protein solution to oil ratio. In all conditions, the droplet size of emulsions prepared with HEC was higher than those prepared with HPI. The stability of the HEC-based emulsions was similar to lower depending on pH and protein concentration in comparison to HPI-based emulsions.

9.5.4.3 Globulins and Albumins

Malomo and Aluko (2015a) compared the emulsifying properties of isolated hemp globulins and albumins at pH 3, 5, 7 and 9 for 1, 2.5 or 5% protein suspensions applying a 83/17 protein solution to oil ratio. Overall within experimental conditions, droplet size and emulsion stability were relatively similar between globulins and albumins except under a few conditions. At 5% protein concentration and at pH 7 and 9, the globulins led to oil droplets of smaller diameter in emulsion than those obtained from albumins. At pH 7 and 1% protein concentration, the emulsion stability was significantly lower with globulins than with albumins.

Teh et al. (2013) have also compared the properties of oil in water emulsions based on alkali soluble hemp globulins or acid soluble ones respectively, to those of emulsions prepared from other hemp or oilseed protein materials. These authors reported similar emulsifying activity and emulsion stability between alkali and acid soluble globulins-based systems at pH 7. Moreover, the emulsifying activity of delipidated hemp cake was higher than that of hemp alkali and acid soluble globulins, itself higher than raw hemp cake. Nevertheless, the raw hemp cake led to more stable emulsions than those formed with delipidated hemp cake and even more than those prepared from acid or alkali soluble hemp globulins. In addition, the emulsifying activity of alkali and acid soluble hemp globulins were similar to their rapeseed counterparts, whereas the emulsifying stability was significantly higher for rapeseed protein-based emulsions.

9.5.5 *Foam*

Food foam results from the stable dispersion of air bubbles in a liquid phase. Generally, protein suspensions have good foamability. The efficiency of proteins to form foam structure is evaluated by foaming capacity (FC) and foaming stability (FS) tests, based on the determination of the foam volume initially generated in the system and its conservation with time respectively.

Several authors have evaluated the foaming properties of hemp protein materials (Table 9.4). Malomo et al. (2014) reported that hemp cakes had lower foaming properties (FC and FS) evaluated at pH 3, 5, 7 and 9 and at 2, 4 or 6% protein concentration than hemp protein isolate produced by isoelectric precipitation (HPI). The FC of the hemp cake increased slightly with pH, whereas a maximum and a minimum were assessed at pH 3 and 5 respectively for HPI. The FS value for HPI was measured close to 100% and was affected nor by pH, nor by protein concentration except at 6% protein concentration-pH 5 condition, where it decreased. The FS value was lower for hemp cakes and depended on pH and protein concentration.

Malomo and Aluko (2015b) compared foaming ability of the same hemp protein isolates obtained by isoelectric point precipitation (HPI) or enzymatic digestion of hemp meal (HEC), as those used to evaluate emulsifying properties as reported in

Sect. 9.5.4. Compared to HPI, HEC led to significantly higher FC values of about three-fold within experimental pH (3–9) and protein concentration (2–6%) ranges. Foams formed from HEC at 2 or 4% protein concentration were generally less stable than those formed from HPI. At 6% protein concentration, similar values of FS were measured for HPI- and HEC-based foams except at pH 5 where the FS value was significantly higher for HEC-based foam.

The enzymatic digestion of HPI by trypsin degraded its foaming properties (FC and FS) at pH 7 (Yin et al. 2008). However, heat treatment of HPI at 95 °C for 10 min increased FS and did not impact FC.

9.6 Biological Activity

9.6.1 Antioxidant Property

According to Pojić et al. (2014), hemp cakes have antioxidant activity evaluated by DDPH (α, α-diphenyl-β-picrylhydrazyl) free radical scavenging method, mainly ascribed to the presence of phenolic compounds. Kim and Lee (2011b) found that hemp protein isolates obtained by isoelectric point precipitation (HPI) had also antioxidant activity. The antioxidant activity of HPI was proportional to protein concentration, with DDPH activity values ranging from 20% at 0.5 mg protein/ml to 80% at 3 mg protein/ml. This antioxidant activity would be partly due to the 11S edestins. Indeed, Kim and Lee (2011a) reported DDPH activity values of edestin ranging from 1% at 0.5 mg protein/ml, to over 20% at 3.5 mg protein/ml.

HPI hydrolysates (HPH) prepared by pancreatin and pepsin had antioxidant activity values measured by the DDPH method, ranging from about 15–65%, and metal chelation rates ranging from about 45–95% depending on the peptide composition (Girgih et al. 2014). Girgih et al. (2010) separated similar HPH into several fractions of <1, 1–3, 3–5 and 5–10 kDa molecular weight. HPH and all peptide fractions had antioxidant activity according to DDPH measurements and displayed effective hydroxyl radical scavenging, metal chelation, ferric reduction and inhibition of linoleic acid oxidation. HPH had the highest metal chelation activity, whereas small and large size peptide fractions had the highest DDPH antioxidant activity and ferric reduction ability respectively. All peptide fractions, regardless of molecular weight, displayed higher hydroxyl radical scavenging activity than HPH. Elsewhere, Girgih et al. (2013b) demonstrated that oxygen radical absorbance capacity (ORAC) and DDPH antioxidant activity of peptides tended to increase with their content in hydrophobic amino acids, whereas hydroxyl radical scavenging decreased.

Tang et al. (2009) generated different hemp protein hydrolysates (HPH) from enzymatic digestions with alkalase, Flavourzyme®, Neutrase®, pepsin, Protamex® or trypsin. By evaluating DDPH antioxidant activity, Fe^{2+} chelation and reducing power for all hydrolysates, the authors highlighted more or less antioxidant efficiency depending on the enzymes used. Wang et al. (2009) also revealed that HPH produced by Neutrase® had significant antioxidant activity determined from similar analytical characterizations.

Both alkali and acid soluble hemp globulins have antioxidant properties (Teh et al. 2016). According to the last authors, the ORAC activity of alkali and acid soluble hemp globulins remained unchanged after their hydrolysis by different proteases, namely AFP 4000 (AFP), HT proteolytic concentrate (HT) or protease G (ProG) from Enzyme Solutions (Australia), actinidin or zingibain. In contrast, DDPH antioxidant activities could increase nearly 5-fold after proteolysis. Proteolysis by HT led to hydrolysates with the highest DDPH antioxidant activity rate reaching up to 73%.

9.6.2 Renal Disease Prevention

Aukema et al. (2011) have studied the impact of the diet of rats having renal polycystic insufficiency, on the symptoms of this pathology including cardiovascular hypertrophy. The rats had different diets based on various protein isolates from hemp (HPI), soybean (SPI) or pea (PPI), or casein (CPI), the last chosen as a reference. SPI and HPI had the same effects for reducing the symptoms of the pathology. Indeed, in comparison with the CPI-based diet, a diet based on HPI or SPI improved the growth of rats, caused a decrease in kidney weight and liquid content, a reduction in the volume of cysts and fibrosis, and led to healthy levels of creatinine and chemokine receptors 2 (CCR2), and heart weights of healthy subjects. On the other hand, the diet based on PPI, compared to that based on CPI, only improved the growth of rats and led to lower creatinine and CRR2 levels. It did not change the volume of fibrosis, and increased kidney weight, renal liquid content and cyst volume.

9.6.3 Antihypertensive Activity

Girgih et al. (2013a) investigated the effect of diets based on hemp meal protein hydrolysates produced by pepsin and pancreatin (HPH), or hemp protein isolates obtained by isoelectric point precipitation (HPI), on rats having spontaneous hypertension. The reference diet was based on casein protein isolates. The HPH- or HPI-based diet reduced systolic blood pressure, through mainly decreases in angiotensin-converting enzyme (ACE) activity and quantitative level of plasma renin. ACE is an enzyme that converts angiotensin 1 to angiotensin 2, which has a strong vasoconstrictor effect. These positive effects were more pronounced for the HPH-based diet than for the HPI-based diet and have been observed during tests in the prevention or treatment phase of hypertension. Contrariwise Girgih et al. (2011) found an increase in systolic blood pressure following hemp-based feeding of rats having spontaneous hypertension. This difference may be due to the different amounts of hemp protein isolates ingested by the rats.

Hemp protein hydrolysates produced by pepsin and pancreatin (HPH) also exhibited antihypertensive abilities, by reducing ACE and renin activity, and systolic blood pressure (Girgih et al. 2014). The involved peptides enabling simultaneously the highest antioxidant (DDPH test), metal chelation and ACE inhibition activities, and the highest systemic blood pressure reduction, have been identified: Proline-Serine-Leucine-Proline-Alanine and Tryptophan-Valine-Tyrosine-Tyrosine. The peptides having the highest combined ACE and renin inhibition rates, and the highest reduction in systolic blood pressure were: Serine-Valine-Tyrosine-Threonine and Isoleucine-Proline-Alanine-Glycine-Valine. According to Girgih et al. (2011), peptides having <1 kDa and 1–3 kDa molecular weights had higher ACE and renin inhibitions than overall HPH. On the other hand, the HPH potential to reduce the systolic blood pressure was 2 times higher than that of <1 kDa and 1–3 kDa peptide fractions.

Orio et al. (2017) performed acid hydrolysis of HPI at 110 °C. The authors obtained protein hydrolysates inhibiting almost 45% of ACE activity and isolated 3 peptides exhibiting ACE inhibition rates higher than 93%: Glycine-Valine-Leucine-Tyrosine, Leucine-Glycine-Valine and Arginine-Valine-Arginine. These 3 peptides were originated from the hydrolysis of the acid subunit of edestin 1 and/or edestin 2.

According to Teh et al. (2016) alkali and acid soluble hemp globulins have limited ACE inhibition properties. These properties can be significantly improved, by nearly 30 times, by the production of enzymatic hydrolysates with AFP 4000 (AFP), HT proteolytic concentrate (HT) or protease G (ProG) (Enzyme Solutions, Australia), actinidin or zingibain. HT hydrolysates showed the highest ACE inhibition rates reaching up to about 77% inhibition.

9.6.4 Hyperglycemia and Inhibition of α-Glucosidase

α-glucosidase is an enzyme that hydrolyses starch and disaccharides to glucose. Enzymatic hydrolysates (HPH) of hemp protein isolates obtained by isoelectric point precipitation (HPI) have been reported to have α-glucosidase inhibition property of interest for the treatment of patients having hyperglycemic diseases (Ren et al. 2016). The last authors found that the HPH obtained by alkalase treatment had much higher α-glucosidase inhibitory activity, reaching almost 60% inhibition rate, than those obtained after enzymatic digestion of HPI with Flavourzyme, Protamex®, neutrase, trypsin or papain, having from 5% to 22% inhibition. From about 10% hydrolysis, the inhibition of α-glucosidase increased in proportion to the increases in hemp protein concentration from 0.1 to 100 mg/ml and hydrolysis degree up to 27%. Inhibition of α-glucosidase was mainly ascribed to 2 peptides: Leucine-Arginine and Proline-Leucine-Methionine-Leucine-Proline.

9.6.5 *Degenerative Brain Disease and Inhibition of Acetylcholinesterase*

Acetycholinesterase (AChE) is an enzyme that converts the neurotransmitter acetylcholine (Ach) into inactive metabolites, i.e. choline and acetate. With aging and dietary changes of humans, less acetylcholine is synthesized, while AChE continues normal activity. Thus, a clear decrease in ACh leads to dysfunctions of the central nervous system, which can in the long run contribute to memory disorders such as dementia or Alzheimer's disease. Malomo and Aluko (2016) showed that hemp protein hydrolysates (HPH) from hemp protein isolate obtained by isoelectric point precipitation (HPI) had AChE inhibition activities that would prevent progression to such troubles. This inhibition did not seem directly related to the degree of hydrolysis of hemp proteins. HPH obtained with 1% pepsin had the greatest AChE inhibition effect, compared with those obtained at other enzyme concentrations varying from 0.5% to 4% or with other types of enzyme such as alkalase, papain, pepsin + pancreatin, thermoase or Flavourzyme®. The authors hypothesized that the higher inhibition for this condition could result from increased synergistic effects between peptides of a wider size range present in the sample.

9.7 Conclusion

Hemp seeds are a good source of plant proteins with a protein content comparable to that of most of legume seeds. The hemp proteins have interesting nutritional value since they are rich in essential amino acids and were demonstrated highly digestible. They present also interesting techno-functional abilities such as emulsifying, foaming and gelling properties that could be exploited in the food industry. In addition, they have revealed a wide range of biological activities, enhanced after partial hydrolysis by various proteases. The most limiting factors in the valorization of hemp proteins in food industry seemed to be their low extraction yield and low solubility.

Previous works have shown that some post-extraction treatments of hemp proteins improved their solubility in the food pH range. However, few have evaluated the impact of these treatments on the general functional properties of hemp proteins.

The challenges for developing hemp protein ingredients are to obtain high extraction yields and highly soluble proteins, while maintaining their nutritional and functional properties. To meet the issue, further studies exploring extraction conditions and post-extraction functionalization treatments of hemp proteins are required.

Acknowledgements This work was supported by the Regional Council of Bourgogne-Franche-Comté through the "plan d'actions régionales pour l'innovation" (PARI) and the European Union through the PO FEDER-FSE Bourgogne 2014/2020 programs (n° 2016-9205AAO033S03076 and BG0009561).

References

Ahrens S, Venkatachalam M, Mistry AM, Lapsley K, Sathe SK (2005) Almond (*Prunus dulcis* L.) protein quality. Plant Foods Hum Nutr 60:123–128. https://doi.org/10.1007/s11130-005-6840-2

Aluko RE (2017) Hemp seed (*Cannabis sativa* L.) proteins: composition, structure, enzymatic modification, and functional or bioactive properties. In: Sustainable protein sources. Elsevier, Oxford, pp 121–132

Amaducci S, Scordia D, Liu FH, Zhang Q, Guo H, Testa G, Cosentino SL (2015) Key cultivation techniques for hemp in Europe and China. Ind Crop Prod 68:2–16. https://doi.org/10.1016/j.indcrop.2014.06.041

Angulo-Bejarano PI, Verdugo-Montoya NM, Cuevas-Rodríguez EO, Milán-Carrillo J, Mora-Escobedo R, Lopez-Valenzuela JA, Garzón-Tiznado JA, Reyes-Moreno C (2008) Tempeh flour from chickpea (*Cicer arietinum* L.) nutritional and physicochemical properties. Food Chem 106:106–112. https://doi.org/10.1016/j.foodchem.2007.05.049

Ash AL (1948) Hemp-production and utilization. Econ Bot 2:158–169. https://doi.org/10.1007/BF02858999

Aukema HM, Gauthier J, Roy M, Jia Y, Li H, Aluko RE (2011) Distinctive effects of plant protein sources on renal disease progression and associated cardiac hypertrophy in experimental kidney disease. Mol Nutr Food Res 55:1044–1051. https://doi.org/10.1002/mnfr.201000558

Balentić JP, Jozinović A, Ačkar Đ, Babić J, Miličević B, Benšić M, Jokić S, Šarić A, Šubarić D (2019) Nutritionally improved third generation snacks produced by supercritical CO2 extrusion I. Physical and sensory properties. J Food Process Eng 42:e12961. https://doi.org/10.1111/jfpe.12961

Bouloc P (2006) Le chanvre industriel: production et utilisations. France Agricole Editions

Callaway JC (2004) Hempseed as a nutritional resource: an overview. Euphytica 140:65–72. https://doi.org/10.1007/s10681-004-4811-6

Carus M, Karst S, Kauffmann A, Hobson J, Bertucelli S (2013) The European hemp industry: cultivation, processing and applications for fibres, shivs, seeds and flowers (Update: March 2017). European Industrial Hemp Association, Hürth (Germany)

Colombo A, Ribotta PD, León AE (2010) Differential scanning calorimetry (DSC) studies on the thermal properties of peanut proteins. J Agric Food Chem 58:4434–4439. https://doi.org/10.1021/jf903426f

Dapčević-Hadnađev T, Hadnađev M, Lazaridou A, Moschakis T, Biliaderis CG (2018) Hempseed meal protein isolates prepared by different isolation techniques. Part II. Gelation properties at different ionic strengths. Food Hydrocoll 81:481–489. https://doi.org/10.1016/j.foodhyd.2018.03.022

Dapčević-Hadnađev T, Dizdar M, Pojić M, Krstonošić V, Zychowski LM, Hadnađev M (2019) Emulsifying properties of hemp proteins: effect of isolation technique. Food Hydrocoll 89:912–920. https://doi.org/10.1016/j.foodhyd.2018.12.002

Docimo T, Caruso I, Ponzoni E, Mattana M, Galasso I (2014) Molecular characterization of edestin gene family in *Cannabis sativa* L. Plant Physiol Biochem 84:142–148. https://doi.org/10.1016/j.plaphy.2014.09.011

Elleuch M, Besbes S, Roiseux O, Blecker C, Attia H (2007) Quality characteristics of sesame seeds and by-products. Food Chem 103:641–650. https://doi.org/10.1016/j.foodchem.2006.09.008

Erbaş M, Certel M, Uslu MK (2005) Some chemical properties of white lupin seeds (*Lupinus albus* L.). Food Chem 89:341–345. https://doi.org/10.1016/j.foodchem.2004.02.040

FAO (2018) The state of food security and nutrition in the world. In: www.fao.org. http://www.fao.org/state-of-food-security-nutrition/en/. Accessed 26 Mar 2019

FAO How to feed the world in 2050. In: FAO. http://www.fao.org/fileadmin/templates/wsfs/docs/expert_paper/How_to_Feed_the_World_in_2050.pdf. Accessed 26 Mar 2019

FAO/WHO, UNU (ed) (2007) Protein and amino acid requirements in human nutrition. WHO, Geneva

Girgih AT, Udenigwe CC, Aluko RE (2010) In vitro antioxidant properties of hemp seed (*Cannabis sativa* L.) protein hydrolysate fractions. J Am Oil Chem Soc 88:381–389. https://doi.org/10.1007/s11746-010-1686-7

Girgih AT, Udenigwe CC, Li H, Adebiyi AP, Aluko RE (2011) Kinetics of enzyme inhibition and antihypertensive effects of hemp seed (*Cannabis sativa* L.) protein hydrolysates. J Am Oil Chem Soc 88:1767–1774. https://doi.org/10.1007/s11746-011-1841-9

Girgih AT, Alashi A, He R, Malomo S, Aluko RE (2013a) Preventive and treatment effects of a hemp seed (*Cannabis sativa* L.) meal protein hydrolysate against high blood pressure in spontaneously hypertensive rats. Eur J Nutr 53:1237–1246. https://doi.org/10.1007/s00394-013-0625-4

Girgih AT, Udenigwe CC, Aluko RE (2013b) Reverse-phase HPLC separation of hemp seed (*Cannabis sativa* L.) protein hydrolysate produced peptide fractions with enhanced antioxidant capacity. Plant Foods Hum Nutr 68:39–46. https://doi.org/10.1007/s11130-013-0340-6

Girgih AT, He R, Malomo S, Offengenden M, Wu J, Aluko RE (2014) Structural and functional characterization of hemp seed (*Cannabis sativa* L.) protein-derived antioxidant and antihypertensive peptides. J Funct Foods 6:384–394. https://doi.org/10.1016/j.jff.2013.11.005

Hadnađev M, Dapčević-Hadnađev T, Lazaridou A, Moschakis T, Michaelidou AM, Popović S, Biliaderis CG (2018) Hempseed meal protein isolates prepared by different isolation techniques. Part I. Physicochemical properties. Food Hydrocoll 79:526–533. https://doi.org/10.1016/j.foodhyd.2017.12.015

He R, He HY, Chao D, Ju X, Aluko R (2013) Effects of high pressure and heat treatments on physicochemical and gelation properties of rapeseed protein isolate. Food Bioprocess Technol 7:1344–1353. https://doi.org/10.1007/s11947-013-1139-z

House JD, Neufeld J, Leson G (2010) Evaluating the quality of protein from hemp seed (*Cannabis sativa* L.) products through the use of the protein digestibility-corrected amino acid score method. J Agric Food Chem 58:11801–11807. https://doi.org/10.1021/jf102636b

Hsu HW, Vavak DL, Satterlee LD, Miller GA (1977) A multienzyme technique for estimating protein digestibility. J Food Sci 42:1269–1273. https://doi.org/10.1111/j.1365-2621.1977.tb14476.x

Jiang J, Chen J, Xiong YL (2009) Structural and emulsifying properties of soy protein isolate subjected to acid and alkaline pH-shifting processes. J Agric Food Chem 57:7576–7583. https://doi.org/10.1021/jf901585n

Karki B, Lamsal BP, Grewell D, Pometto AL, van Leeuwen J, Khanal SK, Jung S (2009) Functional properties of soy protein isolates produced from ultrasonicated defatted soy flakes. J Am Oil Chem Soc 86:1021–1028. https://doi.org/10.1007/s11746-009-1433-0

Khattab RY, Arntfield SD, Nyachoti CM (2009) Nutritional quality of legume seeds as affected by some physical treatments, Part 1: Protein quality evaluation. LWT Food Sci Technol 42:1107–1112. https://doi.org/10.1016/j.lwt.2009.02.008

Kim JJ, Lee MY (2011a) Isolation and characterization of edestin from Cheungsam hempseed. J Appl Biol Chem 54:84–88. https://doi.org/10.3839/jabc.2011.015

Kim JJ, Lee MY (2011b) Characterization of hempseed protein in Cheungsam from Korea. J Korea Acad Ind Coop Soc 12:1763–1769. https://doi.org/10.5762/KAIS.2011.12.4.1763

Lam M, Shen R, Paulsen P, Corredig M (2007) Pectin stabilization of soy protein isolates at low pH. Food Res Int 40:101–110. https://doi.org/10.1016/j.foodres.2006.08.004

Liang HN, Tang CH (2013) pH-dependent emulsifying properties of pea [*Pisum sativum* (L.)] proteins. Food Hydrocoll 33:309–319. https://doi.org/10.1016/j.foodhyd.2013.04.005

Liu F, Tang CH (2015) Soy glycinin as food-grade Pickering stabilizers: Part. I. Structural characteristics, emulsifying properties and adsorption/arrangement at interface. Food Hydrocoll 60:606–619. https://doi.org/10.1016/j.foodhyd.2015.04.025

Malomo SA, Aluko RE (2015a) A comparative study of the structural and functional properties of isolated hemp seed (*Cannabis sativa* L.) albumin and globulin fractions. Food Hydrocoll 43:743–752. https://doi.org/10.1016/j.foodhyd.2014.08.001

Malomo SA, Aluko RE (2015b) Conversion of a low protein hemp seed meal into a functional protein concentrate through enzymatic digestion of fibre coupled with membrane ultrafiltration. Innov Food Sci Emerg Technol 31:151–159. https://doi.org/10.1016/j.ifset.2015.08.004

Malomo SA, Aluko RE (2016) In vitro acetylcholinesterase-inhibitory properties of enzymatic hemp seed protein hydrolysates. J Am Oil Chem Soc 93:411–420. https://doi.org/10.1007/s11746-015-2779-0

Malomo SA, He R, Aluko RE (2014) Structural and functional properties of hemp seed protein products. J Food Sci 79:C1512–C1521. https://doi.org/10.1111/1750-3841.12537

Mamone G, Picariello G, Ramondo A, Nicolai MA, Ferranti P (2019) Production, digestibility and allergenicity of hemp (*Cannabis sativa* L.) protein isolates. Food Res Int 115:562–571. https://doi.org/10.1016/j.foodres.2018.09.017

Mattila P, Mäkinen S, Eurola M, Jalava T, Pihlava JM, Hellström J, Pihlanto A (2018) Nutritional value of commercial protein-rich plant products. Plant Foods Hum Nutr 73:108–115. https://doi.org/10.1007/s11130-018-0660-7

Odani S, Odani S (1998) Isolation and primary structure of a methionine- and cystine-rich seed protein of *Cannabis sativa*. Biosci Biotechnol Biochem 62:650–654. https://doi.org/10.1271/bbb.62.650

Orio LP, Boschin G, Recca T, Morelli CF, Ragona L, Francescato P, Arnoldi A, Speranza G (2017) New ACE-inhibitory peptides from hemp seed (*Cannabis sativa* L.) proteins. J Agric Food Chem 65:10482–10488. https://doi.org/10.1021/acs.jafc.7b04522

Osburn L (1992) Hemp seed: the most nutritionally complete food source in the world. Hemp Line J 1:14–15

Ozanne L (2014) Prospective sur la place des protéines végétales à l'horizon 2030. In the conference: Les légumineuses à graines en alimentation humaine, Paris, France. https://inra-dam-front-resources-cdn.brainsonic.com/ressources/afile/263281-db0a5-resource-prospective-sur-la-place-des-proteines-vegetales-a-horizon-2030.html. Accessed 4 Aug 2019

Paredes-López O, Ordorica-Falomir C, Olivares-Vázquez MR (1991) Chickpea protein isolates: physicochemical, functional and nutritional characterization. J Food Sci 56:726–729. https://doi.org/10.1111/j.1365-2621.1991.tb05367.x

Park SK, Seo JB, Lee MY (2012) Proteomic profiling of hempseed proteins from Cheungsam. Biochim Biophys Acta BBA Proteins Proteomics 1824:374–382. https://doi.org/10.1016/j.bbapap.2011.10.005

Patel S, Cudney R, McPherson A (1994) Crystallographic characterization and molecular symmetry of edestin, a legumin from hemp. J Mol Biol 235:361–363. https://doi.org/10.1016/S0022-2836(05)80040-3

Pihlanto A, Mattila P, Mäkinen S, Pajari AM (2017) Bioactivities of alternative protein sources and their potential health benefits. Food Funct 8:3443–3458. https://doi.org/10.1039/C7FO00302A

Pimentel D, Pimentel M (2003) Sustainability of meat-based and plant-based diets and the environment. Am J Clin Nutr 78:660S–663S. https://doi.org/10.1093/ajcn/78.3.660S

Pojić M, Mišan A, Sakač M, Dapčević Hadnađev T, Šarić B, Milovanović I, Hadnađev M (2014) Characterization of byproducts originating from hemp oil processing. J Agric Food Chem 62:12436–12442. https://doi.org/10.1021/jf5044426

Ponzoni E, Brambilla IM, Galasso I (2018) Genome-wide identification and organization of seed storage protein genes of *Cannabis sativa*. Biol Plant 62:693–702. https://doi.org/10.1007/s10535-018-0810-7

Ranalli P, Venturi G (2004) Hemp as a raw material for industrial applications. Euphytica 140:1–6. https://doi.org/10.1007/s10681-004-4749-8

Ren Y, Liang K, Jin Y, Zhang M, Chen Y, Wu H, Lai F (2016) Identification and characterization of two novel α-glucosidase inhibitory oligopeptides from hemp (*Cannabis sativa* L.) seed protein. J Funct Foods 26:439–450. https://doi.org/10.1016/j.jff.2016.07.024

Rodrigues IM, Coelho JFJ, Carvalho MGVS (2012) Isolation and valorisation of vegetable proteins from oilseed plants: methods, limitations and potential. J Food Eng 109:337–346. https://doi.org/10.1016/j.jfoodeng.2011.10.027

Rutherfurd SM, Fanning AC, Miller BJ, Moughan PJ (2015) Protein digestibility-corrected amino acid scores and digestible indispensable amino acid scores differentially describe protein quality in growing male rats. J Nutr 145:372–379. https://doi.org/10.3945/jn.114.195438

Salgado PR, Ortiz SEM, Petruccelli S, Mauri AN (2011) Sunflower protein concentrates and isolates prepared from oil cakes have high water solubility and antioxidant capacity. J Am Oil Chem Soc 88:351–360. https://doi.org/10.1007/s11746-010-1673-z

Sari YW, Mulder WJ, Sanders JPM, Bruins ME (2015) Towards plant protein refinery: review on protein extraction using alkali and potential enzymatic assistance. Biotechnol J 10:1138–1157. https://doi.org/10.1002/biot.201400569

Sarv V (2017) A comparative study of camelina, canola and hemp seed processing and products. Master degree thesis, University of Toronto

Sarwar G (1997) The protein digestibility–corrected amino acid score method overestimates quality of proteins containing antinutritional factors and of poorly digestible proteins supplemented with limiting amino acids in rats. J Nutr 127:758–764. https://doi.org/10.1093/jn/127.5.758

Sarwar G, Peace RW (1986) Comparisons between true digestibility of total nitrogen and limiting amino acids in vegetable proteins fed to rats. J Nutr 116:1172–1184. https://doi.org/10.1093/jn/116.7.1172

Schaafsma G (2000) The protein digestibility–corrected amino acid score. J Nutr 130:1865S–1867S. https://doi.org/10.1093/jn/130.7.1865S

Shen L, Tang CH (2012) Microfluidization as a potential technique to modify surface properties of soy protein isolate. Food Res Int 48:108–118. https://doi.org/10.1016/j.foodres.2012.03.006

Sorgentini DA, Wagner JR (1999) Comparative study of structural characteristics and thermal behavior of whey and isolate soybean proteins. J Food Biochem 23:489–507. https://doi.org/10.1111/j.1745-4514.1999.tb00033.x

Tang CH, Ten Z, Wang XS, Yang XQ (2006) Physicochemical and functional properties of hemp (*Cannabis sativa* L.) protein isolate. J Agric Food Chem 54:8945–8950. https://doi.org/10.1021/jf0619176

Tang CH, Wang XS, Yang XQ (2009) Enzymatic hydrolysis of hemp (*Cannabis sativa* L.) protein isolate by various proteases and antioxidant properties of the resulting hydrolysates. Food Chem 114:1484–1490. https://doi.org/10.1016/j.foodchem.2008.11.049

Teh SS, Bekhit AED, Carne A, Birch J (2013) Effect of the defatting process, acid and alkali extraction on the physicochemical and functional properties of hemp, flax and canola seed cake protein isolates. J Food Meas Charact 8:92–104. https://doi.org/10.1007/s11694-013-9168-x

Teh SS, Bekhit AEDA, Carne A, Birch J (2016) Antioxidant and ACE-inhibitory activities of hemp (*Cannabis sativa* L.) protein hydrolysates produced by the proteases AFP, HT, pro-G, actinidin and zingibain. Food Chem 203:199–206. https://doi.org/10.1016/j.foodchem.2016.02.057

Wang XS, Tang CH, Yang XQ, Gao WR (2008) Characterization, amino acid composition and in vitro digestibility of hemp (*Cannabis sativa* L.) proteins. Food Chem 107:11–18. https://doi.org/10.1016/j.foodchem.2007.06.064

Wang XS, Tang CH, Chen L, Yang XQ (2009) Characterization and antioxidant properties of hemp protein hydrolysates obtained with Neutrase®. FTB J 47(4):428–434

Wang Q, Jin Y, Xiong YL (2018) Heating-aided pH shifting modifies hemp seed protein structure, cross-linking, and emulsifying properties. J Agric Food Chem 66:10827–10834. https://doi.org/10.1021/acs.jafc.8b03901

Yin SW, Tang CH, Wen QB, Yang XQ (2007) Properties of cast films from hemp (*Cannabis sativa* L.) and soy protein isolates. A comparative study. J Agric Food Chem 55:7399–7404. https://doi.org/10.1021/jf071117a

Yin SW, Tang CH, Cao JS, Hu EK, Wen QB, Yang XQ (2008) Effects of limited enzymatic hydrolysis with trypsin on the functional properties of hemp (*Cannabis sativa* L.) protein isolate. Food Chem 106:1004–1013. https://doi.org/10.1016/j.foodchem.2007.07.030

Yin SW, Tang CH, Wen QB, Yang XQ (2009) Functional and structural properties and in vitro digestibility of acylated hemp (*Cannabis sativa* L.) protein isolates. Int J Food Sci Technol 44:2653–2661. https://doi.org/10.1111/j.1365-2621.2009.02098.x

Chapter 10
Hemp Fibers for Wastewater Treatment

Lavinia Tofan, Carmen Paduraru, and Carmen Teodosiu

Abstract Sorption by low-cost lignocellulosic materials is the method receiving the utmost attention in searching for the best green approach in water pollution remediation. Among all the forms under which lignocellulosic materials are available for wastewater treatment purposes, a special position is occupied by lignocellulosic fibers such as hemp fibers. The plant fibers provides ecological and economical conveniences of high practical significance. This chapter reports relevant published researches dealing with new sorbents based on hemp fibers applied in batch and dynamic systems for the removal of pollutants from environmental aqueous media. It is divided into the following sections: distinctive features of hemp fibers as sorption media; sorption removal of pollutants from aqueous solutions by different types of hemp fibers; the applicability of sorbents based on hemp fibers for water and wastewater treatment; comparison of sorbents based on hemp fibers with other green sorbents for wastewater treatment. The results of all studies systematized and interpreted in this chapter strongly recommend the use of hemp fibers as a sustainable solution to ameliorate the quality of wastewaters.

Keywords Sorption · Fibrous sorbents · Natural hemp fibers · Loaded hemp fibers · Chemically modified hemp fibers · Carbonized hemp fibers · Heavy metal ions · Organic pollutants · Wastewaters · Removal

Abbreviations

FT-IR Fourier transform infra-red spectrophotometry
SEM Scanning electron microscopy

L. Tofan (✉) · C. Paduraru · C. Teodosiu
Department of Environmental Engineering and Management, "Cristofor Simionescu" Faculty of Chemical Engineering and Environmental Protection, "Gheorghe Asachi" Technical University of Iasi, Iasi, Romania
e-mail: cteo@ch.tuiasi.ro

G. Crini, E. Lichtfouse (eds.), *Sustainable Agriculture Reviews 42*, Sustainable Agriculture Reviews 42, https://doi.org/10.1007/978-3-030-41384-2_10

10.1 Introduction

The ecological disequilibria and health hazards from water pollution are significant parts of the price paid by the contemporary society for its technological advances. Realizing the issue gravity, the research for the development of sustainable technology for water pollution remediation is steadily increasing. An in- depth literature survey revealed that the sorption by sorbents is one of the preferred method for heavy metals and organic pollutants removal from aquatic environments (Fu and Wang 2011; Barakat 2011; Oller et al. 2011; Rivera-Utirilla et al. 2013; Amin et al. 2014; Zhao et al. 2016; Chandrakant et al. 2016; Femina Carolina et al. 2017; Saravan et al. 2017; Nidheesh et al. 2018). The key factor of the sorption popularity is versatility of the material sorbents which are available in an ever increasing range of structures and properties from natural, synthetic and waste sources. They may have different origins (mineral, organic, biological), and can be used under different physical (insoluble beads, gels, sponges, and fibers) and chemical (raw or functionalized) forms (Pan et al. 2009; Crini and Badot 2010; Renu et al. 2016; Burakov et al. 2018). The need to avoid the disadvantages of the conventional material sorbents such as the cost and difficulties associated with the regeneration of activated carbon or the expensive and restricted environmental applicability of synthetic polymers has stimulated the increase of usage of sorption on the so–called low- cost sorbents, which today occupy an important position from the perspective of sustainability. The popularity of these non- conventional sorbents is ensured mainly by their low- cost, wide availability and multiple recyclability (Table 10.1). Taking into account the chemical composition, the natural substances or agro-industrial by- products and wastes that can act as low- cost sorbents in water and wastewater treatment, these can be classified as follows: (i) lignocellulosic materials and (ii) non- cellulosic materials (Tran et al. 2015; Malik et al. 2017). Low–cost sorbents from waste materials can belong to one of the following categories: (i) agricultural and household wastes; (ii) industrial by – products; (iii) sludge; (iv) sea materials; (v) soil and ore minerals (Ali et al. 2012; De Gisi et al. 2016; Afroze and Sen 2018; Dai et al. 2018; Silva et al. 2018). Among these classes of materials, the favorable cost- potential ratio, abundance and renewability requirements are best meet by the lignocellulosic sorbents that have the highest applicability (Sud et al. 2008; Hubbe et al. 2011, 2012, 2013, 2014).

Table 10.1 The main characteristics of non-conventional sorbents

Non-conventional sorbents		
Cheapness	Availability	Renewable and recyclable character
Natural materials;	Cellulosic and non-cellulosic materials	"zero-waste" concept
By-products and wastes of industrial and agricultural processes	Untreated;	
	Physically modified;	
	Chemically modified	

The attractiveness of lignocellulosic materials for aquatic pollutants removal is due to the corroboration of their particular features such as: ubiquity; biodegradability; hydrophilicity (meaning fast kinetics, biocompatibility and great potential of functionalization); high surface area; significant porosity; good mechanical strength; favorable degree of swelling. The multitude of the forms under which the cellulosic materials can work as sorbents is also remarkable: husk, shell, straws and bran, leaves, fibers, barks, grasses, pulps, seeds, others (Abdolali et al. 2014; Varghese et al. 2018; Neris et al. 2019). Of all these forms, a special attention has been devoted to the sorbents based on lignocellulosic fibers. On the basis of their origin, natural lignocellulosic fibers can be classified as follows: (i) bast fibers (flax, hemp, jute, ramie, kenaf); (ii) fruit fibers (coir, oil palm); (iii) leaf fibers (pineapple, sisal, banana, date palm); (iv) seed fibers (cotton); (v) grass fibers (bagasse, bamboo) (Manna et al. 2017). Regardless of their origin, the gained credibility of the plant fibers is due to the fact that unlike other low– cost sorbents, their use brings some advantages, namely the possibility of operation at low values of pH, resistance to the toxicity levels of metal ions and higher robustness from the chemical and physical points of view (Table 10.2). Also, their ecological and economical convenience for filter production is highly significant (Rojas et al. 2017).

The performances of plant fibers for the removal of aquatic pollutants by sorption are strongly correlated with their variable cellulose content. One of the types of fibers obtained from the hemp plant, namely the bast fibers has the highest content of cellulose (75–80%) (Hokkannen et al. 2016). The sorption potential of hemp fibers has been explored in older and recent researches. The results of all studies suggest that hemp fibers sorption has potential for implementation in wastewater decontamination. This chapter emphasizes recent advances and trends, highlighting the pollutants removal abilities by the sorbents based on: natural hemp fibers, loaded hemp fibers, chemically modified hemp fibers and carbonized hemp fibers, considering also the original contribution of the authors on this topics. By framing the data reported by the authors for the first time in 1999 in the current scientific context and providing recent information, this study attempts to be a platform in describing the synthesis, characterization and applicability of different types of sorbents based on hemp fibers for pollutants removal, as well as the comparison of these sorbents with other sorbents effective in such processes.

10.2 Distinctive Features of Hemp Fibers as Sorption Media

Hemp (*Cannavis sativa* L.) is a fast growing plant with a unique and versatile potential that is suggestively described as the plant of thousand and one molecules (Andre et al. 2018). Among the multi – purpose applications of the cellulosic and woody fibers obtained from the hemp stem, this one is of special interest because it is based on the valorization of their sorption properties for the efficient removal of different pollutants from aquatic media (Crini and Lichtfouse 2018).

Table 10.2 Applicability of lignocellulosic fibers for remediation purposes

Vegetal fiber sorbent	Targeted pollutant	Study type	References
Jute fibers			
Unmodified	Azo dyes	Equilibrium and dynamic studies	Roy et al. (2013)
Grafted with acrylic acid	Hg(II), Pb (II)	Equilibrium and kinetic studies	Hassan and Zohdy (2018)
Carboxyl modified	Pb(II), Cd (II), Cu(II)	Equilibrium, kinetic and desorption studies	Du et al. (2016)
Pyromellitic modified dianhydride	Aniline	Characterization and batch studies	Gao et al. (2015)
		Fixed – bed column studies	Jiang et al. (2016)
Kenaf fibers			
Unmodified	Cu(II), Ni (II)	Batch and fixed bed column studies in single and binary solutions	Hasfalina et al. (2010, 2012)
Citric acid modified	Methylene blue	Equilibrium, kinetic and thermodynamic studies	Sajab et al. (2011)
Coconut fiber			
Unmodified	Ag(I)	Equilibrium, kinetic, thermodynamic and desorption studies	Staroń et al. (2017)
Pristine and NaOH-treated	Hg(II)	Equilibrium and kinetic studies	Johari et al. (2014)
Agave americana fibers	Cd(II), Pb (II)	Equilibrium, kinetic and thermodynamic studies	Ben Hamissa et al. 2010
	Malachite green dye	Equilibrium, thermodynamic and kinetic studies	Altinisik et al. (2015)
	Methylene blue dye	Fractal kinetics and regeneration studies.	Ben Hamissa et al. (2013)
Sisal fibers			
Unmodified	Pb(II), Cd (II)	Characterization and equilibrium studies	dos Santos et al. (2011)
Coated with polyaniline	Pb(II)	Equilibrium and kinetic studies	Teklu et al. (2018)

Hemp fibers carry important features which predetermine their use as pollutants sorbent, namely:

- excellent mechanical strength, high fiber length (Table 10.3);
- high stiffness and roughness;
- high porosity, low elongation at break, high elasticity (Kabir et al. 2017);
- high hygroscopicity and absorbency, good thermal and electric properties, UV protection properties (Kostic et al. 2010);
- moisture content 12%; significant degree of swelling;
- good chemical stability:

Table 10.3 Physical and mechanical characterization of hemp fibers (Shahzad 2012)

Property	Value	Property	Value
Length	8.3–14 mm	Tensile strength	310–750 MPa
Diameter	17–23 μm	Specific tensile strength	210–510 MPa
Aspect ratio (length/diameter)	549	Young's module	30–60 GPa
Specific apparent density	1500	Specific Young's module	20–41 GPa

- the action of the oxidant agents (inorganic and organic peroxides, chlorine compounds) underlying the bleaching process, after which the hemp becomes golden yellow (Tofan and Paduraru 1999);
- the hydrolytic action of the alkalis (especially 18%NaOH) leads to the increase of the hemp fibers hydrophilicity (Paduraru and Tofan 2002);
- under the action of mineral acids (HNO_3, HCl, H_2SO_4) the fibers of hemp loss their resistance;

• tolerance to biological structures, easiness in functionalizations (Tofan and Paduraru 2000).

The above variable properties of hemp fibers are determined by their chemical composition (Table 10.4) that is very heterogeneous due to the combined action of many factors, such as: source, geographical region, climatic and environmental conditions, age, technique of retting and separation etc. (Shahzad 2012). Thus, like other lignocellulosic materials, hemp fibers contain variable amounts of cellulose and accompanying substances: hemicelluloses, lignins, pectins, proteins, minerals, fats, waxes.

The main component of hemp fibers is cellulose that consists of thousands of glucose molecular units joined in long chains and has the functional role of frame substance. The linear macromolecule of cellulose consists of D – anhydroglucose ($C_6H_{11}O_5$) repeating units joined by β-1,4 – glycosidic linkages. Each repeating unit contains three hydroxyl groups, able to form intramolecular and intermolecular hydrogen bonds (Petroudy 2017). A significant part of the cellulose is crystalline (Table 10.4) and with a high degree of orientation, being inaccessible for water and other solvent molecules, and the remaining part of cellulose consists of less oriented chains (Abdolali et al. 2014). The second basic unit is represented by the hemicellulose, a polymer mainly derived from the chains of pentose sugars and acts as cementing material Typical components of hemicelluloses (d-xylose, d-mannose, d-glucose, d-galactose, l-arabinose, l-rhamnose, 4-O-methyl-d-glucuronic acid, d-glucuronic acid and d-galacturonic acid) may be linked together either in a linear or branched arrangement (Anwar et al. 2014). The third component of the fibers, lignin is a cross-linked aromatic polymer with strong covalent bonds disposed as a 3D-network. Defined as a polymer of phenylpropane units with three different aromatic units (*p*-hydroxyphenyl, guaiacyl, and syrin-gyl), lignin plays the role of encrusting substance (Leite and Pereira 2017). The fiber construction is based on microfibrilis, a cellulose chains network formed by inter- and intramolecular hydrogen bonds (Fig. 10.1a). Through the hemicellulose and lignin that surrounds them,

Table 10.4 Systematization of the properties of the hemp fibers selected for the studies concerning the sorption removal of pollutants

Feature	Description			References
Chemical composition	Cellulose (%) 74–75	Ash (%) 0.82		Paduraru and Tofan (2008)
	Hemicelluloses (%) 18.4–15.4	Xylans (%) 3.0–7.0		
	Lignin (%) 3.7	Proteins (%) 0.5–1		
	Waxs (%) 4.04	Pectins (%) 4.0–8.0		
	Celullose (%) 78.15	Fats and waxes (%) 0.69		Pejic et al. (2009)
	Hemicelluloses (%) 0.72	Pectin (%) 1.39		
	Lignin (%) 6.06			
	Water solubles (%) 1.5			
	Cellulose (%) 57–77	Lignin (%) 5–9		Placet et al. (2017)
	Hemicelluloses (%) 9–14			
	Cellulose (%) 75	Lignin (%) 3		Loiacono et al. (2018)
	Hemicellulose (%) 15	Pectin (%) 5		
Chemical structure (main IR transitions for a hemp fiber of reference)	**Wavenumber (cm^{-1})**	**Vibration**	**Source**	
	3336	OH stretching	Cellulose, hemicellulose	Dai and Fan (2010)
	2887	C–H symmetrical stretching	Cellulose, hemicellulose	
	1729	C=O stretching vibration	Pectin, waxes	
	1623	OH bending of absorbed water	Water	
	1423	HCH and OCH in – plane bending vibration	Cellulose	
	1202, 1155	C–O–C asymmetrical stretching	Cellulose, hemicellulose	
	1048, 1019	C–C, C–OH, C–H ring and side group vibrations	Cellulose, hemicellulose	
Crystallinity	Crystallinity index: 0.82; percentage of crystallinity: 84.39%			Kyzas et al. (2015)
Surface characteristics	Amount of functional groups:Q(COOH) = 0.535 mmol/g; Q(CHO) = 0.052 mmol/g;			Vukcevic et al. (2014a), Balintova et al. (2014), and Kostic et al. (2018)
	Specific surface area: $S_{BET} = 0.236\ m^2/g$; $0.99\ m^2/g$			
	Point of zero charge pH: $pH_{PZC} = 4.4$			
	Accessibility of the surface area to aqueous solution:			
	Value of water retention (WRV) = 60%;			
	Iodine sorption value (ISV) = $130 mgI_2/g$ of fibers			

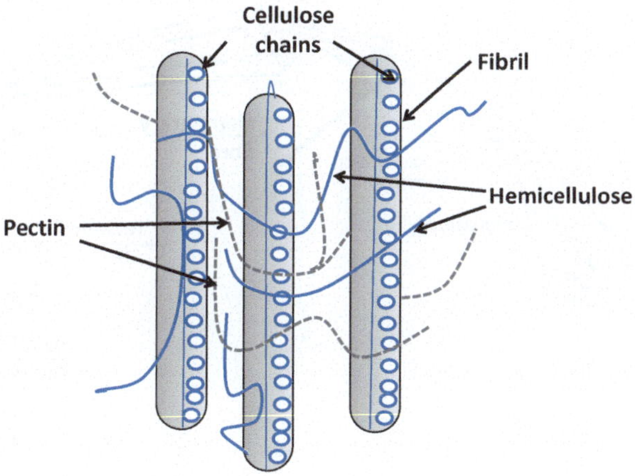

Fig. 10.1 The role of cellulose, hemicellulose and lignin in hemp fiber construction (**a**) – fibril of cellulose. (Adapted from Sedan et al. 2008). (**b**) – architecture of hemp fiber. (Adapted from Kaczmar et al. 2011)

microfibrils are binded together and form a matrix structure of fibril and subsequently, fiber (Fig. 10.1b). The fibers are connected by pectins and encrusted with lignin in bundles.

Cellulose is the main responsible for some fundamental features such as: the strength, stability and hydrophilicity of the fibers. Hemicellulose has a major impact on the moisture adsorption and thermal stability of the hemp fibers, while lignin strongly influences the UV degradation of fibers (Sedan et al. 2008). In an attempt to explain the individual influences of hemicelluloses and lignin on the properties of hemp fibers, some researchers have concluded that hemicelluloses removal increases, while lignin removal decreases the moisture and iodine sorption of hemp fibers (Kostic et al. 2010). According to the results of another significant study, hemicellulose removal determines the increase of the degree of fiber swelling and decrease of water retention value of hemp fibers, while the effect of lignin removal is the decrease of the degree of fiber swelling and increase of the water retention ability of hemp fibers (Pejic et al. 2009). It has also been shown that the removal of both hemicellulose and lignin leads to the peeling of hemp fiber surface (Kostic et al. 2018).

The heterogeneity of hemp fibers chemical composition is associated with a very complicated structure, corresponding to their multi-cellular constitution (Fig. 10.2). Hemp fibers are made up of bundles of individual cells (elementary fibers), cemented between them by the medial lamella, with a predominantly lignin content. Morphologically, an individual cell consists of the primary wall (disorganized arrangements of cellulose fibrils embedded in an organic matrix of hemicelluloses, lignin and proteins), the second wall (three layers of cellulose fibrils with different axial orientations, bonding by hemicelluloses), the tertiary wall (one layer of cellulose) and the lumen (Morin-Crini et al. 2019).

The feasibility and suitability of hemp fibers in the sorption removal of pollutants, from aqueous media can be considered as a corollary of these specific features. For this purpose the major role is played by the functional groups incorporated in the structure of hemp fibers. The carboxylic, phenolic, hydroxyl and carbonyl groups existing in the basic components of these biofibres (Table 10.4), able to form strong bondings, especially with the pollutant metal ions, are at the basis of the sorption performances. The high accessibility of the cell wall components to the aqueous

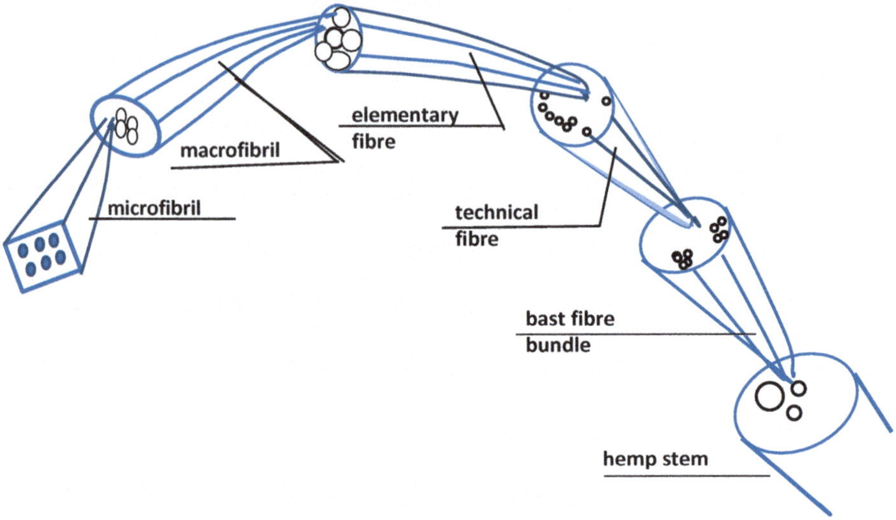

Fig. 10.2 Scanning electron microscopy photographs describing the multi-cellular structure of natural hemp fibers (Paduraru and Tofan 2008)

Table 10.5 Highlighting of natural hemp fibers capability to enhance their sorption performance

Treatment method applied to natural hemp fibers	Type of sorbents based on modified hemp fibers with improved capacity of pollutant retention	References
Physical methods	Loaded hemp fibers	Tofan and Paduraru (1999); Tofan et al. (2001)
Carbonization	Carbonized hemp fibers	Yang et al. (2011), Vukcevic et al. (2012, 2014b, 2015, 2017), and Lupul et al. (2015b)
Chemical methods	Chemically modified hemp fibers	Tofan and Paduraru (2004), Cassano et al. (2013), Pejic et al. (2009, 2011), Vukcevic et al. (2014a), Balintova et al. (2014), Kyzas et al. (2015), Loiacono et al. (2017a), and Kostic et al. (2018)

solutions must also be under consideration (Table 10.4). A group of researchers have studied the influence of the chemical composition and surface characteristics and found that sorption properties of short hemp fibers are predominantly influenced by surface acidity and the amount of functional groups, while fiber structure and specific surface area have a secondary role in the sorption of Pb(II), Cd(II) and Zn (II) ions (Pejic et al. 2009, 2011; Vukcevic et al. 2014a).

The adjustability of their surface chemistry is very beneficial for enhancing the hemp fibers abilities for the pollutants sorption. Thus, through the adequate methods of physical and chemical pretreatments, modified hemp fibers can be obtained with significantly improved sorption performance both in terms of sorption capacity as well as selectivity (Table 10.5).

These particular features of hemp fibers can be valorized for the simultaneous or sequential removal of pollutants from wastewaters, in batch and dynamic systems of sorption.

10.3 Removal of Pollutants from Aqueous Media by Different Types of Hemp Fibers

Hemp fibers can be used in aquatic pollutants removal without any treatment (natural hemp fibers) or after having undergone physical (loaded hemp fibers) or chemical (chemically modified hemp fibers) treatments. Furthermore, the chemical composition of hemp fibers enables them to act as precursor for biocarbon based sorbents.

The batch studies on this topics reported in the literature so far addressed issues related to the setting of optimum conditions of sorption and understanding of the sorption mechanism by:

1. studies regarding the influence of different process variables
   - initial pH;
   - initial pollutant concentration of the solution;
   - sorbent dosage;
   - contact time;
   - temperature
2. isothermic modelling
   - Langmuir equation (Langmuir 1916);
   - Freundlich equation (Freundlich 1906)
3. kinetic modelling
   - pseudo-first order model (Lagergren 1898);
   - pseudo-second order model (Ho and McKay 1999)
4. thermodynamic parameters (Li and Liu 2008)
   - free energy change (ΔG);
   - enthalpy change (ΔH);
   - entropy change (ΔS)

10.3.1 Natural Hemp Fibers

The idea of the authors of this work to use natural hemp fibers as heavy metal sorbent was born in the '90s with the aim of reaching a double goal: to develop the basis of an efficient technology for polluted water treatment by using industrial wastes and to provide a possible solution for the industrial wastes complex and expensive management problem, by converting them into value- added, environmentally sustainable sorbents. This approach is very close of the circular economy concept, which nowadays is approached by many researchers in the field of environmental science and engineering. In this context, the hemp thick and rigid fibers (resulting as wastes from a textile factory in the North–East Region of Romania) were explored for the first time as sorbent for the removal of some toxic and polluting metal ions from aqueous monometallic solutions in batch conditions. For this purpose, the following succession of stages has been performed:

1. *sorbent preparation*: boiling for 4 h in a solution containing soap and soda ash → washing several times with water → rinsing with bidistilled water → drying at 45 °C;
2. *batch sorption experiments*: samples of about 0.25 g hemp were equilibrated with 50 mL of aqueous solution containing a defined amount of the targeted metal ion, at desired temperature and pH → hemp fibers were removed from aqueous solution by filtration → the metal ion concentration in the final solution has been determined by atomic absorption spectrometry;
3. *calculation of the parameters* characteristic to the metal ions sorption by natural hemp:

   - retention percentage, R (%) = $[(C_0 - C)/C_0].100$;
   - retained amount of metal ion, q = $[(C_0 - C)/G]$

where C_0 is the initial concentration of metal ion (mg/L), C is the cation concentration after sorption (mg/L), V is the volume of solution (L); and *G* is the weight of hemp fibers (g).

The original results giving the evidence of the possible benefits of using the hemp fibers waste for the sorptive removal of heavy metals from aqueous media were first published by Tofan and Paduraru (1999). The data referring to the sorption characteristics of Cu(II), Cd(II), Zn(II) and Pb(II) on natural hemp fibers were selected and systematized in Table 10.6. The choice of these heavy metal ions was dictated by their potential pollution impact and frequent presence in industrial effluents.

Our studies data are in good agreement with those subsequently published by Pejic et al. (2009), Vukcevic et al. (2014a), and Kostic et al. (2018) for Cd^{2+}, Zn^{2+} and Pb^{2+}, respectively, Balintova et al. (2014) and Rozumova and Legatova (2018) for Cu^{2+} ions sorption on unmodified hemp fibers. The sorption performances of hemp fibers were explained on the basis of functional groups, micropores and microcracks incorporated in the structure of hemp fibers (Pejic et al. 2009). The sorption of the heavy metal ions on hemp fibers was described as an endothermic process of chemical nature which follows a mechanism of ion exchange on acidic functional groups and occurs through fast surface adsorption, intraparticle diffusion and final equilibrium stage (Paduraru and Tofan 2008; Tofan et al. 2010a, c; Vuckevic et al. 2014a, b).

Recent studies are focused on the decontamination of individual and poly-metallic aqueous solutions by raw felted fibers of hemp in batch systems. The increased efficiency of the felted fibers of hemp as compared to that of the loose fibers in the simultaneous removal of Al^{3+}, Co^{2+}, Cr^{3+}, Cu^{2+}, Ni^{2+} and Zn^{2+} from poly-metallic aqueous solutions has been pointed out (Bugnet et al. 2017). The maximum sorption capacities of a hemp felt made of 100% fiber for Cd^{2+} (27.47 mg/g), Cu^{2+} (14.64 mg/g of hemp), Zn^{2+} (10.59 mg/g of hemp), Co^{2+} (7.99 mg/g of

Table 10.6 Systematization of the authors' results of the batch studies concerning the Cu(II), Cd(II), Zn(II) and Pb(II) sorption on natural hemp fibers

Effect of process parameters			
Parameter	Findings	Explanation	
Initial pH (Paduraru and Tofan 2008; Tofan et al. 2010b)	This dependence was investigated in solutions with initial pH in the range of 2–5 where the metals exist in their double positively charged ionic forms (Zn^{2+}, Pb^{2+}) and their precipitation as metal hydroxides is avoided. The sorption for both metal ions under study was minimum at pH = 2 and then increased, reaching the maximum at initial pH = 5 from unbufferd solutions	Low pH → Protonation of the superficial functional groups → positive charge of hemp fibers → repulsive forces that limit the approach of the ions	Increasing pH → Dissociation of the functional groups (hydroxyl, carboxyl) → surface of hemp fibers is negatively charged→ strong attraction of the ions in solution → increased uptake
Initial metal concentration (Paduraru and Tofan 2008; Tofan et al. 2010b)	Increasing metal ion concentration → increase of the amount of heavy metal ion retained on the hemp fibers → decrease of the removal efficiency	This behavior is due to the initial metal concentration action of driving force	
Hemp dose (Tofan et al. 2010b)	Sorption percentage values increased with increasing hemp dose. At the maximum dose of 40 mg hemp/L, the percentage of Pb(II) retention reached a value of 96%	This trend is in good agreement with the higher number of available sites of the natural hemp for Pb(II) binding	
Contact time (Paduraru and Tofan 2008; Tofan et al. 2010c)	Increasing contact time of the phases→ sharp increase of the heavy metal amounts sorbed on the hemp → reaching values that remained almost constant	Lack of available active sites required for high initial concentration of metal ions → reduction in immediate metal ions sorption (Chaiyasith et al. 2006)	
Temperature (Paduraru and Tofan 2008; Tofan et al. 2010a, c)	The capacity of sorption and sorption intensity are enhanced at higher temperature	The change in temperature affects not only the diffusion rate of metal ions, but also the solubility of metals (Park et al. 2010)	

(continued)

Table 10.6 (continued)

Characterization of the sorption systems based on natural hemp fibers								
			Kinetics		Thermodynamics			
Cation	T (K)	q_0 (mg/g)	Pseudo-first order k_1 (min^{-1})	Pseudo-second order k_2 (g/mg min)	ΔG (KJ/mol)	ΔH (KJ/mol)	ΔS (J/mol K)	References
Cu(II)	293	9.0735			−5.79			Tofan and Paduraru (2000)
Cd(II)	275	2.5909	$5.42 \cdot 10^{-2}$		−6.771	10.92	0.0636	Tofan et al. (2010a)
	293				−8.069			
	323				−9.646			
Zn(II)	277	16.54	$6.678 \cdot 10^{-3}$		−14.183	16.98	0.112	Paduraru and Tofan (2008)
	293	21.047			−16.066			
	323	24.401			−19.350			
Pb(II)	277	20.700	$2.53 \cdot 10^{-3}$	$8.84 \cdot 10^{-4}$	−20.035	14.21	0.122	Tofan et al. (2010c)
	293	25.047			−22.046			
	333	28.359			−25.292			

hemp), Fe^{2+} (7.85 mg/g of hemp), Ni^{2+} (7.87 mg/g of hemp), Cr^{3+} (6.53 mg/g of hemp), Al^{3+} (6.38 mg/g of hemp) and Mn^{2+}(4.55 mg/g of hemp) are very promising for real poly-metallic effluents detoxification (Loiacono et al. 2018).

10.3.2 Loaded Hemp Fibers

During the last years, a special interest was given for finding the most effective method to enhance the affinities of low-cost sorbents towards the removal of aquatic pollutants. The authors of this work proposed a simple approach, namely loaded hemp fibers, that, on one hand, can be fibrous sorbents with improved sorption capacity and selectivity, and on the other hand, open a possible horizon to develop a process where the hemp fibers can be simultaneously applied to reduce water pollution produced by organic compounds and heavy metal ions.

The features of the loaded materials prepared by the impregnation of natural hemp with α- benzoinoxime and alizarin S point out that, due to their special properties, hemp fibers are a good support for the immobilization by physical adsorption of different organic compounds (Table 10.7) (Tofan and Paduraru (1999); Tofan et al. (2001)). To the best of our knowledge, there exists no other available study reporting the potential of natural hemp fibers for organic pollutants sorption. In our studies, α- benzoinoxime and alizarin S were chosen as modifying agents for hemp fibers on the basis of the selectivity function performed in any sorbent by the chelating functional groups, possessing donor atoms (O, N, S) able to form coordination bonds with metal ions (Fig. 10.3). The behavior of these materials

Table 10.7 Loaded hemp fibers for the selective removal of Cu(II) and Cr(III) ions

Hemp loaded with **α- benzoinoxime – Cu(II)**

				Thermodynamics			
Sorption capacity of the organic reagent	pH	T (K)	q_0 (mg Cu/g of hemp)	ΔG (kJ/mol)	ΔH (kJ/mol)	ΔS (kJ/kmol)	Reference
0.074 mmol	8	278	11.7612	−7.378			Tofan and Paduraru (1999); Tofan et al. (2001)
α-benzoinoxime/g of hemp		291	13.8072	−7.949	+3.829	0.00401	
		313	15.1225	−8.610			

Hemp loaded with **alizarin S – Cr(III)**

Langmuir constants			Freundlich constants			
Temperature (K)	q_0 (mg Cr/g of hemp)	K_L (ml/mmol)	K_F	n	ΔH (kJ/mol)	Reference
277	8.632	0.264	2.34	4.54	−5.25	Tofan et al. (2001)
293	6.338	0.249	1.85	4.33		
313	4.726	0.203	1.79	2.08		

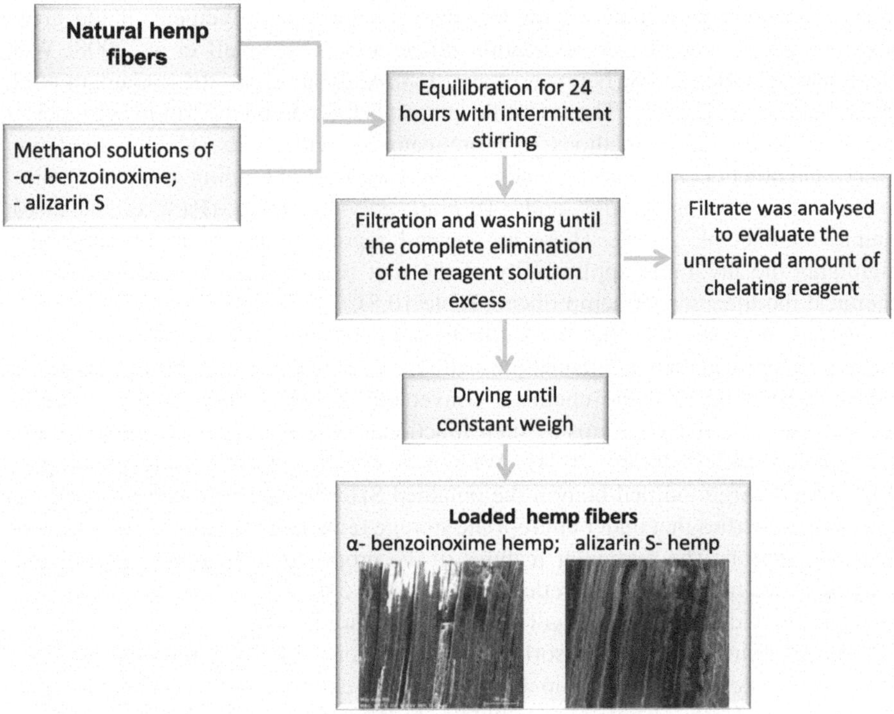

Fig. 10.3 Scheme of hemp loading

prepared by the "sorption" of chelating reagents on natural hemp fibers was assessed for the selective removal of Cu(II) and Cr(III) ions from aqueous solutions. The main characteristics of the hemp fibers- α- benzoinoxime – Cu(II) and hemp fibers – alizarin S- Cr(III) batch sorption systems are presented in Table 10.7. As expected, the physical impregnation of hemp fibers resulted in significant enhancements of uptakes of Cu(II) and Cr(III), respectively. Thus, there is a sensitive difference between Cu(II) sorption capacity of hemp – α- benzoinoxime (13.8072 mg/g of hemp) and natural hemp fibers (9.0735 mg/g of hemp). Also, alizarin–S impregnated hemp exhibited higher Cr(III) sorption capacity (6.340 mg/g impregnated hemp) than natural hemp fibers (4.006 mg/g).

According to Table 10.7, the favorable behavior of loaded hemp fibers provides some new attractive possibilities, such as:

- removal of organic pollutants;
- selective removal of heavy metal ions from aqueous effluents;
- removal of mixtures of heavy metal ions and organic pollutants.

10.3.3 Chemically Modified Hemp Fibers

The most appropriate methods for removal efficiency improvement of the green sorbents are the chemical modification procedures (O'Connell et al. 2008; Wan Ngah and Hanafiah 2008; Nguyen et al. 2013; Abdolali et al. 2014; Gautam et al. 2014; Salman et al. 2015). The scientific interest is focused on the following types of chemical modification methods: (1) pretreatment with acids, alkalis, oxidizing agents and other chemical compounds; (2) enhancement of binding groups; (3) elimination of inhibiting groups and (4) graft polymerization (Park et al. 2010; Ramrakhiani et al. 2016). There are several reports available in literature that highlights the favorable applicability of the first two of these procedures for the chemical modification of hemp fibers (Table 10.8).

Starting from the fact that the sulfur ligands are more selective towards heavy metal ions removal than their analogues with nitrogen and oxygen, the authors of this study reported in 2004 the chemical conversion of hemp fibers into a chelating sorbent with sulphydryl groups by their functionalization with β-mercaptopropionic acid (Tofan and Paduraru 2004). With a S content of 44%(corresponding to 0.1375mmoleS/g modified hemp), the obtained SH – hemp fibers had sorption and selective properties that under appropriate storage remained unchanged for 3 months after preparation. The particular features of the proposed sorbent based on chemically modified hemp fibers are reflected in 8.8- and 5.41-fold enhancements uptakes of Ag(I) and Cd(II), respectively(Tofan and Paduraru 2004).

Another potential chelating sorbent candidate for Cd(II) ions removal was synthesized via esterification of hemp with 2-benzyl- 4-chlorophenol, a germicide agent, that was covalently coupled to cellulose backbone of hydrophilic fibers by a heterogeneous synthesis (Cassano et al. 2013).

Table 10.8 Chemically modified hemp fibers

Type and source of hemp fiber	Targeted pollutant	Chemical treatment	Effect of hemp fibers chemical modification	q_0, mg/g	References
Waste hemp fibers of hemp from a textile factory in the North–East Region of Romania	Ag(I),	Purification of hemp fibers → preparation of the reaction mixture(-β-mercaptopropionic, acid, acetic anhydride, acetic acid, sulphuric acid) → soaking purified hemp fibers in the reaction mixture for 3–4 days at 40 °C	Introducing sulphydryl chelating groups → improved sorption performance	10.786	Tofan and Paduraru (2004)
	Cd(II)			14.051	
	Pb(II)			22.999	
Short and entangled hemp fibers obtained from ITES Odzaci, Serbia	Cd(II)	Treating the fiber sample with 17.5% NaOH at room temperature → neutralization with 1% acid acetic → washing → drying	Progressive removal of hemicellulose → structural and morphological changes → increased sorption capacity	8.767	Pejic et al. (2009)
	Pb(II)			16.161	
	Zn(II) from single and ternary metal ion solutions			5.099	
	Zn(II)			7.4–8.1	Vukcevic et al. (2014a)
Short and entangled hemp fibers obtained from ITES Odzaci, Serbia	Cd(II)	Treating the fiber sample with 0.7% $NaClO_2$ at pH = 4 and boiling temperature→ washing →drying	Progressive removal of lignin → structural and morphological changes → improved sorption capacity	4.383	Pejic et al. (2009, 2011)
	Pb(II)			7.666	
	Zn(II) from single and ternary metal ion solutions			2.484	
	Zn(II)			7.1–8.3	Vukcevic et al. (2014a)

(continued)

Table 10.8 (continued)

Type and source of hemp fiber	Targeted pollutant	Chemical treatment	Effect of hemp fibers chemical modification	q_0, mg/g	References
Hemp shives from the Hemp Flax company	Cu(II)	Treatment with 1.6 M NaOH solution at room temperature for 48 h → neutralisation with 1% acetic acid → washing → drying	Gradual removal of hemicellulose or lignin → enhanced sorption capacity	4.45	Balintova et al. (2014)
Hemp fibers (HF) and shives (HS) from the Hemp Flax company	Ni(II)	Soaking in 5%NaOH solution →stirring at room temperature for 1 h → soaking in 12.5% aqueous solution of citric acid →stirring at room temperature for 30 min → drying under vacuum at 60 °C → rinsing	Removal of any impurities → adding of extra carboxyl groups → improved sorption performance	HF ⇩ 242 HS ⇩ 237	Kyzas et al. (2015)
Hemp felted fibers from a hemp processing company in France	Cd(II)	Washing hemp felt → treatment with a mixture of maltodextrin and 1,2,3,4 – butanetetracarboxylic acid → washing → drying	Providing ion – exchange properties by introducing carboxylic groups → increased sorption capacity	129.87	Loiacono et al. (2017a)
	Co(II)			37.88	
	Cu(II)			63.32	
	Mn(II)			44.25	
	Ni(II)			40.98	
	Zn(II) from aqueous mixtures			68.50	
Short and entangled hemp fibers obtained from ITES Odzaci, Serbia	Pb(II)	Treating the fiber sample with 17.5% NaOH for 5 and 45 min, respectively	Hemicelluloses removal → interfibrillar regions of fibers less dense and rigid → fibrils more capable to rearrange	15.54 16.16	Kostic et al. (2018)
		Treating the fiber sample with 0.7% $NaClO_2$ (5 and 60 min, respectively)	Removal of lignin → more homogenous middle lamella due to the gradual elimination of micropores and the less rigid cell wall	16.16 16.37	

The enhanced sorption abilities of the chemically modified hemp fibers by alkaline (sodium hydroxide) and oxidative (sodium chlorite) treatments (Table 10.8) has been explained by the domination of sorption at outer surfaces of fibers that increased the roughness of hemp fiber surfaces and induced new capillary spaces in inter-surfacial layer between completely or partially separated fibers due to the removal of lignin and hemicelluloses (Pejic et al. 2009). Strong sorption capacities of the felted fibers of hemp coated with a maltodextrin-1,2,3,4-butane tetracarboxylic polymer (Table 10.8) were attributed to a chemisorption mechanism involving complexation and ion exchange (Loiacono et al. 2017a, b).

The improved sorption characteristics of the chemically modified hemp fibers can be minimized by the main disadvantages of the chemical modification methods, namely the increased cost of treatment and the secondary pollution. Thus, for the full exploitation of the beneficial effects of the chemical pretreatments, the research needs to focus on the improvement of the current modification methods and development of innovative methods.

10.3.4 Carbonized Hemp Fibers

There is unanimous recognition of the feasibility and suitability of activated carbon for the removal of pollutants in water and wastewater treatment. In this context, a topic of high interest is the exploitation of the raw biomass as inexpensive and efficient precursor for the development of activated carbon (Dias et al. 2007; Yahya et al. 2015; Jain et al. 2016a, b; Luka et al. 2018). The great interest especially addressing the lignocellulosic biomass for the preparation of activated carbons was justified mainly on the basis of unique features of this biomass in terms of porosities and surface functional groups (Altenor et al. 2009; Chawdhury et al. 2013).

Hemp fibers have already been proven as promising alternative precursors for preparation of activated carbons. Thus, the extremely favorable properties of the activated carbon fibers prepared by the physical activation with steam and chemical activation with zinc chloride and phosphoric acid, respectively were emphasized in literature (Williams and Reed 2003, 2004; Rosas et al. 2009; Lupul et al. 2015a). A recent study concerning the preparation and characterization of bast and leaf fibers–derived mesoporous activated carbon sorbents shows that hemp fibers derived activated carbon has the biggest BET surface area and the highest carbon content as compared to the flax fiber– derived activated carbon and sisal fiber–derived activated carbons (Dizbay-Onat et al. 2018).

The sorptive performances of carbon materials from natural hemp fibers in water and wastewater treatment are highlighted in Table 10.9. As it can be seen from Table 10.9, unlike natural, loaded and chemically modified hemp fibers, biocarbon fibers prepared by carbonization and chemical activation of hemp are preferable for

Table 10.9 Preparation and characterization of carbonized hemp fibers with applications in wastewater treatment

Synthesis procedure of activated carbon fibers	Resultant ACF	Textural properties of ACF		Pollutant sorption properties of ACF		References
		S_{BET} (m^2/g)	V_{tot} (cm^3/g)	Targeted pollutant	Remarks	
Bast fibers of hemp from Yunnan Province in China → impregnation with 2.5 M H_3PO_4 → boiling at 110 °C → filtration → activation under continuous N_2 flow at 400, 450, 500, 550, 600 °C → washing with hot distilled water → drying	P400	986.3	0.5567	Acid blue 9 dye	Langmuir maximum capacity of sorption $q_0 = 28.75$ mg dye/g of P450	Yang et al. (2011)
	P450	1142.4	0.6663			
	P500	1069.6	0.6162			
	P550	1058.3	0.6017			
	P600	1038.6	0.5084			
Short hemp fibers from ITES Odzaci, Serbia → chemical treatment	Ch 1	518.5	0.291	15 pesticides from different chemical classes	Carbonized and activated hemp fibers could act as a good solid – phase sorbent for the pesticide analysis in water samples	Vukcevic et al. (2012)
	ChH5	425.9	0.207			
	ChH45	573.5	0.290			
NaOH; $NaClO_2$	ChL5	428.6	0.208			
⇩	ChL60	388.6	0.194			
Carbonization at 1000 °C → **carbonized hemp fibers** → chemical activation with KOH at 900 °C → **activated hemp fibers**	Ach1	673.0	0.403			
	Ach2	2192	1.203			
Short hemp fibers from ITES Odzaci, Serbia → chemical treatment	Ch1	518	0.290		Experimental values of the equilibrium uptake, $q_{e,exp}$ ranged between 15.57 and 23.59 mg Pb/g of carbonized hemp fiber	Vukcevic et al. (2014b)
	Ch L5	429	0.208	Pb(II)		
	ChL60	389	0.193			
NaOH; $NaClO_2$	Ch H5	426	0.206			
⇩	ChH45	574	0.289			
Carbonization at 1000 °C → **carbonized hemp fibers**						

Short hemp fibers → carbonization at 700 and 1000 °C under constant nitrogen flow → **carbonized hemp fibers** → chemical activation with KOH at 700 and 900 °C → **activated hemp fibers**	ACh119	673		Pesticides	Values of the Langmuir maximum capacity of sorption, q_0, ranged between for 12.2–19.3, 11.8–14.7, 11.6–19.5, 13–15.7, 14.5–15.5 mg pesticide/g of carbonized fiber, for Ac, D, N, C and At, respectively	Vukcevic et al. (2015)
	ACh127	1178		Ac		
	ACh129	2192		D		
	ACh149	1858		N		
	ACh719	352		C		
	ACh729	1210		At		
Dry hemp stems → carbonization at 600 °C → mixture with HNO_3 → heating at 700 °C under nitrogen flow → washing (HAC)	HAC	2135	0.919	Atrazine	Values of the Langmuir maximum capacity of sorption, q_0, ranged between for 227, 263, 179, 169 mg pesticide/g for HAC, $HACN_2$, $HACNH_3$ and $HACHNO_3$, respectively	Lupul et al. (2015b)
HAC → oxidation with HNO_3 → washing (HAC HNO_3)	$HACHNO_3$	2067	0.910			
HAC → treatment with gaseous NH_3 → heating → cooling (HAC NH_3)	HAC NH_3	2088	0.927			
HAC → heating at 700 °C under nitrogen flow (HAC N_2)	HAC N_2	2213	0.953			
Waste hemp fibers → alkaline treatment (NaOH)	Ch1	519	0.291	Pb(II)	Values of the Langmuir maximum capacity of sorption,q_0, ranged between for 106.7, 103.1, 113.3, 116.3, 114.9 mg Pb/g for Ch1, ChL5, ChL60, ChH5 and ChH45, respectively	Vukcevic et al. (2017)
	ChL5	429	0.208			
Waste hemp fibers → oxidative treatment ($NaClO_2$)	ChL60	389	0.194			
Chemically modified hemp fibers → carbonization at 1000 °C under constant N_2 flow	ChH5	426	0.207			
	ChH45	573	0.290			

Note: ACF = activated carbon fibers; S_{BET} = specific surface area obtained by BET method; V_{tot} = total pore volume; Ac = acetapirimid; D = dimethoate; N = nicosulfuron; C- carbofuran; At – atrazine; ITES = name of hemp provider from Serbia the removal of organic pollutants (pesticides and dyes) by sorption. These carbonized hemp fibers applicability, in good agreement with that of other activated carbons, is possible due to their very high surface area and high pore volume (Table 10.9). It was suggested that these outstanding features of the hemp – based activated carbon fibers are due to the beneficial effect of the hemp fibers complex structure and heterogeneous chemical composition (Vukcevic et al. 2012, 2014b). Kostic's research group recently studied the Pb(II) removal from aqueous solutions by biocarbons prepared by the carbonization of chemically modified hemp fibers and established that the Pb(II) sorptive ability of the biocarbons can be significantly improved by increasing the amount of biocarbons surface oxygen groups and enhanceding the fibrillation, meaning facilitated penetration of Pb(II) ions into the biocarbon structure (Kostic et al. 2018).

The eligibility of hemp fibers for the preparation of efficient activated carbon sorbents can be regarded as a contribution to the implementation of sustainable development in both hemp production and environmental protection (Yang et al. 2011).

10.4 Applicability of Sorbents Based on Hemp Fibers in Wastewater Treatment

Although still very limited as number, the studies performed in order to improve pollutant sorption on hemp fibers as a green technical solution for full scale applications had very promising results. The studies on this topic were conducted on simulated as well as real industrial effluents belonging to the following categories differentiated in the scientific literature: (i) bearing low concentration of pollutants in large volume (mining wastewaters); (ii) with high values of total dissolved solids in small volumes (wastewaters from electroplating industry) (Vijayaraghavan and Yun 2008). The high effectiveness of the treatment is conditioned by strong affinity of the sorbent for the first category of wastewaters and high capacity of sorption in the second case (Atkinson et al. 1998). In good agreement with these considerations, the reported sorption performances of hemp fibers in both cases of real industrial conditions are extremely encouraging.

The first estimation of the hemp fibers practical value targeting environmental aspects (removal of toxicity from aqueous media and safe disposal of hemp waste) and economic (metal recovery) was for the first time reported in 2009 (Table 10.10). For this purpose, the concentration of trace amounts of Cd(II) from large amounts of aqueous solutions on sulphydryl hemp fibers and the subsequent desorption of Cd (II) from the hemp loaded fibers were carried out, the significant results being shown in Table 10.11 (Tofan et al. 2009). It is obvious from Table 10.8 that the use of sulphydryl hemp fibers could lead to the decrease of the Cd(II) content from large

Table 10.10 Scheme of Cd(II) sorption – desorption in batch conditions

First step – sorption	Second step – desorption/calcination
Contacting different volumes of solution (200 mL, 500 mL and 1000 mL) containing the same amount of Cd(II) (112 μg) with 25 g of sulphydryl hemp fibers(24 h);	Contacting samples of 25 g of Cd(II) loaded SH-hemp with 25 mL of HNO_3 (24 h) → filtration of mixture;
Intermittent stirring;	Calcination of Cd(II) loaded SH-hemp at 800 °C → dissolution of the obtained residues with 25 mL of HCl 1:1;
Filtration	Determination of the concentration of Cd(II) in solution by atomic absorption spectrometry

Table 10.11 Concentration of Cd(II) in traces by hemp – SH(V_k = 25 mL) (Tofan et al. 2009)

Experiment	V_{sample} (mL)	Cd(II) initial amount (μg)	Cd (II) found μg	Concentration factor[a]	Remarks
1.	200	112	88.1	6.29	Desorption with HNO_3 2 M
2.	200	112	109.4	7.81	Calcining SH–
3.	500	112	96.8	17.29	Hemp and dissolution of the
4.	1000	112	87.7	30.27	Obtained residue with HCl 1:1

Note: [a]Concentration factor $= \frac{q_k \cdot V_{sample}}{q_{sample} \cdot V_k}$, where q_k and q_{sample} are the absolute amounts of Cd(II) in concentrate and sample, respectively; V_k is the volume of the concentrate; V_{sample} is the volume of the sample

volumes of wastewaters below allowable discharge limits. On the other hand, Table 10.11 points out the almost quantitative recovery of the retained Cd (II) (97.67%) by calcination of Cd loaded sulphydryl hemp fibers at 800 °C and the obtained residues dissolution with HCl 1:1.

In recent years, attempts were made to evaluate the capacity of some hemp – based materials for the simultaneous removal of pollutants from real polymetallic effluents (Table 10.12). As it can be seen from Table 10.12, the sorption potential of the hemp was practically unaffected by the complex nature of real effluents, appreciable levels of metallic ion removal being attained. Moreover, the significant sorption ability of hemp has been proven again by ecotoxicological tests whose results indicated a drastic decrease of effluent toxicity (Loiacono et al. 2018).

These sorption performance of hemp in treatment of real wastewaters must be confirmed by other extensive and varied studies.

Table 10.12 The compatibility of sorbents based on hemp fibers with real wastewaters

Hemp based sorbent	Treated wastewater	Targeted pollutant	Removal efficiency	Working batch conditions	References
Hemp shives	Acid mine drainage from Smolnik mine	Fe(II)	100%	Initial pH ≈ 4; 1 g/100 mL; 24 h of contact	Balintova et al. (2014)
Unmodified Treated with NaOH		Cu(II)	62.65%; 75.90%		
Hemp felt made from 90% fiber and 10% added polypropylene	Three representative samples of effluents from a metal finishing factory	Cd(II)	100%	2 g of hemp; 100 mL of solution; 60 min of contact	Loiacono et al. (2017b)
		Co(II)	13–68%		
		Cu(II)	9–93%		
		Mn(II)	12–21%		
		Ni(II)	39–84%		
		Zn(II)	64–91%		
Hemp felt (100% fiber)	Eight representative samples of effluents from a metal finishing factory	Cu(II)	~100%	1.5 g of hemp; 100 mL solution; pH ~5	Loiacono et al. (2018)
		Pb(II)	~100%		
		Cu(II)	~100%		
		Ni(II)	~100%		
		Fe(II)	85%		
		Zn(II)	84%		
		Co(II)	69%		
		Cr(III)	68%		
		Al(III)	43%		
		Mn(II)	23%		

10.5 Comparison of Sorbents Based on Hemp Fibers with Other Green Sorbents for Wastewater Treatment

It should be mentioned from the very beginning that the significant differences between the sources, operating conditions and technical solutions make very difficult and subjective the direct comparison between sorption performances of various materials.

A literature survey reveals that by comparison with other sorptive materials, the use of fibrous sorbents such as hemp fibers for the removal of aquatic pollutants can provide some special benefits, as follows:

- towards the conventional sorbents in beads form, the fibers can be applied in the form of felt or fabrics, i.e. high surface area during the contact with the contaminated aquatic environment → faster rates of the sorption process → enhanced efficiency (Jain et al. 2016a, b);
- fibrous sorbents ensure improved kinetics and easiness of handling when compared with powdered and granular sorbents (Suzuki 1994);
- unlike powders which require time- consuming techniques for removal from aqueous solutions, due to their macroscopic nature, hemp fibers can be easily retrieved from water by a simple filtration (Kyzas et al. 2015).

As a confirmation of these findings, the interesting results of a study comparing the performances of some commercial and low- cost sorbents demonstrate the ability of hemp shives to compete with synthetically prepared sorbents for Cu(II) removal from acidic media as well as the sodium hydroxide modified hemp superiority to peat at the use with the same purpose (Balintova et al. 2014). Furthermore, a recent study emphasized the significant value of the sorption capacity of short hemp fibers for Pb (II) ions which has been found to be higher than that of activated carbon in the granular or powder form or comparable with that of nanotubes (Kostic et al. 2018).

From another perspective, the maximum sorption capacities of hemp fibers vis- à-vis other sorbents based on natural and modified cellulosic fibers for Cu(II), Cd(II), Zn(II) and Pb(II) removal are presented in Table 10.13.

Table 10.13 Comparison of maximum sorption sorption capacity of various fibrous sorbents for Cu(II), Cd(II), Zn(II) and Pb(II) removal from aqueous solutions

	Maximum capacity of sorption (mg/g)				
Fibrous sorbent	Cu(II)	Cd(II)	Zn(II)	Pb(II)	References
Unmodified *juniper* fibers		9.18			Min et al. (2004)
Base–treated *juniper* fibers		29.54			
Unmodified *jute* fibers	4.23		3.55		Shukla and Pai (2005)
Dye loaded *jute* fibers	8.40		5.95		
Oxidized *jute* fibers	7.73		8.02		
Unmodified *coir* fibers			1.83		Shukla et al. (2006)
Oxidized *coir* fibers			7.88		Shukla and Shukla (2013)
Alkali treated coir fibers	9.43		29.41		
Palm kernel fiber	20.605				Ofomaja (2010)
Agave sisalana(sisal) fiber		1.34		1.35	dos Santos et al. (2011)
Palm fibers				18.622	Al-Haidary et al. (2011)
Coconut Mucifera fiber	0.08		0.09		Asiagwu et al. (2013)
Natural cotton fiber chemically modified with citric acid	6.12	8.22	4.53	21.62	Paulino et al. (2014)
Kenaf bast fiber	33.56		13.68	50.00	Shamsudin et al. (2016)
Flax fibers	9.921		8.453	10.741	Abbar et al. (2017)
Natural hemp fibers	9.073	2.59	21.047	25.047	Tofan and Paduraru (2000); Paduraru and Tofan (2008); Tofan et al. (2010c)
Hemp-based material in felt	2.14	1.09			Morin-Crini et al. (2019)
Fiber hemp modified with α-benzoin oxime	13.8				Tofan and Paduraru (1999)
Sulfhydryl hemp fibers		14		22.999	Tofan and Paduraru (2004)
Unmodified hemp fibers				14.14	Kostic et al. (2018)

It can be observed from Table 10.13 that hemp fibers are competitive against other fibrous sorbents, being adequate for the use as sorption filter material in wastewater treatment.

10.6 Conclusion

This chapter presents an integrated approach of the synthesis, sorption behaviour, applicability and comparison of different types of hemp fibers for the removal of various pollutants from aqueous effluents. The sorption performances of hemp in wastewater treatment is based on its remarkable fundamental features: low cost, availability, high mechanical strength and porosity, hydrophilic character, fast sorption, tolerance to biological structures, adjustability of the surface chemistry, possibility of being used as fibers and filters. The hemp fibers can be used for the inorganic and organic pollutants removal without any treatment (natural hemp fibers), or after the application of physical (loaded hemp fibers) or chemical (chemically modified hemp fibers) treatments. The chemical composition of hemp fibers provides their ability to act as precursor for biocarbon based sorbents. The results of all batch studies systematized in this chapter strongly recommend the sorption by natural hemp fibers as a sustainable solution to ameliorate the quality of wastewater contaminated with heavy metal ions. The favorable behavior of loaded hemp fibers provides the opportunity of hemp material reuse for the sequential removal of organic pollutants and heavy metal from wastewaters. Besides the sorption capacity increase, the proper chemical modifications of hemp fibers also can ensure the selectivity of heavy metal ions sorption that can be raised to the level of specificity. Unlike natural, loaded and chemically modified hemp fibers, biocarbon fibers prepared by carbonization and chemical activation of hemp are preferable for the sorption removal of organic pollutants (pesticides and dyes). Although still very limited as number, but with promissing results, studies to recognize the pollutants sorption by hemp fibers as a green technical solution are still needed for full scale applications in wastewater treatment. The heavy metal uptake capabilities of hemp fibers are similar to or even higher that that of other sorbents based on natural and modified vegetable fibers. The researches on these topics must be continued and deepened from the following perspectives:

- the assessment of the chemism and mechanisms of the retention processes of heavy metal ions on hemp fibers;
- the subsequent valorization of the results already obtained in order to find the most efficient solutions for the sorption performances of the hemp fibers improvement, by different methods of physical and chemical treatments;
- the assessment of the performance of hemp fibers in multi– pollutant solution systems prior to their industrial applications;
- comprehensive studies on the regeneration and reuse of the hemp fibers in the context of circular economy;

- novel achievement procedures of the pollutant retention on the proposed sorbents, as improvement variants of the performances of the actual sorption process;
- comprehensive dynamic studies in order to use the hemp fibers for treating complex industrial wastewaters and also for the simultaneous removal of heavy metal ions and organic substances;
- studies under pilot scale applications are needed to complete the aspects concerning the removal efficiencies with technical and economic factors that influence process scale-up.

References

Abbar B, Alem A, Marcotte S, Pantel A, Ahfir N-D, Bizet L, Duriatti D (2017) Experimental investigation on the removal of heavy metals(Cu^{2+}, Pb^{2+} and Zn^{2+}) from aqueous solutions by flax fibers. Process Saf Environ 109:639–647. https://doi.org/10.1016/j.psep.2017.05.012

Abdolali A, Guo WS, Ngo HH, Chen SS, Nguyen NC, Tung KL (2014) Typical lignocellulosic wastes and by-products for biosorption process in water and wastewater treatment: a critical review. Bioresour Technol 160:57–66. https://doi.org/10.1016/j.biortech.2013.12.037

Afroze S, Sen T (2018) A review of heavy metals and dye adsorption from water by agricultural solid waste adsorbents. Water Air Soil Pollut 229:225. https://doi.org/10.1007/s11270-018-3869-z

Al-Haidary AM, Zanganah FHH, Al-Alzawi SRF, Khalili FI, Al-Dujaili AH (2011) A study on using date palm fibers and leaf base of palm as adsorbents for Pb(II) ions from its aqueous solution. Water Air Soil Pollut 214:73–82. https://doi.org/10.1007/s11270-010-0405-1

Ali I, Asim M, Khan TA (2012) Low cost adsorbents for the removal of organic pollutants from water. J Environ Manag 113:170–183. https://doi.org/10.1016/j.jenvman.2012.08.028

Altenor S, Carene-Melane B, Gaspard S (2009) Activated carbons from lignocellulosic waste materials for water treatment: a review. Int J Environ Tech Manag 10:308–323. https://doi.org/10.1504/IJETM.2009.023737

Altinisik A, Seki Y, Ertas S, Akar E (2015) Evaluating of *Agave americana* fibers for biosorption of dye from aqueous solution. Fiber Polym 16:370–377. https://doi.org/10.1007/s12221-015-0370-9

Amin MT, Alazba AA, Manzoor U (2014) A review of pollutants from wastewater using different types of nanomaterials. Adv Mater Sci Eng 2014:1–24. https://doi.org/10.1155/2014/825910

Andre CM, Hausman JF, Guerriero G (2018) *Canabis sativa*: the plant of thousand and one molecules. Front Plant Sci 7:1–17. https://doi.org/10.3389/fpls.2016.00019

Anwar Z, Gulfraz G, Irshad M (2014) Agro- industrial lignocellulosic biomass a key to unlock future bio-energy – a brief review. J Radiat Res Appl Sci 7:163–173. https://doi.org/10.1016/j.jrras.2014.02.003

Asiagwu AK, Owamah HI, Cristopher IO (2013) Kinetic model for the sorption of Ni(II), Cu (II) and Zn(II) onto *Cocos nucifera* fiber waste biomass from aqueous solution. J Chem Eng Process Technol 4:1000149. https://doi.org/10.4172/2157-7048.100014

Atkinson BW, Bux T, Kasan HC (1998) Considerations for application of biosorption technology to remediate metal – contaminated industrial effluents. Water SA 24:129–135

Balintova M, Holub M, Stevulova N, Cigasova J, Tesarcikova M (2014) Sorption in acidic environment—biosorbents in comparison with commercial adsorbents. Chem Eng Trans 39:625–630. https://doi.org/10.3303/CET1439105

Barakat MA (2011) New trends in removing heavy metals from industrial wastewater. Arab J Chem 4:361–377. https://doi.org/10.1016/j.arabjc.2010.07.019

Ben Hamissa AM, Lodi A, Seffen M, Finocchio E, Botter R, Converti A (2010) Sorption of Cd (II) and Pb(II) from aqueous solutions onto Agave americana fibers. Chem Eng J 159:67–74. https://doi.org/10.1016/j.cej.2010.02.036

Ben Hamissa AM, Brouers F, Nibi MC, Seffen M (2013) Kinetic modeling study on methylene blue sorption onto Agave americana fibers: fractal kinetics and regeneration studies. Sep Sci Technol 48:2834–2842. https://doi.org/10.1080/01496395.2013.809104

Bugnet J, Morin-Crini N, Cosentino C, Chanet G, Winterton P, Crini G (2017) Hemp decontamination of poly-metallic aqueous solutions. Environ Eng Manag J 16:535–542

Burakov AE, Galunin EV, Burakova IV, Kucherova AE, Agarwal S, Tkachev AG, Gupta VK (2018) Adsorption of heavy metals on conventional and nanostructured materials for wastewater treatment purposes: a review. Ecotoxicol Environ Saf 148:702–712. https://doi.org/10.1016/j.ecoenv.2017.11.11034

Cassano R, Trombino S, Ferrarelli T, Nicoletta FP, Mauro MV, Giraldi C, Picci N (2013) Hemp fiber (*Cannabis sativa* L.) derivatives with antibacterial and chelating properties. Cellulose 20:547–557. https://doi.org/10.1007/s10570-012-9804-3

Chaiyasith S, Chaiyasith P, Septhum C (2006) Removal of cadmium and nickel from aqueous solutions by adsorption onto treated fly ash from Thailand. Thammasat Int J Sci Tech 11:13–20

Chandrakant RH, Ananda JJ, Dipak VP, Naresh MM, Anirundaha BP (2016) A critical review on textile wastewater treatments: possible approaches. J Environ Manag 182:351–366. https://doi.org/10.1016/j.jenvman.2016.07.090

Chawdhury ZZ, Hamid SBA, Das R, Hasan Md R, Zain SM, Khalid K, Uddin MN (2013) Preparation of carbonaceous adsorbents from lignocellulosic biomass and their use in removal of contaminants from aqueous solutions. Bioresources 8:6523–6555. https://doi.org/10.15376/biores.8.4.6523-6555

Crini G, Badot P (eds) (2010) Sorption processes and pollution control. Conventional and non-conventional sorbents for pollutant removal from wastewater. Presses Univ Franche-Comté. ISBN:978-2-84867-304-2

Crini G, Lichtfouse E (eds) (2018) Green adsorbents for pollutant removal. Innovative materials. Environmental chemistry for a sustainable world. Springer. https://doi.org/10.1007/978-3-319-92162-4

Dai D, Fan M (2010) Characteristic and performance of elementary hemp fiber. Mater Sci Appl 1:336–342. https://doi.org/10.4236/msa.2010.16049

Dai Y, Sun Q, Wang W, Lu L (2018) Utilization of agricultural waste as adsorbents for the removal of contaminants: a review. Chemosphere 211:235–253. https://doi.org/10.1016/j.chemosphere.2018.06.79

De Gisi S, Lofrano G, Grassi M, Notarnicola M (2016) Characteristics and adsorption capacities of low-cost sorbents for wastewater treatment: a review. Sustain Mater Technol 9:10–40. https://doi.org/10.1016/j.susmat.2016.06.002

Dias J, Alvim-Ferraz M, Almeida M, Rivera-Utrilla J, Sanchez-Polo M (2007) Waste materials for activated carbon preparation and its use in aqueous- phase treatment: a review. J Environ Manag 85:833–846. https://doi.org/10.1016/j.jenvman.2007.07.031

Dizbay-Onat M, Vaidya U, Balanay JA, Lungu C (2018) Preparation and characterization of flax, hemp and sisal fibers derived mesoporous activated carbon adsorbents. Adsorpt Sci Technol 36:441–457. https://doi.org/10.1177/0263617417700635

dos Santos WNL, Cavalcante DD, da EGP S, das Virgens CF, de Souza Dias F (2011) Biosorption of Pb(II) and Cd(II) ions by *Agave sisalana* (sisal fibers). Microchem J 97:269–273. https://doi.org/10.1016/j.microc.2010.09.04

Du Z, Zheng T, Wang P, Hao L, Wang Y (2016) Fast microwave–assisted preparation of a low-cost and recyclable carbonyl modified lignocellulose- biomass jute fiber for enhanced heavy metal removal from water. Bioresour Technol 201:41–49. https://doi.org/10.1016/j.biortech.2015.11.009

Femina Carolina C, Senthil Kumar P, Saravan A, Janet Joshiba G, Naushad M (2017) Efficient techniques for the removal of toxic heavy metals from aquatic environment: a review. J Environ Chem Eng 5:2782–2799. https://doi.org/10.1016/j.jece.2017.05.029

Freundlich HM (1906) Uber die adsorption in lusungen. Z Phys Chem 57:85–470. https://doi.org/10.1515/zpch-1907-5723

Fu F, Wang Q (2011) Removal of heavy metal ions from wastewaters: a review. J Environ Manag 92:407–418. https://doi.org/10.1016/j.jenvman.2010.11.011

Gao D-W, Hu Q, Pan H, Jiang J, Wang P (2015) High–capacity adsorption of aniline using surface modification of lignocellulosic – biomass jute fibers. Bioresour Technol 193:507–512. https://doi.org/10.1016/j.biortech.2015.06.138

Gautam RK, Mudhov A, Lofrano G, Chattopadhyaya WC (2014) Biomass- derived biosorbents for metal ion sequestration: adsorbent modification and activation methods and adsorbent regeneration. J Environ Chem Eng 2:239–259. https://doi.org/10.1016/j.jece.2013.12.019

Hasfalina CM, Maryam RZ, Luquan CA, Rashid M (2010) The potential use of kenaf bioadsorbent for the removal of copper and nickel from single and binary solution. J Nat Fibers:267–275. https://doi.org/10.1080/15440478.2010.527508

Hasfalina CM, Maryam RZ, Luquan CA, Rashid M (2012) Adsorption of copper(II) from aqueous medium in a fixed bed column by kenaf fibers. APCBEE Proc 3:255–263. https://doi.org/10.1016/j.apcbee.2012.06.079

Hassan MS, Zohdy MH (2018) Removal of toxic heavy metal ions from aqueous solutions using jute fibers grafted with acrylic acid by gamma irradiation. J Vinyl Addit Technol 24:39–346. https://doi.org/10.1002/vnl.21603

Ho YS, McKay G (1999) Pseudo–second order for the sorption process. Process Biochem 34:451–465. https://doi.org/10.1016/S0032-9592(98)00112-5

Hokkannen S, Bhatnagar A, Sillanpää M (2016) A review on modified methods to cellulose – based adsorbents to improve adsorption capacity. Water Res 91:156–173. https://doi.org/10.1016/j.watres.2016.01.008

Hubbe M, Hasan SH, Ducoste JJ (2011) Cellulosic substrates for removal of pollutants from aqueous systems: a review. 1. Metals. Bioresources 6:2161–2287

Hubbe M, Beck KR, O'Neal W, Sharma YC (2012) Cellulosic substrates for removal of pollutants from aqueous systems: a review. 2. Dyes. Bioresources 7:2592–2687. https://doi.org/10.15376/biores.7.2.2592-2687

Hubbe M, Rojas O, Fingas M, Gupta BS (2013) Cellulosic substrates for removal of pollutants from aqueous systems: a review. Spilled oils and emulsified organic liquids. Bioresources 8:3038–3097. https://doi.org/10.15376/biores.8.2.3038-3097

Hubbe M, Park J, Park S (2014) Cellulosic substrates for removal of pollutants from aqueous systems: a review. Dissolved petrochemical compounds. BioResorces 9:7782–7925. https://doi.org/10.15376/biores.9.4.7782-7925

Jain A, Balasubramanian R, Srinivasan MP (2016a) Hydrothermal conversion of biomass waste to activated carbon with high porosity: a review. Chem Eng J 283:789–805. https://doi.org/10.1016/j.cej.2015.08.014

Jain CK, Malik DS, Yadav AK (2016b) Applicability of plant based biosorbents in the removal of heavy metals: a review. Environ Process 3:495–523. https://doi.org/10.1007/s40710-016-0143-5

Jiang J, Pan H, Gao DW (2016) Column adsorption of aniline by a surface modified jute fibers and its regeneration property. J Environ Chem Eng 4:2243–2249. https://doi.org/10.1016/j.jece.2016.03.022

Johari K, Saman N, Song J, Heng YY, Mat H (2014) Study of Hg(II) removal from aqueous solutions using lignocellulosic coconut fibers biosorbent: equilibrium and kinetic evaluation. Chem Eng Commun 201:1198–1220. https://doi.org/10.10180/00986445.2013.806311

Kabir S, Kamir L, Islam K (2017) A comparative study of dyeing properties of hemp and cotton fibers. Eur Sci J 13:378–389. https://doi.org/10.19044/esj.2017.v13n33p378

Kaczmar JW, Pach J, Burgstaller C (2011) The chemically treated hemp fibers to reinforce polymers. Polymer 56:817–822. https://doi.org/10.14314/polymery.2011.817

Kostic M, Pejic P, Asanovic KA, Aleksic V, Skundric P (2010) Effect of hemicellulose and lignin on the sorption and electric properties of hemp fibers. Ind Crop Prod 32:169–174. https://doi.org/10.1016/j.indcrop.2010.04.014

Kostic M, Pejic P, Vukcevic M (2018) Waste hemp (*Cannabis sativa*) fibers as a biosorbent and a precursor for biocarbon sorbents: influence of their chemical composition on Pb(II) removal. In: Stevanovic T (ed) Chemistry of lignocellulosics: current trends. Taylor & Francis, Boca Raton, pp 3–21

Kyzas G, Terzopoulu Z, Nikolaidis V, Alexopoulu E, Bikiaris D (2015) Low cost hemp biomaterials for nickel ions removal from aqueous solutions. J Mol Liq 209:209–218. https://doi.org/10.1016/j.molliq.2015.05.060

Lagergren S (1898) About the theory of so-called adsorption of solid substance. Hand 24:1–39

Langmuir I (1916) The constitution and fundamental properties of solids and liquids. J Am Chem Soc 38:2221–2295. https://doi.org/10.1021/ja02268a002

Leite C, Pereira H (2017) Cork–containing bark – a review. Front Mater 3:163. https://doi.org/10.3389/fmats.2016.00063

Li Y, Liu Y-J (2008) Biosorption isotherms, kinetics and thermodynamics. Sep Purif Technol 61:229–242. https://doi.org/10.1016/j.seppur.2007.10.002

Loiacono S, Crini G, Martel B, Chanet G, Cosentino C, Raschetti M, Placet V, Torri G, Morin-Crini N (2017a) Simultaneous removal of Cd Co, Cu, Mn, Ni and Zn from synthetic solutions on a hemp-based felt. 2: chemical modification. J Appl Polym Sci 134:1–16. https://doi.org/10.1002/app.45138

Loiacono S, Morin-Crini N, Cosentino C, Torri G, Chanet G, Winterton P, Crini G (2017b) Simultaneous removal of Cd, Co, Cu, Mn, Ni and Zn from synthetic solution on a hemp – based felt: experimental design. J Appl Polym Sci 134:1–11. https://doi.org/10.1002/app.44422

Loiacono S, Crini G, Chanet G, Raschetti M, Placet V, Morin- Crini N (2018) Metals in aqueous solutions and real effluents: biosorption behavior of a hemp-based felt. J Chem Technol Biotechnol. https://doi.org/10.1002/jctb.561

Luka Y, Highina BK, Zubairu A (2018) The promising precursors for development of agricultural waste materials: a review. Int J Adv Sci Res Eng 4:46–59. https://doi.org/10.7324/IJASRE.2018.32615

Lupul I, Yperman J, Carleer R, Gryglewicz G (2015a) Tailoring of porous structure of hemp stem – based activated carbon produced by phosphoric acid activation in steam atmosphere. J Porous Mater 22:283–289. https://doi.org/10.1007/s10934-014-9894-4

Lupul I, Yperman J, Carleer R, Gryglewicz G (2015b) Adsorption of atrazine on hemp stem-based activated carbons with different surface chemistry. Adsorpt J Int Adsorpt Soc 21:489–498. https://doi.org/10.1007/s10450-015-9689-1

Malik S, Jain CK, Yadav AK (2017) Removal of heavy metals from emerging low-cost adsorbents: a review. Appl Water Sci 7:2113–2136. https://doi.org/10.1007/s13201-016-0401-8

Manna S, Saha P, Chudhury S, Thomas S (2017) Chapter 3: Alkali treatment to improve physical, mechanical and chemical properties of lignocellulosic natural fibers fibers for use in various applications. In: Kuila A, Sharma V (eds) Lignocellulosic biomass production and industrial applications. Scrivener Publishing Willey, pp 42–64

Min SH, Han JS, Shin EW, Park JK (2004) Improvement of cadmium ion removal by base treatment of juniper fiber. Water Res 38:1289–1295. https://doi.org/10.1016/j.watres.2003.11.016

Morin-Crini N, Loiacono S, Placet V, Torri G, Bradu C, Kostic M, Cosentino C, Chanet G, Martel B, Lichtfouse E, Crini G (2019) Hemp – based adsorbents for sequestration of metals – a review. Environ Chem Lett 17:393–408. https://doi.org/10.1007/s10311-018-0812-x

Neris JB, Luzardo FHM, Silva EGP, Velasco FG (2019) Evaluation of adsorption process of metal ions in multi –element aqueous systems by lignocellulosic adsorbents applying different isotherms: a critical review. Chem Eng J 307:404–420. https://doi.org/10.1016/j.cej.2018.09.125

Nguyen TAH, Ngo HH, Guo WS, Zhang J, Liang S, Yue QY, Li Q, Nguyen TV (2013) Applicability of agricultural wastes and by – products for adsorptive removal of heavy metals from wastewater. Bioresour Technol 148:574–585. https://doi.org/10.1016/j.biortech.2013.08.124

Nidheesh PV, Zhou M, Oturan MA (2018) An overview on the removal of synthetic dyes from water by electrochemical advanced oxidation processes. Chemosphere 197:210–227. https://doi.org/10.1016/j.chemosphere.2017.12.195

O'Connell DW, Birkinshaw C, O'Dwyer TF (2008) Heavy metal adsorbents prepared from the modification of cellulose: a review. Bioresour Technol 99:6709–6724. https://doi.org/10.1016/j.biortech.2013.12.037

Ofomaja AE (2010) Equilibrium studies for copper ion adsorption onto palm kernel fibre. J Environ Manag 91:1491–1499. https://doi.org/10.1016/j.jenvman.2010.02.029

Oller I, Malato S, Sanchez-Perez JA (2011) Combination of advanced oxidation processes and biological treatments for wastewaters decontamination – a review. Sci Total Environ 409:4141–4166. https://doi.org/10.1016/j.scitotenv.2010.08.061

Paduraru C, Tofan L (2002) Equilibrium studies for the sorption of metal ions onto hemp. Cellul Chem Technol 36:375–380

Paduraru C, Tofan L (2008) Investigations on the possibility of natural hemp fibres use for Zn (II) removal from wastewaters. Environ Eng Manag J 7:687–693

Pan B, Pan B, Zhang W, Lv L, Zhang Q, Zheng S (2009) Development of polymeric and polymer–based hybrid adsorbents for pollutant removal from waters. Chem Eng J 151:15–29. https://doi.org/10.1016/j.cej.2009.02.036

Park D, Yun YS, Park JM (2010) The past, present and future trends of biosorption. Biotechnol Bioprocess Eng 15:86–102. https://doi.org/10.1007/s12257-009-0199-4

Paulino AG, da Ciuha AJ, da Silva Alfaya RV (2014) Chemically modified natural cotton fiber: a low-cost biosorbent for the removal of Cu(II), Zn(II), Cd(II) and Pb(II) from natural water. Desalin Water Treat 52:4223–4233. https://doi.org/10.1080/19443994.2013.804451

Pejic B, Kucevic M, Kostic M, Skundric P (2009) Biosorption of heavy metal ions from aqueous solutions by short hemp fibers: effect of chemical composition. J Hazard Mater 164:146–153. https://doi.org/10.1016/j.jhazmat.2008.07.139

Pejic B, Kucevic M, Pajic– Lijakovic I, Lausevic M, Kostic M (2011) Mathematical modeling of heavy metal ions (Cd^{2+}, Zn^{2+} and Pb^{2+}) biosorption by chemically modified hemp fibers. Chem Eng J 172:354–360. https://doi.org/10.1016/j.cej.2011.06.016

Petroudy SRD (2017) Physical and mechanical properties of natural fibers. Woodhead Publishing Series in Composite Science and Engineering 74:59–83. https://doi.org/10.1016/B978-0-08-100411-1.00003-0

Placet V, Day A, Beaugrand J (2017) The influence of unintended field retting on the physico-chemical and mechanical properties of hemp bast fibers. J Mater Sci 52:5759–5777. https://doi.org/10.1007/s10853-017-0811-5

Ramrakhiani L, Ghosh S, Majumdar S (2016) Surface modification of naturally available biomass for enhancement of heavy metal removal: efficiency, upscaling prospects and management aspects of spent biosorbents: a review. Appl Biochem Biotechnol 180:41–78. https://doi.org/10.1007/s12010-016-2083-y

Renu, Agarwal M, Singh K (2016) Heavy metal removal from wastewater using various sorbents. J Water Reuse Desal 7:387–419. https://doi.org/10.2166/wrd.2016.104

Rivera-Utrilla J, Sanchez-Polo M, Ferro-Garcia MA, Prado-Joya G, Ocampo-Perez R (2013) Pharmaceuticals as emerging contaminants and their removal from water. A review. Chemosphere 93:1268–1287. https://doi.org/10.1016/j.chemosphere.2013.07.059

Rojas CMM, Rivera Velasquez MF, Tavalaro A, Molinari A, Fallico C (2017) Use of vegetable fibers for PRB to remove heavy metals from contaminated aquifers – comparisons among cabuya fibers, broom fibers and ZVI. Int J Environ Res Public Health 14:684–702. https://doi.org/10.3390/ijerph14070684

Rosas JM, Bedia J, Rodriguez-Mirasol J, Cordero T (2009) Hemp derived activated carbon fibres by chemical activation with phosphoric acid. Fuel 88:19–26. https://doi.org/10.1016/j.fuel.2008.08.004
Roy A, Chakraborthy S, Kundu SP, Adhikari B, Majumder SB (2013) Lignocellulosic jute fiber as a bioadsorbent for the removal of azo dye from its aqueous solution: batch and column studies. J Appl Polym 129:15–27
Rozumova L, Legatova B (2018) Biosorbent from industrial hemp hurds for copper ions removal. Int J Mater Sci Eng 6:114–125. https://doi.org/10.17706/ijmse.2018.6.4.114-125
Sajab MS, Chia CH, Zakaria S, Jani SM, Ayob MK, Chee KLC, Khiew PS, Chiu KS (2011) Citric acid modified kenaf fiber for removal of methylene blue from aqueous solution. Bioresour Technol 102:7237–7243. https://doi.org/10.1016/j.biortech.2011.05.011
Salman M, Athar M, Farooq U (2015) Biosorption of heavy metals from aqueous solutions using indigeneous and modified lignocellulosic materials. Rev Environ Sci Biotechnol 14:211–228. https://doi.org/10.1007/s11157-015-9362-x
Saravan A, Janet Joshiba G, Naushad M (2017) Efficient techniques for the removal of toxic heavy metals from aquatic environment. A review. J Environ Chem Eng 5:2782–2799. https://doi.org/10.1016/j.jece.2017.05.029
Sedan D, Pagnoux C, Smith A, Chotard T (2008) Mechanical properties of the hemp reinforced cement: influence of the fiber/matrix interaction. J Eur Ceram Soc 28:183–192. https://doi.org/10.1016/j.jeurceramsoc.2007.05.019
Shahzad A (2012) Hemp fiber and its composites – a review. J Compos Mater 46:973–986. https://doi.org/10.1177/0021998311413623
Shamsudin R, Abdullah H, Kanari A (2016) Application of kenaf bast fiber to adsorb Cu(II), Pb (II) and Zn(II) in aqueous solutions: single and multi–metal systems. Int J Environ Sci Develop 7:715–723. https://doi.org/10.18178/ijesd.2016.7.10.868
Shukla SR, Pai RS (2005) Adsorption of Cu(II), Ni(II) and Zn(II) on modified jute fibers. Bioresour Technol 96:1430–1439. https://doi.org/10.1016/j.biortech.2004.12.010
Shukla P, Shukla SR (2013) Biosorption of Cu(II), Pb(II), Ni(II) and Fe(II) on alkali treated coir fibers. Sep Sci Technol 48:421–428. https://doi.org/10.1080/01496395.2012.691933
Shukla SR, Pai R, Shendarkar AD (2006) Adsorption of Ni(II), Zn(II) and Fe(II) on modified coir fibers. Sep Purif Technol 47:141–147. https://doi.org/10.1016/j.seppur.2005.06.014
Silva CP, Jamia G, Otero M, Esteves VI, Calisto V (2018) Waste–based alternative adsorbents for the remediation of pharmaceutical contaminated waters: has a step forward already been taken? Bioresour Technol 250:888–901. https://doi.org/10.1016/j.biortechn.2017.11.102
Staroń P, Chwastowski J, Banach M (2017) Sorption and desorption studies on silver ions from aqueous solution by coconut fiber. J Clean Prod 149:290–301. https://doi.org/10.1016/j.jclepro.2017.02.116
Sud D, Majayan G, Kaur MP (2008) Agricultural waste material as potential adsorbent for sequestering heavy metal ions from aqueous solutions – a review. Bioresour Technol 99:6017–6027. https://doi.org/10.1016/j.biortechn.2007.11.064
Suzuki M (1994) Activated carbon fibers: fundamentals and applications. Carbon 32:577–586. https://doi.org/10.1016/0008-6223(94)90075-2
Teklu T, Wangatia L, Alemayelu E (2018) Removal of Pb(II) from aqueous media using adsorption onto polyaniline coated sisal fibers. J Vinyl Addit Technol. https://doi.org/10.1002/vnl.21652
Tofan L, Păduraru C (1999) Removal of copper(II) ions in traces from waste waters by sorption on hemp modified with α-benzoinoxime. J Balkan Ecology 2:106–112
Tofan L, Păduraru C (2000) The hemp – a new fibrous natural unconventional sorbent. Buletinul Institutului Politechnic Iasi, Series of Chemistry and Chemical Engineering, Tome XLVI (L) fascicule 3–4:113–118
Tofan L, Păduraru C (2004) Sorption studies of Ag (I), Cd(II) and Pb(II) ions on sulphydryl hemp fibres. Croat Chem Acta 77:581–586

Tofan L, Păduraru C, Toma O (2001) Hemp fibres impregnated with alizarine S as unconventional material for Cr(III) ions in technological systems. Scientific Annals of the State University of Moldavia Chisinau, Chisinau, pp 204–208

Tofan L, Păduraru C, Volf I (2009) Concentration of cadmium (II) trace amounts from large volumes of aqueous samples of chemically modified hemp fibres. Sci Pap J Agron Ser 52:506–511

Tofan L, Păduraru C, Volf I (2010a) Comparative study concerning the retention thermodynamics of some heavy metal ions on hemp fibres. Buletinul Institului Politechnic Iasi LVI(LX):137–145

Tofan L, Păduraru C, Volf I, Balan C (2010b) Removal of lead(II) from aqueous solution by sorption by on natural hemp fibres. Sci Pap J Agron Ser 53:150–153

Tofan L, Păduraru C, Volf I, Balan C (2010c) Kinetic and thermodynamic profile of Pb(II) sorption by untreated hemp fibres. Sci Pap J Agron Ser 53:146–150

Tran VS, Ngo HH, Guo W, Zhang J, Liang S, Ton-That C, Zhang X (2015) Typical low- cost biosorbents for adsorptive removal of specific organic pollutant from water. Bioresour Technol 182:353–363. https://doi.org/10.1016/j.biortech.2015.02.003

Varghese AG, Paul SA, Latha MS (2018) Remediation of heavy metals and dyes from wastewater using cellulose based adsorbents. Environ Chem Lett:1–11. https://doi.org/10.1007/s10311-018-00843-z

Vijayaraghavan K, Yun LS (2008) Bacterial biosorbents and biosorption. Biotechnol Adv 26:266–291. https://doi.org/10.1016/j.biotechadv.2008.02.002

Vukčević M, Kalijadis A, Radisic M, Pejić B, Kostić M, Laušević Z, Laušević M (2012) Application of carbonized hemp fibres as a new solid-phase extraction sorbent for analysis of pesticides in water samples. Chem Eng J 211:224–232. https://doi.org/10.1016/j.cej.2012.09.059

Vukčević M, Pejic B, Lansevic M, Pajic-Lijakovic I, Kostic M (2014a) Influence of chemically modified short hemp fiber structure on biosorption process of Zn^{2+} from wastewater. Fiber Polym 15:687–697. https://doi.org/10.1007/s12221-014-0687-9

Vukčević M, Pejić B, Kalijadis A, Pajić-Lijaković I, Kostić M, Laušević Z, Laušević M (2014b) Carbon materials from waste short hemp fibres as a sorbent for heavy metal ions—mathematical modeling of sorbent structure and ions transport. Chem Eng J 235:284–292. https://doi.org/10.1016/j.cej.2013.09.047

Vukčević M, Kalijadis AM, Vasiljevic TM, Babic BM, Laušević ZV, Laušević MD (2015) Production of activated carbon derived from waste hemp (*Cannabis sativa*) fibres and its performance in pesticide adsorption. Microporous Mesoporous Mater 214:156–165. https://doi.org/10.1016/j.micromeso.2015.05.012

Vukcevic M, Pejic BM, Pejic- Lijakovic I, Kalijadis A, Lausevic Z, Lausevic M (2017) Influence of precursor chemical composition on the heavy metal adsorption properties of hemp (*Cannabis Sativa*) fibers based biocarbons. J Serb Chem Soc 82:1–16. https://doi.org/10.2298/JSC170310080V

Wan Ngah WS, Hanafiah MAKM (2008) Removal of heavy metal ions from wastewaters by chemically modified plant wastes as adsorbents: a review. Bioresour Technol 99:3935–3948. https://doi.org/10.1016/j.biortech.2007.06.011

Williams PT, Reed AR (2003) Pre- formed activated carbon matting derived from the pyrolysis of biomass natural fiber textile waste. J Anal Appl Pyrolysis 70:563–577. https://doi.org/10.1016/S0165-2370(03)00026-3

Williams PT, Reed AR (2004) High grade activated carbon matting derived from the chemical activation of pyrolysis of natural fiber textile waste. J Anal Appl Pyrolysis 71:971–986. https://doi.org/10.1016/j.jaap.2003.12.007

Yahya MA, Al-Qudah Z, Zanariah Ngah CW (2015) Agricultural bio-waste materials as a potential sustainable precursors for activated carbon production: a review. Renew Sust Energ Rev 46:218–235. https://doi.org/10.1016/j.rser.2015.02.051

Yang R, Liu GQ, Xu XH, Li M, Zhang JC, Hao XM (2011) Surface texture, chemistry and adsorption properties of acid blue 9 of hemp (*Cannabis sativa* L.) bast-based activated carbon fibres prepared by phosphoric acid activation. Biomass Bioenergy 35:437–445. https://doi.org/10.1016/j.biombioe.2010.08.061

Zhao M, Xu Y, Zhang C, Rong H, Zeng G (2016) New trends in removing heavy metals from wastewater. Appl Microbiol Technol 100:6509–6519. https://doi.org/10.1007/s00253-016-7646-x

The manufacturer's authorised representative in the EU is Springer Nature Customer Service Centre GmbH, Europaplatz 3, 69115 Heidelberg, Germany. If you have any concerns regarding our products, please contact ProductSafety@springernature.com

Printed and bound by CPI Group (UK) Ltd, Croydon, CR0 4YY
30/09/2026
02225697-0009